PRACTICAL APPLICATION OF DEPENDABILITY ENGINEERING

An Effective Approach to Managing Dependability in Technological and Evolving Systems

By

Thomas Van Hardeveld

David Kiang

ASME PRESS

Library of Congress Cataloging-in-Publication Data

Van Hardeveld, Thomas.
Practical application of dependability engineering: an effective approach to managing dependability in technological and evoving systems / by Thomas Van Hardeveld and David Kiang.
pages cm
Includes bibliographical references and index.
ISBN 978-0-7918-6001-4 (alk. paper)
1. Reliability (Engineering) I. Kiang, David, 1935- II. Title.

TA169.V36 2012
620'.00452—dc23

2012027460

DEDICATION

This book is dedicated to the memory of the past Chairman of IEC/TC56 Dependability

Mr. Geoffrey C. Alstead

who *inspired experts from many countries in the development of International Standards on Dependability, including the authors of this book.*

PREFACE

Our modern technological society expects everything to work as expected, that is, without interruption or failure. Our memories are short, whereas the history of innovation started with the earliest human quests for survival. It is only very recently that we have reached the currently high levels of dependability that we now demand in our daily lives without much thought. But the path toward even higher levels of dependability continues. It is becoming increasingly difficult to address dependability issues as system sophistication and technological advances move forward.

Many people call this general area of endeavor "reliability," although, in fact, it actually consists of a number of interrelated characteristics, primarily availability, reliability, maintainability and supportability. In this book, we use the umbrella term "dependability," which is the one that is known in international standardization.

Dependability has negative connotations since it naturally deals with failures. Although we know that failures are what we must analyze, we want to also portray dependability in a positive light as building trust and successful achievement of objectives.

Dependability is a multi-faceted and varied field of engineering, but one that is critical to industry and society as a whole. The fundamentals of dependability are well understood and documented, so there is no pressing need to reinvent the foundations of dependability engineering (e.g., statistics) and duplicate the rich vein of literature in this area. It is more important to discuss how to manage for successful dependability performance in rapidly changing times. It is necessary to tailor the application of dependability since it varies substantially across different industries, types of assets and application of technologies.

This book is intended for engineers and practitioners who need to solve problems and find answers to achieve dependability performance of technological and evolving systems. It provides current industry best practices and practical knowledge and guidance to address dependability management issues and engineering processes from a system life cycle perspective. The topics presented herein highlight the essence of life cycle management practices and systematic cost-effective solutions, focusing on dependability performance characteristics for project risk avoidance and failure prevention.

The rationale for the need to achieve dependability performance is presented from the viewpoint of the causative effects and probable risk exposures of system malfunction. Recommendations are provided to assist and devise practical means for cost-effective solutions to support the decision process for mitigation of negative consequences. The starting point for resolution of dependability issues is to exercise prudent engineering judgement by fully understanding the problem at hand. The intention is to avoid using cookbook answers for expediency, which often lead to misinterpretation of the theoretical assumptions to fit the case in point.

The development of technological systems involves hardware, software and human aspects of dependability in order to realize system performance requirements. The interaction of hardware and software to achieve interoperability, survivability and serviceability in system performance is addressed. Dependability issues of evolving systems, where merging of the "old" and the "new" coexists in system integration to meet performance needs, are discussed. Applicable dependability methods for assessment and assurance are provided to facilitate confidence in reaching project acceptance and delivery targets.

This book is organized in a logical sequence where individual chapters present specific dependability topics to highlight key dependability management and engineering activities that can be readily referenced. The main framework of the book is the life cycle and its various stages of concept definition, design and development, realization, operation and maintenance, and disposal or retirement. It is crucial to recognize the wide spread in life cycles, from very short where little time is allowed for achieving high dependability, such as many consumer products, to very long life cycles over many decades, such as much of our infrastructures, which stress the continual achievement and improvement that are expected.

We have tried to recognize that there is a very wide diversity of applications of dependability. In most of our examples, however, there is an acknowledgement of the rich legacy of the American Society of Mechanical Engineers (ASME), and thus the focus is mainly on industries related to power generation, monitoring and control, oil and gas, and the energy industry in general. This of course leaves out many other industries where dependability is equally important, but it is the view of the authors that the principles and methods are easily applicable elsewhere.

Thomas Van Hardeveld, P. Eng.
David Kiang, P. Eng.

ACKNOWLEDGEMENTS

We earnestly wish to share our collective years of experience as engineers serving global clients in a broad range of industry sectors and business enterprises. We are privileged to have been actively involved for many years in contributing to the development of standards for the International Electrotechnical Commission (IEC) Technical Committee (TC56) on Dependability. To this committee, we are thankful for the opportunity to share our technical expertise. The knowledge and experience captured in this book reflects the dependability requirements and technology advancement cultivated in the international market place. Dependability applies to all industry sectors in product development and service provision, technological and evolving systems, and legacy assets and equipment.

We sincerely wish to thank all our colleagues who have shared tremendous insights and knowledge over the many years of our involvement in IEC/TC56 Dependability. We wish to dedicate this book to our future generation of dependability engineers and practitioners, who are and will be continuing our endeavor to challenge technology evolution from a dependability perspective with confidence and trust for dependability value creation. The authors wish to express sincere thanks and acknowledge the contributions of the following colleagues in the preparation of this manuscript:

Cathy Hunsley, London Overground Infrastructure, IEC/TC56
Peter Kesselyak, IEC/TC56
Valter Loll, Chairman, IEC/TC56
Eric Mahy, IEC/TC56
Gerald Taranto, Taranto Consultancy, IEC/TC56

The authors thank the IEC for permission to reproduce information from its International Publication IEC 62628 ed 1.0 (2012).

We are immensely grateful to the ASME for agreeing to publish this book and their dedication to the field of dependability in their conferences and publications. We are especially grateful to the staff of ASME Press, Mary Grace Stefanchik and Tara Smith Collins, for their support, patience and encouragement in making this book a reality.

Finally, we would be remiss if we did not mention the support and indulgence of our spouses, Marilyn Ewald and Alice Kiang, who had to suffer our intense dedication to realizing one of our dreams.

FOREWORD

In our modern society, we are highly dependent on technological systems and innovative products, and often not by choice but by the need to use public facilities for essential services. We enjoy the conveniences of the Internet, e-commerce and wireless services. We rely on our electricity supply. We have cultivated renewable energy, but we still need the oil and gas to support our transportation systems. We are fortunate to live in a world at a time of immense progress, but we are often confronted with century-old infrastructure that requires maintenance and extensive support costs for their upkeep. This is where dependability engineering principles and practices are put into action.

The technical committee TC56 under the auspices of the International Electrotechnical Commission (IEC) has defined the term "dependability" as including reliability, availability, maintenance, maintainability and technological risk. Dependability thus encompasses the major aspects of our primary concerns of being dependent on such technological systems and utilization of available products and services for safety and security reasons. Dependability has become an indispensible notion for the survival of human species.

As the Chairman of IEC TC56, I welcome the book by Tom Van Hardeveld and David Kiang. Both Tom and David have long been associated with IEC dependability standards development work and have both received the prestigious IEC 1906 Award for their distinguished services and significant contributions to IEC/TC56. The book describes dependability aspects of large technological and evolving systems. It emphasizes the need for integration of hardware, software and human aspects for system development and implementation. The life cycle process goes beyond development, operation and maintenance until the system is retired from service. It also addresses our safety concerns, avoidance of risk and furtherance of a sustainable environment. To achieve an optimum balance between cost, schedule and performance, input from dependability specialists is needed. Industry at large will appreciate and realize the value of dependability in return on investment.

New systems are complicated and challenging. A first step is to describe the problem and put it into a proper perspective. The book contains many practical examples and good checklists. I hope that you will feel challenged and inspired while reading it.

Valter Loll, M.Sc., B.Com.
Chairman of IEC TC56 Dependability

ABOUT THE AUTHORS

Thomas Van Hardeveld has 38 years (as of 2012) experience in all aspects of the operation and maintenance of gas turbines, compressors and other gas transmission and process equipment. He is a specialist in maintenance management and condition monitoring of all types of equipment as well as reliability techniques and risk and integrity management. For more than 20 years, he has been involved in international consulting and training activities in Pakistan, New Zealand, Kuwait, Thailand, Argentina, Trinidad, Mexico, Malaysia, Kazakhstan and the Middle East. For the past 16 years he has presented training courses on maintenance management, maintenance planning and scheduling, Reliability Centered Maintenance, condition monitoring, compressors, gas turbines, reliability of rotating equipment and pipeline operation and maintenance. He has conducted maintenance assessments, performed Reliability Centered Maintenance analyses, implemented condition monitoring systems and consulted on various aspects of rotating equipment operation and maintenance. He is actively involved in standardization activities with the IEC/TC56 Committee on Dependability as a convenor and project leader and was recognized in 2010 by the International Electrotechnical Commission with the IEC 1906 Award. He has been an instructor for SAIT Polytechnic both in Calgary and internationally. He has co-authored a book on Pipeline Operation and Maintenance: A Practical Approach (now the second edition) and more recently one on Pipeline Pumping and Compression Systems, both published by ASME Press. He has a M.Sc. in Mechanical Engineering from the University of Calgary and is a registered Professional Engineer in the province of Alberta.

David Kiang has worked in the information and communications technology (ICT) industry for many years. He has dedicated his knowledge and experience to International Electrotechnical Commission (IEC) standardization work over the past two decades. David has contributed extensively in the development of IEC/TC56 standards on *Dependability*. He has coordinated the IEC/TC56 dependability seminars since 2006 and facilitated workshops in several countries on dependability of technological systems and related methodologies by international experts. These outreached programs had helped engineers and practitioners better understand the concept of dependability and enabling them to apply appropriate methods and tools to effectively manage dependability projects and resolve technical problems. David has provided quality and dependability engineering consulting services to multinational corporations and system design organizations on life cycle management, dependability development and implementation processes, and conformity assessment of products and services involving hardware, software and human aspects of dependability. David was appointed in 2002 by IEC/TC56 as Convener of *System aspects of dependability* responsible for development of relevant dependability standards to meet current market needs for industry applications. In 2004 David received the prestigious IEC 1906 Award for contribution to IEC dependability management standards. David Kiang is a Professional Engineer registered in Ontario, Canada. He is the Principal Consultant of *T.D. Kiang & Associates* that he founded in 1989. He retired as Senior Technical Adviser from *Nortel Networks* after 28 years of service. David graduated from the University of Toronto with a B.A.Sc. degree in Electrical Engineering. He received a Diploma in Management Sciences and a master degree (MBA) in Operations Research from the University of Ottawa. David is a Life Senior Member of IEEE.

CONTENTS

Chapter 1

INTRODUCTION TO DEPENDABILITY

1.1 WHAT IS DEPENDABILITY

1.1.1 Defining Dependability

Dependability is the ability of a system to perform as and when required [1]. Dependability is an inherent system property applying to any system, product, process or service involving hardware, software and human aspects. It is crucial to consider dependability in the development and application of technological and evolving systems. In today's global business environment, dependability is a key decision factor for evaluation and acceptance of successful system performance. Dependability represents customer objectives and values and determines critical system performance to earn user trust and achieve customer satisfaction.

Systems vary in form, fit and function, with performance complexity determined by user needs and often identified by types for application. Technological systems such as automated monitoring and control systems engage the design incorporation of interacting hardware and software elements to deliver specific system functions. In some cases, human interactions are involved and form part of the performance process in system operation. Functions are the building blocks of a system. They are arranged and configured to form the system architecture that facilitates interaction of relevant functions in order to deliver the required system output. A simple function is an elementary operation performed by the system to accomplish a task. A complex function can consist of several simple functions working together to accomplish a more complicated task or a sequence of tasks.

Evolving systems such as communications networks, power generation and pipelines have to adapt to changes with time as necessitated by technology advancements, new features and customer service demands associated with market competition. By their nature, evolving systems will contain legacy systems, which may involve mature technologies from previous generations. A non-evolving system may also at some point turn into a legacy system due to obsolescence. Dependability issues should be considered in all types of systems where maintenance and upgrades are necessary to sustain an acceptable level of performance for ongoing service.

The prime objective of a system is to deliver the required capability or performance. Dependability is a collective set of time-related performance characteristics achieved by design and implementation into a system. It coexists with other desirable characteristics such as quality, safety and immunity and is incorporated into a system to create added value by enhancing system performance for specific applications.

For technological and evolving systems, availability and its associated characteristics focus on hardware/software technology issues as they relate to dependability performance. Credibility of data and its associated characteristics focus on information throughput delivered dependably where applicable and required. Dependability as a performance attribute consists of two separate but interdependent measures [2], one related to hardware/software and the other to information integrity.

The main dependability characteristics focusing on hardware/software technology include:

- *availability* for readiness of operation;
- *reliability* for continuity of service provision;
- *maintainability* for ease of preventive and corrective maintenance actions;
- *supportability* for provision of maintenance support and logistics to perform maintenance tasks.

The advent of software technology and diverse information technology applications in products and services has precipitated industry needs to address dependability issues concerning data integrity and credibility of information throughput. The urgency is driven by e-commence and rapid adaptation of communications services for social networking. Privacy and security issues for dependency and usage of the information media have become paramount in the minds of people overwhelmed with the craze for virtual multimedia connections and communications. Additional application-specific dependability characteristics are needed to address information technology involved in system performance.

The dependability characteristics focusing on information technology include:

- *credibility* for identification and recognition of the system state to withstand incorrect inputs or unauthorized access;
- *data integrity* and credibility of throughput;
- *security* for protection from intrusion in application and use.

Figure 1.1 shows the relationships of dependability characteristics of technological and evolving systems.

Other application-specific dependability characteristics include, but are not limited to, the following:

- *recoverability* for system restoration following a failure with or without external actions;
- *safety* for prevention of harm in application and use;
- *durability* for long service life;
- *operability* for ease of user control and successful operation of system functions;
- *serviceability* for access and retention of system functions in application.

Not every system exhibits all possible dependability characteristics. Some characteristics are more dominant than others by design intent and incorporation. For example, a control system for a fire alarm should have high availability for activation upon sensing abnormal conditions with a low false alarm rate, and high reliability to sustain successful operation for the duration after activation, while maintainability and supportability are less important because the entire control system is designed as a throwaway item that can be readily replaced at low cost. The general observation that "dependability means many

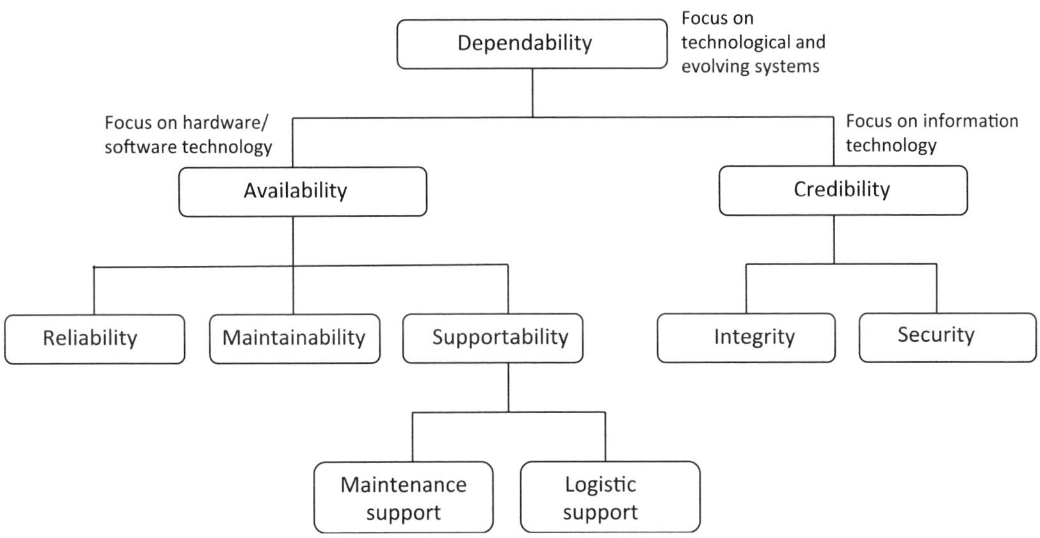

Figure 1.1 Dependability characteristics of technological and evolving systems

things to many people" is due to different expectations and diverse perceptions concerning the interpretation of dependability performance. This leads to the need for dependability standardization.

1.1.2 The Application of Dependability

The principles and practices of dependability can be applied to a wide range of products, systems, processes and services. The examples in this book focus on industries related to power generation, transportation, oil and gas, petrochemicals and telecommunications in which hardware, software and human aspects are involved, utilized or deployed at different levels of complexity. For simplicity, the application of dependability is addressed in three major categories: components, systems and networks, although what is considered to be in each category will depend on one's perspective. This is to facilitate description of the dependability as applied at three key levels of the system hierarchy. Dependability application often needs to address upward and downward, as well as forward and backward compatibility, and interoperability and interface issues.

Components are generally used as building blocks for system construction. Examples are electronic devices, smaller valves, control switches, compressor impellers or blades, and software application programs and module units. Major dependability characteristics are reliability, durability and maintainability to facilitate ease of access and assembly in component applications. Maintenance is typically carried out by component replacement or software update. Component technology is relatively mature. The focus of this book addresses issues concerning technology selection, supply-chain management and assurance of off-the-shelf products, including software bundles.

Systems are generally more complex with combinations of components assembled and configured to perform multiple tasks or functions. Applicable dependability characteristics entail availability, reliability, maintainability and supportability for hardware and software technology application. Human aspects can be addressed as an internal system function in performing a task, or as an external user of the system in service. For specific systems involving data transport or for financial transactions, the applicable dependability characteristics include credibility, integrity and security for information technology applications.

A network is a combination of various interconnecting and collaborating systems that facilitate a common objective in delivering a specific set of defined services. The major technological challenge is to achieve interoperability and connectivity of the systems. The service area may extend beyond physical boundaries, such as wireless communications via satellite, and may spread over wide geographical regions for delivery of transportation service, such as oil and gas pipelines. The systems that collaborate in the network evolve with time. The network may change in both structure and configuration with new systems being added to legacy systems. This may involve adapting physical assets and infrastructure to connect the legacy and new systems while at the same time sustaining operation during their incorporation to meet business needs. Applicable dependability characteristics include availability, reliability, durability and supportability, which are key to sustain continuous operation without major interruptions.

1.1.3 Industry Views on Dependability

There are major industrial sectors that advocate the importance of dependability and respond to technological changes in a global competitive environment. The application scenarios differentiate on how dependability is strategically deployed, effectively engaged and efficiently used in society, and identify specific industries where dependability is crucial to their innovation and business success. Recognizing these specific scenarios assists in understanding the nature of the different industry approaches and appreciating the value of dependability in addressing specific dependability application issues.

(a) Consumer electronics industry
The consumer electronics industry is a highly competitive and volatile market. Technology-savvy consumers are hungry for the latest widgets such as smart phones, laptops, tablets and a multitude of electronic gadgetry. The Annual International Consumer Electronics show attracts both technology professionals and potential product users to examine the latest gadgets. The advancement in semiconductor fabrication technology has drastically driven down the production cost for chip manufacturing and packaging, permitting chip incorporation for a wide range of applications in consumer electronic products. The advent of software, Internet and web applications allows new online social networking services to accommodate some sophisticated multimedia content in social communications. Consumer electronic products can be characterized by the need for ease of use, convenience, social acceptance and a relatively short life. New versions of personal computers are available on the market every two to six months with new features and attractions. The consumer electronics industry strategy is either to be a product leader to gather the premium price for the new innovation or to be a product follower to offer a much lower price for a similar product within a time lapse of two to six months for those who can wait. Because of product miniaturization, packaging size and weight limitations, most portable consumer electronic products are not designed for maintainability, with some exceptions to allow adding memory chips or inserts for plug and play connections. Products in the lower price range can be considered throwaway items. More expensive products may

cost more to repair than purchasing a new one. The standard manufacturer warranty for defective product return is one year; however, protection plans at a cost are offered by the product retailers to exchange a product for a new replacement in case of product damage. A secondary resale market has also been developed and is thriving in refabricated products for reuse. Product brand value is highly regarded as an indication of a popular brand that could fetch a premium. Product dependability from product user experience is closely linked to such purchase decisions.

(b) Transportation service industry

The public has come to expect very high performance levels from transportation services such as railways, air travel, mass transit systems in cities, transport of goods by trucks and rail lines, courier services and bus lines. For transportation services, there are two points of view for dependability: one by the customer and the other for the operator of the service.

The customer perceives dependability of service in performance as probably more important than the quality of service. This usually translates into on-time departure and arrival. It is now becoming common for transportation services such as airlines and trains to inform the customer, and even provide a guarantee, that certain reliability performance targets will be met. Performance may be described as achievement of punctuality within a window of acceptable time delay for a certain percentage of journeys. Customer satisfaction is a general measure, with dependability performance as a significant contributing factor. The very high levels of safety, now common in most transportation networks, mean that most travelers give little thought to how travel safety is achieved until an accident occurs.

Dependability has a much wider and complicated meaning for the operator of the transportation service. Here it applies to the many transportation-related assets that need to operate successfully in order to meet the needs of the customer. Using air travel as one of our essential services, on-time and safe travel requires the cooperation and interaction of not only the airline operator but also the supporting ground-based functions of the airport, air traffic control and even other aircraft that travel the same routes. The resultant complexity is a testament to the very high levels of dependability that can now be achieved and that all of us expect. In addition to having the same performance objectives as those of the customers, the operator has another significant objective, that of minimizing operating costs and being profitable. This objective is quite dependent on maximizing dependability in that reliability and availability will enable higher capacity to generate additional revenue. A complicating factor is that unwitting and ill-conceived cost reductions such as using less expensive equipment can easily impact dependability negatively in terms of downtimes. The operator of transportation services is responsible not only to the customers and their shareholders but also needs to comply with government regulations and transportation safety rules.

(c) Manufacturing industry

The manufacturing industry covers a wide range of industries from extracting mineral resources to the manufacture of products or goods in mass production. The reliability of the equipment and machinery used for the manufacturing process is linked to both production capacity and the quality of the manufactured product. Maintenance support of the equipment and machinery will enhance the reliability and availability of the entire production process. The manufacturing strategy is to sustain production operation with well-maintained equipment and machinery to protect capital investment and minimize sunk costs, such as maintaining idle unused equipment and keeping excess inventory of spare parts.

Loss of production is of major concern in the manufacturing industry because it links directly to production capability and capacity for revenue generation. The production quality process controls the product yields and limits the number of rejects. Customer liaison is essential to minimize the unnecessary customer complaints on failed product returns under warranty or in service operation. Dependability performance in production advocates control of product reliability and minimization of product returns for any reason in the context of customer satisfaction.

(d) Petroleum industry

The petroleum industry is a long chain of activities from the wellhead to processing to petrochemicals for both gas and liquid production and products. It includes the producing plants and facilities as well as their interconnecting transportation links, which may consist of pipelines and other means such as rail, sea or trucking for fuel transport. Dependability involvement in the petroleum industry needs to be considered from a network point of view as well as from an individual plant/facility/equipment perspective.

The major factor for dependability consideration in petroleum-related industries is production availability because this is tied directly to revenue generation. There are two different business objectives from a production perspective: maximizing production where production is capacity-constrained, and emphasizing production efficiency and cost minimization where it is market-constrained. In the first case, the temptation will be to maximize production, but with the risk that exceeding capacity could lead to increased overload failures. In the second case, cost control measures may jeopardize the production process reliability, which could result in increased maintenance support costs.

The safety of workers and the public in the petroleum industry is of prime importance. The link between safety and dependability performance is to ensure that high reliability is a primary requirement for safe operation.

(e) Power generation and the electrical grid

Power generation and the electrical grid are considered to be an essential service in a modern society highly dependent on technological innovation and up-to-date service facilities. Loss of electrical supply is not only a major inconvenience but can have drastic economic consequences.

The very nature of power demand is that there are peak times that come close to maximum available capacity. In a deregulated market, the individual power producer can charge customers a higher premium for electricity use at peak periods of energy demands and lower rates during non-peak periods. Since much of base load capacity is centralized close to the source of energy being used, the electrical grid is a crucial component of the delivery system. Measurement of the availability and reliability of the generating capacity and load or capacity planning is critical for the individual power producer and the grid operator. Dependability of service is crucial for sustaining customer quality of service.

(f) Major equipment supplies and services

Major equipment supplies and services are specialized equipment vendors providing industry with original equipment manufacturer (OEM) products and services. These equipment vendors have a fundamental stake in ensuring superior dependability in their products and services, since their continuing economic existence depends on such a business strategy. Their main role is to ensure that the vendor's equipment is on the industry procurer's preferred list of suppliers to increase equipment sales. To provide sustainable dependability of service, vendors have become much more aggressive and closely involved with their industry users or partners by means of long-term service agreements. The agreements include increased equipment monitoring of dependability performance and guarantees in provision of maintenance support services. Such agreements may have mutual benefits in a supply chain management process.

(g) General infrastructure and utility services
General infrastructure and utility services include public services such as water supply, storage and distribution systems, telecommunications and the Internet, and roads and bridges. As society becomes more dependent on such infrastructure and utility services in our daily lives, long-range planning and public consultations are often assembled and debated in an attempt to address and justify multi-year cost investments of public funds and to evaluate alternate services for possible trade-offs. The emphasis on the dependability of such infrastructure and utility services also increases as the public becomes more aware of the value of long-term dependability investments.

1.1.4 The Importance of Dependability

There are at least four major reasons why dependability is important to technological systems and products.

1. Technological systems are sophisticated and often engaged in mission-critical applications. Dependability plays a leading role in enabling technology selection, design incorporation and service enhancement to achieve system performance with integrity, safety and security protection while delivering functional capability and dependable services.
2. Today's technology is fast moving. Products incorporated with technologies have a much shorter life cycle in the consumer market than ever before. New innovations and features tend to appeal to product users. Dependability provides the essential technical approaches and methodology to entice business investments in innovation to adapt to changes and sustain a longer product life prior to technological obsolescence.
3. Customers expect value for money. Technological products acquired or purchased are expected to deliver dependability of service. Dependability sustains product value and permits new features and upgrades to maintain serviceability and operability in a changing market environment.
4. Technological systems are complex. The cost of ownership could carry a heavy burden. Unexpected failures are often viewed as potential revenue losses. The dependability process involves designing for system performance durability and reliability for fault avoidance and prevention. Dependability provides efficient supportability and reduces ongoing operation and maintenance costs.

The importance of dependability from an industry application perspective can be summarized as follows:

- Dependability is a key decision factor in business contracts and material procurements.

- Dependability provides a strategic framework for life cycle process management.
- Dependability policy drives technology advancements and innovation.
- Dependability promotes green technology for product realization.
- Dependability principles and practices support sustainable environments.
- Dependability methodology supports risk assessment processes.
- Dependability ensures safety, security and integrity in operation and maintenance.
- Dependability influences system performance achievements, resource allocations and life cycle cost implications.
- Dependability expedites the maturity processes in project implementation.
- Dependability advocates experience database capture and instigates knowledge acquisition and enhancement.
- Dependability supports the brand value of products.
- Dependability earns user trust and gains customer satisfaction.
- Dependability is a key enabler of asset management.

1.1.5 The History of Dependability

The genesis of *dependability* evolves from the first use of the word "reliability" by Samuel Taylor Coleridge, who bestowed the word on his friend, the poet Robert Southey, to praise his steadfastness [3]. From this seemingly insignificant usage of the term, reliability has grown enormously to a broadly accepted, if not entirely understood, property that everyone expects for a wide range of situations. Searches for "reliability" and related terms result in thousands of references in papers and manuscripts and literally millions of hits on the Internet.

The main pillars of reliability are the concepts of probability and statistics, which emerged earlier from the work of two Frenchmen, Blaise Pascal and Pierre de Fermat. The emergence of the need for quality became apparent with mass production [3], and this evolved into statistical quality control and later into statistical process control in the 1920s.

Reliability principles and practices became active as an engineering discipline around the 1950's, with a catalyst being the vacuum tube and the many failures that occurred. A key moment was the initiation of the Advisory Group on Reliability of Electronic Equipment (AGREE), jointly established in 1952 between the Department of Defense and the American electronics industry. The AGREE Report of June 4, 1957, provided all the armed services with the assurance that reliability could be specified, allocated and demonstrated. The reliability engineering discipline had thus come to existence. The first conference on quality control and reliability (of electronics) was held in 1954, and its proceedings evolved into a journal that is still being published as the IEEE Transactions on Reliability. Another important development was the work of Wallodi Weibull, who pioneered the flexible statistical distribution that now carries his name.

Reliability came into further prominence in the 1960's when many Mil-standards and specifications were developed to meet the needs of design and implementation for defense production in the USA. Worldwide industry acceptance of the Mil-standards was noted as the leading source of the reliability knowledge database. The most well-known reference is the Mil-HDBK-217 *Reliability prediction of electronic equipment* [4], which has been adopted in many countries and is used by industry organizations [5] as the framework methodology and basis for failure rate estimation. Other methods on testing, reliability growth and reliability analysis have their origins in the MIL-standards.

Since systems do fail during operation, maintenance and maintenance support are needed in order to sustain continual operation. Integrated Logistic Support is practiced in some industry sectors to provide system readiness and enhance availability performance in system operation.

The advent of software technology in the 1980's for system integration precipitated the need for more rigorous processes to be established in capability maturity models [6, 7] and software reliability engineering [8]. Human factors engineering [9] and ergonomics [10] recognized that human–machine interfaces are issues of concern and probable causes of system malfunctions. The events of system failures and the resultant service impact have extensive influence on dependability performance if not appropriately addressed from a system life cycle perspective.

Reliability engineering now encompasses statistical methods, techniques such as FMEA and Fault Tree Analysis, physics of failure, hardware, software and human reliability, probabilistic or quantitative risk assessment and reliability growth and prediction, to name only a few. Databases of information have been widely established and their use has increased dramatically. Practically every engineering discipline has a focus on these aspects as a key component of business success.

The term "reliability" now has a much broader meaning and includes not only the specific meaning of reliability as the probability that something may fail but also related concepts of availability, maintainability, supportability, safety, integrity and a host of other terms. This has led to a proliferation of aggregate terms such as R&M (Reliability and Maintainability), RAM (Reliability, Availability and Maintainability), RAMS (where the additional "S" is safety or sometimes supportability) and Dependability, which is used by International Standards.

On the international scene, the IEC (International Electro-technical Commission) established Technical Committee 56 (TC56) in 1965 to address reliability standardization in response to a German proposal in 1962 that was approved by the IEC Committee of Action in 1964 [11]. The first meeting of IEC/TC56 was held in Tokyo, October 1965, under the Chairmanship of France and the Secretariat held by the USA. The title of IEC/TC56 was initially "Reliability of Electronic Components and Equipment." Different countries have since hosted the annual IEC/TC56 meetings, with the participation of many international delegates to develop reliability standards. In 1980, the title was amended to "Reliability and Maintainability" to address reliability and associated maintainability characteristics applicable to products. In 1989, the title was changed to "Dependability" to better reflect the technological evolution and business needs of a broader scope of applications based on the umbrella of dependability. In 1990, following consultations with ISO (International Organization for Standardization) , it was agreed that the scope of IEC/TC56's work should no longer be limited to the electro-technical field but should address generic dependability issues across all disciplines and for all types of equipment and systems. The scope of the IEC/TC56 Strategic Business Plan covers the generic aspects on dependability program management, testing and analytical techniques, software and system dependability, technical risk assessment and life cycle costing. The IEC/TC56 dependability standards published or under preparation as of 2012 include an extensive list of core standards addressing dependability management; process standards on dependability engineering guidance for systems and network application, reliability-centered maintenance and integrated logistic support; and support standards covering test methods, analysis and prediction techniques and dependability assurance.

The IEEE-CS Technical Committee on Fault-Tolerant Computing was formed in 1970 and laid the foundation for development of dependability concepts and terminology for the computer industry [12]. The dependability concepts of fault-tolerant computing systems consist of three parts: the first part on attributes addresses availability, reliability, safety, confidentiality, integrity and maintainability; the second part on threats includes identification of faults, errors and failures; and the third part concerning means for mitigation presents methods for fault prevention, fault tolerance, fault removal and fault forecasting.

The dependability concepts of fault-tolerant computing systems have been adopted by the computing and software industry for incorporation in software engineering practices.

1.1.6 The Evolution of Dependability

The evolution of dependability is highly influenced by technology advancements and is fostered by the rapid adaptation of new technology in industry applications. Consumers are eager for dependable new products.

Figure 1.2 shows the technology trends that influence dependability in its evolution process.

Technology comes in waves as new innovations and novel processes are discovered and adopted for industry applications. These waves are mirrored in advances in dependability techniques. The first wave shows the gradual maturity of hardware reliability from 1950 to 1990 due to improved manufacturing processes and quality control methods. The second wave shows the steady climb of system dependability from 1980 to 2000 due to software integration to enhance the performance of combined hardware-software systems. The third wave shows the rapid expansion of network dependability from 2000 onwards as technology convergence and infusion becomes realizable and practicable for e-commerce and different forms of industrial and social communications. These networks encompass

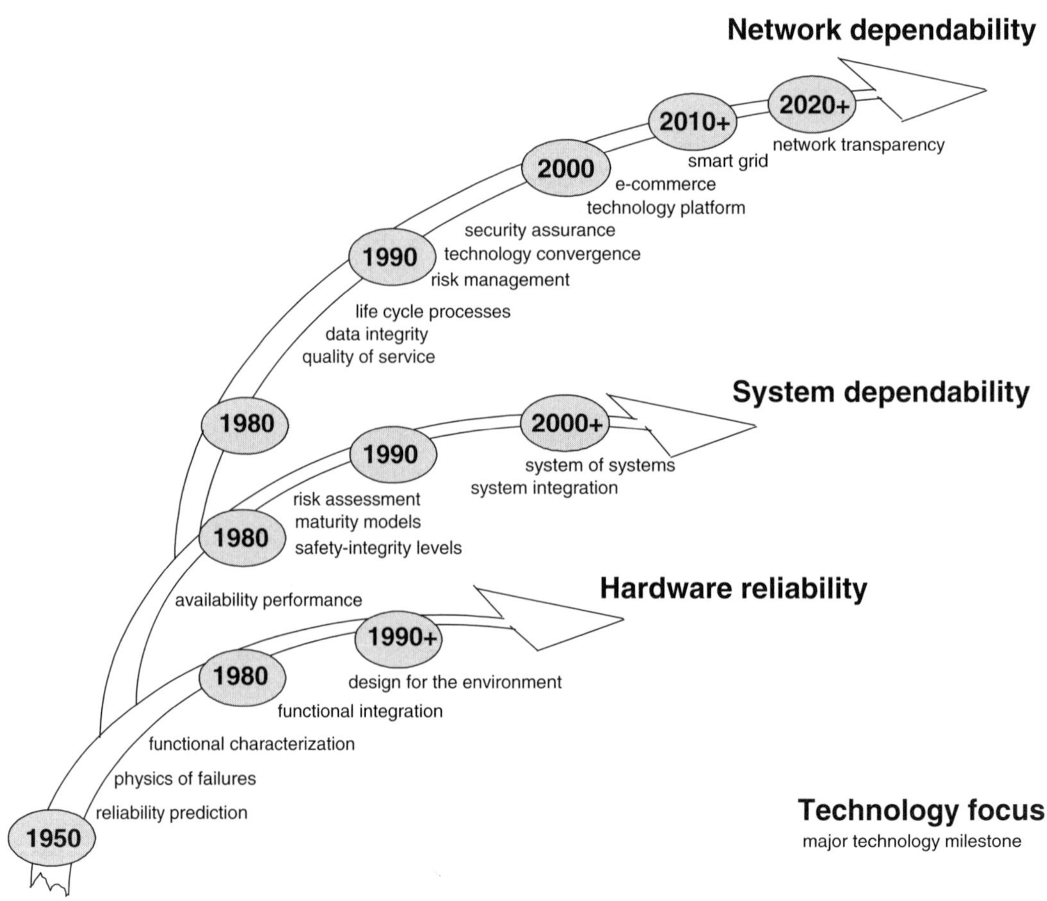

Figure 1.2 Dependability technology trends

both physical entities such as energy transportation and virtual ones such as communications and data transfer and access.

The use of dependability in engineering practice has matured over the past decades. Dependability has evolved into a formal technical discipline for industry adaptation and engineering standardization. Because of the broad scope of the dependability methodology, it facilitates the applications in new technology innovation for implementation of diverse systems such as the Smart Grid initiatives around 2007 to address issues concerning dependability of interacting systems and interoperability of networks. The Smart Grid is an international collaboration of industries and governments to modernize the existing electric grids due to aging infrastructure and intermittent generation by current energy sources.

The Smart Grid utilizes technological products and services to:

- facilitate network connectivity and interoperability of equipment;
- allow consumers to optimize their own electricity consumption;
- enhance communications with intelligent monitoring and control systems and self-healing technologies.

The Smart Grid is aimed at:

- providing sustainable and economical generation and supply of electricity;
- delivering enhanced levels of dependability performance and security of service;
- reducing significantly the environmental impact of the whole electricity supply system.

Smart Grid development involves the coordination and cooperation of IEC, NIST (National Institute of Standards and Technology), IEEE (Institute of Electrical and Electronics Engineers) [13, 14] and other organizations to establish and implement applicable Smart Grid technology standards.

In today's global business environment, network developers and service providers are intensely involved in volatile market competitions. This leads to frequent repositioning of their business ventures with innovative gadgets to satisfy the ever-growing service demands. For example, the technological products' life cycle for some personal communications and computing devices are relatively short, and they require early planning for their obsolescence and succession. Dependability of service is at the top of every business agenda to sustain reliable operation.

1.2 DEPENDABILITY CONCEPTS

1.2.1 Dependability Principles and Practices

The concepts of dependability from an application perspective can be presented as a set of underlying principles that cover a collection of broad-based values, rules and assumptions. These underlying principles lay the foundation for the dependability engineering discipline and practices. Table 1.1 presents an example of a set of dependability principles and practices. They may be adapted in different ways to fit actual system requirements.

The driving mechanism to achieve superior dependability performance is derived from a set of relevant dependability characteristics that are selected from both the main dependability characteristics and the application-specific dependability characteristics.

The main dependability characteristics are intrinsic in the design process and are related to aspects such as material composition, manufacturing processes and performance

TABLE 1.1 Dependability principles and practices

Dependability Principles	Dependability Practices
1. Dependability consists of unique performance characteristics represented by intrinsic main characteristics and application-specific characteristics	1. Dependability characteristics specific to application needs are selected and incorporated into an application to achieve the desired performance outcome
2. Dependability is time-dependent and adaptive to operational changes and environmental conditions	2. Dependability characteristics relevant to system performance requirements, such as robustness and long life, are designed and evaluated to meet actual life cycle applications
3. Dependability is a technical discipline and is managed by sound business principles in a risk-based competitive environment	3. Dependability embraces the system approach for design and implementation of technical processes for optimal return on investments
4. Dependability epitomizes value and trustworthiness	4. Dependability promotes the creation of value to establish performance, integrity and user confidence
5. Dependability embodies technology solutions to alleviate performance problems	5. Dependability methodology instigates appropriate techniques for fault avoidance, fault tolerance and fault/failure forecasting
6. Dependability promotes safety	6. Dependability enhances safety during all phases of the life cycle, including design, installation, operation and maintenance
7. Dependability advocates "green" technology in design selection and application	7. Dependability activities promote a reduce, reuse and recycle policy in designing for the environment

requirements. They are based on hardware product development concepts and can be extended for use in system applications with some limitations.

Application-specific dependability characteristics are oriented toward achieving specific performance objectives such as safety, security and integrity, which are incorporated into a system along with the primary performance characteristics derived from the system requirements. In practice, system requirements consist of a combination of functional and non-functional requirements that identify specific dependability characteristics incorporated into a system by the nature of its design and implementation.

Functional requirements specify the desired design outcomes, such as system performance, product features, process capability and deliverables. These identify what a system is supposed to do by defining specific functions or tasks. The plan for implementing functional requirements is addressed in the system design. The functional requirements drive the application aspects in terms of the actions the system must perform.

Non-functional requirements specify overall system characteristics such as availability, serviceability and supportability. These non-functional requirements establish the criteria that can be used to determine or evaluate the adequacy of system performance [15]. The system architecture has significant impact on fostering the dependability of the system and should be considered in the dependability plan. The non-functional requirements drive the technological aspects on how well the system must perform.

Non-functional requirements are often referred to as "qualities," "quality of service," "application-specific" or the "ilities" of a system. Dependability is a time-dependent characteristic among the extensive list of "ilities" definitions [16]. The modifier term "non-functional" is often misunderstood to imply a negative connotation. In the context of dependability engineering, the term "application-specific" is used to present a positive image. For technological systems application, the examples of "application-specific" characteristics such as safety, security and usability are classified as "execution qualities," which are observable during system operation. Characteristics such as maintainability, testability and scalability are classified as "evolution qualities," which are embodied in the static structure of the system. They change only when the system configuration is updated or the architecture has evolved.

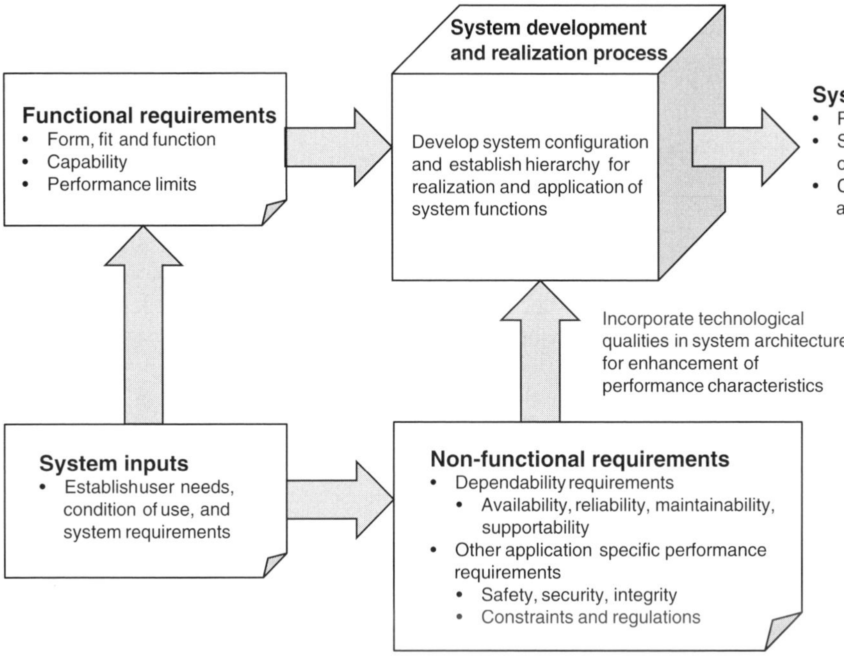

Figure 1.3 Functional and non-functional requirements for the system development and realization process

Figure 1.3 illustrates the relationships and process of functional and non-functional requirements for system development and realization. The dependability requirements are highlighted as non-functional requirements.

The black box represents the system as a tangible deliverable item. Functional requirements drive the functions as per the design and as realized by its implementation. These functions interact according to a hierarchical structure and the system configuration to perform specific tasks, which deliver the required system output. The tangible output delivered by the system is the product or service required to meet customer needs.

Non-functional requirements drive the technology selected for incorporation into system functions. The process results in the achievement of specific desirable performance characteristics such as reliability and maintainability and verification and validation of acceptable system performance. The intangible output delivered by the system is how well the system performs as experienced by customer.

If the system outputs as perceived or experienced by the users or customers meet or exceed their expectation, then the result of the process is that customer satisfaction has been achieved.

1.2.2 Dependability Concept Realization

The conceptual idea of dependability needs to be harnessed and developed before it can become useful for practical applications. Transforming dependability principles into practice requires decisive actions and the implementation of appropriate technical processes. A successful transformation process is guided by a strategic approach and fostered by sound business management practices.

The transformation process involves four basic steps:

(a) *requirements definition*: identifying dependability needs associated with the system application in terms of operating environment, technology constraints, failure consequences and user expectations;
(b) *requirements analysis*: determining operational scenarios related to expected dependability performance by defining system failures, probable risk exposures, maintenance support strategies, project delivery constraints and budget limitations;
(c) *design configuration*: identifying feasible make-up of system components, structure and interfaces for allocation of performance functions, dependability apportionment, system partitioning, protocol establishment, maintenance access and environmental impact;
(d) *functional evaluation*: analyzing and evaluating functional design to determine hardware/software composition, realization of functions, make-buy decision process, test verification and evaluation and replacement and restoration schemes.

The results of the transformation define the scope and limitations for system development and application. The transformation process ensures that relevant dependability characteristics are embedded in the system as part of its performance and functions.

1.2.3 Dependability Knowledge Base

Dependability is technology driven. The dependability knowledge base is built on experience and creativity in a broad range of technical applications. In some applications, dependability delves into the breadth and depth of technology innovation to seek optimal solutions through incorporation in system performance. Dependability should add value to a system when correctly implemented. If wrongly applied it would be a waste of time and effort, some with negative consequences. In practice, it is essential that the nature of the application, the time and budget constraints, the availability of resources and other influencing factors should be taken into consideration.

In all cases, the definition of failures associated with system performance should be established. Failure symptoms should be observed and recorded where possible. By understanding failures and their consequences, it will greatly enlighten the failure diagnostic process to find suitable solutions for dependability improvement. For example, the knowledge of physics of failure will help produce a more reliable integrated circuit chip; the understanding of manifestation of software faults will lead to fault-tolerant fail-safe design; the incorporation of robustness and redundancy techniques in system configuration will result in higher availability performance; and field performance studies will provide the rationale to justify NFF (no fault found) ratios on field returns for logistic support improvement. Human factors have to be considered with respect to failure mitigation. The task for dependability assessment of system performance demands technical know-how and timely effort to achieve successful results. The dependability analysis data must be meaningful to support technical recommendations.

1.3 THE SYSTEM APPROACH TO DEPENDABILITY

1.3.1 Definition of a System

A system is a bounded physical/virtual entity consisting of interacting elements operating in an environment to achieve a defined objective. The system concept is illustrated in Figure 1.4.

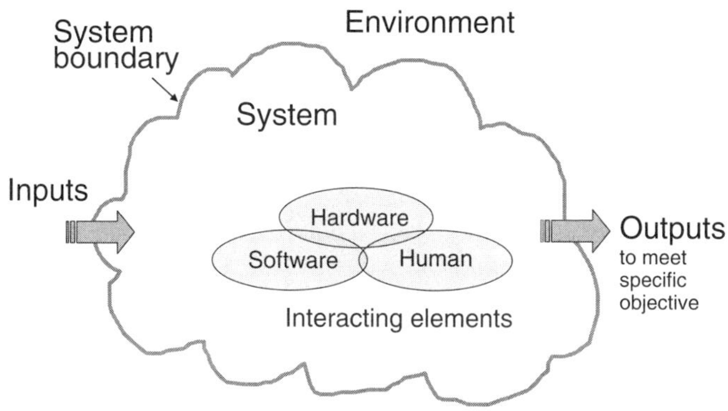

Figure 1.4 Illustration of the system concept

A system may consist of combined hardware and software with human interactions to realize the delivery of desirable performance in a specific application. The performance of the system will exhibit dependability characteristics designed into the system. For example, designing for high reliability in a safety critical system is desirable, but inadequate maintenance and software upgrades may trigger undesirable control system false alarms for equipment protection. A system may also represent a set of software-controlled processes deployed in various hardware equipment and subsystems in order to enhance human operation for delivery of the required functions.

A system may be simple, such as a digital timer, or complex, such as a distributed functional safety control system. The majority of systems in use are generally repairable throughout their life cycle for economic reasons and practical applications. Inexpensive simple systems such as COTS (commercial-off-the-shelf) products may be non-repairable throwaway items by design to meet market and business requirements.

A system can link to other systems to form a network. The boundaries separating a product from a system, and a system from a network, can be distinguished by defining the application of the entity of interest. For example, a digital timer as a product can be used to synchronize the operation of a computer; the computer as a system can be linked to other computers via the Internet to form a network. The environment surrounding the system boundary represents the application conditions of system operation. The environment may change according to the operational scenarios for completion of the mission or upon entering different stages of the system life cycle process.

Open systems are systems with permeable boundaries that permit external interactions with other compatible systems in the network to achieve interoperability and accomplish mission-critical performance objectives. Examples of open systems include distributive networks, e-commerce, Cloud computing services, and Smart Grid. Open systems present new challenges for dependability engineering to address availability performance, integrity impairment, security violations, service interference and service agreement disputes associated with open-system networks.

An example is that of gas transmission as illustrated in Figure 1.5. Pipelines transport natural gas from the wellhead through a gathering system to a mainline transmission pipeline for delivery to customers. The prime dependability characteristic for a pipeline is long-term availability and structural integrity. A small group of pipelines may be considered to be a system. Various pipeline systems will be interconnected to form a network from the original

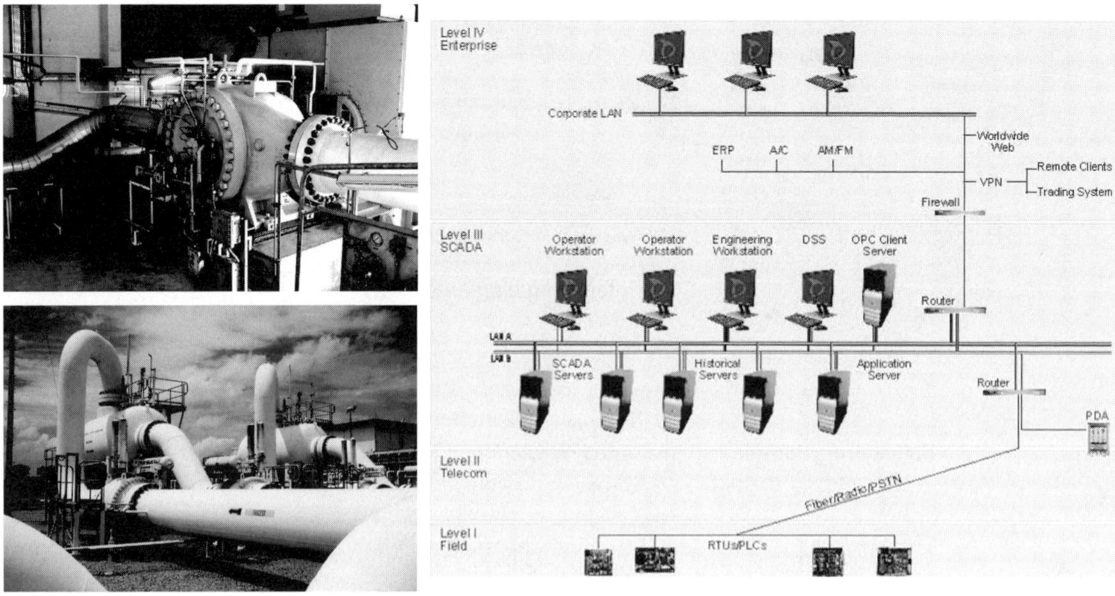

Figure 1.5 Example of a compressor system and gas transmission network

supply to the final consumer. Each pipeline system is normally controlled and monitored by a control center that is heavily dependent on information technology and communications networks for its operation. Much of the control is automatic, but the human aspect is still crucial. Cooperation between these control centers is critical. A continuous and dependable supply of gas is the main objective of both the gas transmission companies and the end users.

Along the way, compressors are utilized to reduce the volume of the gas in order to minimize the size of the pipeline and to overcome friction losses. Many hardware and software technologies converge for a compressor unit, which consists of a driver, such as a gas turbine and centrifugal compressor, and their auxiliary systems. Most of the operation is automatic, but manual operation is still required at times. High availability and reliability are paramount and are very dependent on excellent maintainability and supportability. The compressor unit may be considered to be a system with many sub-systems and lower levels of equipment and components.

This is but one example of the extensive infrastructure on which our society depends. This example can easily be transferred to the electrical grid and power generation, water supply systems and any of a multitude of transportation systems.

1.3.2 Life Cycle Approach to Dependability

Taking a life cycle approach to dependability is essential to achieving dependability. The precursor to successful operation is proper design and implementation. Dependability characteristics are affected differently at the various stages of the life cycle and have to be managed appropriately.

A generic system life cycle process is depicted in Figure 1.6.

The system life cycle follows a sequence of process stages from concept/definition, through design/development, realization/implementation, operation/maintenance and enhancement, and finally to system retirement or termination of service. The transition

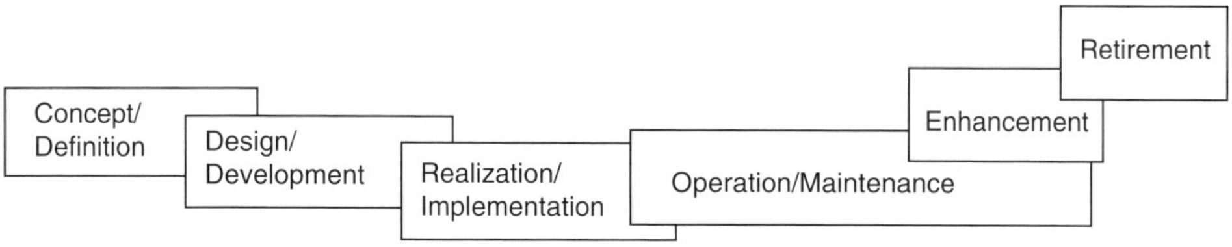

Figure 1.6 System life cycle process

from one stage to the next presents an opportunity for management reviews to support the decision-making process in order to advance the project forward. There are no cookbook answers on how to select dependability characteristics for system incorporation, or suitable engineering templates that offer precise dependability solutions. System dependability is achieved by due diligence and prudent application of relevant technical disciplines, skills and resource allocation, and product knowledge supported by practical experience.

Some major dependability programs, such as reliability growth, field tracking, spares provisioning and integrated logistic support, require extensive dependability participation. These programs are created to support major project life cycle management objectives with long-term resource and facility commitments. Many of the project activities require close coordination with suppliers and customers to deal with contract compliance, warranty conditions, incentive and penalty resolution and third-party support requirements. The cooperation effort often requires information sharing and activity collaboration through negotiated processes. These dependability program activities should be recognized in the planning of dependability strategy and incorporated in the management framework to assess their cost-effective operations.

1.3.3 Methodology and Process Application

The selection and application of processes for engineering dependability into a specific system are carried out by the project tailoring process and guided by management direction. The selection of applicable methods can proceed with the knowledge of criteria and the understanding of their significance for achieving system dependability. The objective is to utilize these methods to incorporate relevant dependability characteristics into system functions. The methodology for implementing dependability into system functions can be viewed from two perspectives:

- a top-down approach to synthesize system dependability based on specified system requirements and market information to develop the system architecture;
- a bottom-up approach to build dependability into system functions based on dependability design rules for simplification, fault tolerance and risk reduction and mitigation.

Both approaches involve the identification of dependability characteristics and determination of their values. Dependability characteristics are the fundamental measures for assessment and achievement of system dependability.

Most system functions in technological systems utilize combined hardware and software elements in system designs. They offer a broad range of design features for diverse applications. Dependability of system functions is achieved by incorporation of design rules and established processes for applications. Design trade-off can be attained by proper combination of technologies suitable to meet specific application needs. Economic values can be gained through modular packaging and standardization for mass-scale production. System functions can be automated for self-checking to improve performance effectiveness by means of built-in tests or other monitoring schemes. Software-driven system functions can provide an added advantage by reducing direct human involvement in system operation. In this respect, human intervention in system functions may only be necessitated by safety and security regulations.

There are three generic approaches to determine that system dependability has been or can be achieved. They serve different purposes with varying degrees of engineering rigor. In practice, a combination of these approaches is likely to be used.

(a) *Demonstration* – this is achieved by means of actual system operation in an application environment over a scheduled time period to demonstrate dependability performance.
(b) *Inference* – this is achieved by means of statistical methods using observed data of constituent system functions based on established criteria and assumptions to arrive at quantitative values representing system dependability performance characteristics.
(c) *Progressive evidence* – this is achieved by progressive accomplishment of project milestones with auditable arguments to support objective evidence.

It should be noted that all testing programs are labor intensive. Pass-fail criteria and test conditions should be well established prior to test initiation. The test objectives should be clearly defined and understood for each test case. Test cases are developed to simulate actual system field operating conditions in which specific interest areas or potential problems may be encountered. A test case is a set of test inputs, execution conditions and expected results developed for a particular testing objective. Test case selection should be based on test performance accuracy to represent usage or application conditions. Test case execution should be able to establish the impact of failure contributions relevant to the test performance objectives. In some cases, it would be prudent to run pilot test projects to verify their test effectiveness prior to full-scale test program implementation.

For software-intensive systems, the dependability assessment process should focus on software assurance activities, which are systematically deployed for dependability analysis and testing during the software development life cycle. The objective of process implementation is to ensure software system maturity and dependability achievement by test evaluation to improve reliability growth. The assessment process is the enabling mechanism to ensure verification of software requirements and validation of software dependability results.

1.3.4 Hardware Aspects

Hardware constitutes the building blocks of the system's physical structure and is often dictated by size and weight limitations. Hardware elements consist of mechanical, electrical, electronic, optical and other physical components that are used in various configurations to realize hardware functions. Most hardware products built today are relatively mature in terms of their technology. Design rules and standards are well established. Hardware products will exhibit consistency in dependability when there is a controlled manufacturing process environment. Product quality and dependability can be ascertained by appropriate assurance programs. There is ample practical previous operating experience and failure rate data to support reliability performance of these hardware-based products.

However, some products with active electronic components are sensitive to varying application environments. The physics of failure of these components dominates the hardware failure and infant mortality phenomenon. Proper reliability design, packaging and screening can help to significantly reduce early failures. Some hardware elements may wear out due to operation or extensive use, while others may have a limited shelf life. These inherent reliability problems can be resolved by implementation of preventive maintenance efforts. Hardware system structure is hierarchical. Maintenance support can be assisted by proper functional design and packaging strategy of the lowest replaceable assembly or unit. This facilitates maintainability design and logistics support activities to improve system availability performance.

Hardware assemblies can be designed in-house, outsourced or purchased as COTS products. They are used extensively as building blocks for practical design and engineering expediency. This leads to the need for supply-chain management that affects the make-buy decisions, outsourcing and subcontracting schemes, verification and validation procedures and documentation, and monitoring and assurance processes. The significance of supply-chain management is based on purchaser-supplier cooperation and sharing of relevant information in the procurement and acquisition process. The supply chain provides the necessary linkage for tracking important information. The impact on business is expediency in the administrative process, reduction in provisioning costs and incentives for delivery of quality products and services.

1.3.5 Software Aspects

The increasing trend of software applications in the service industry has resulted in the rapid growth of Internet services and Web development. Standardized interfaces and protocols have enabled the use of third-party software functionality over the Internet to permit cross-platform, cross-provider and cross-domain applications. Software has become a driving mechanism for realizing complex system operations and enabling the achievement of viable e-businesses based on seamless integration and enterprise process management. Software design has assumed the primary function in data processing, safety monitoring, security protection, and communication links in network services. This paradigm shift has put the global business community in a situation of relying heavily on the software systems to sustain business operations. Software dependability plays a dominant role by influencing success in system performance and data integrity.

Software is a virtual entity. Software refers to procedures, programs, codes, data and instructions for system control and information processing. A software system consists of an integrated collection of software items such as computer programs, procedures and executable codes that are incorporated into the physical host of the processing and control hardware to realize system operation and deliver performance functions. The hierarchy of the software system can be viewed as a structure representing the system architecture and consisting of subsystem software programs and lower-level software units. A software unit can be tested as specified in the design of a program. In some cases, two or more software units are required to perform a software function. The system encompasses both hardware and software elements interacting to provide useful functions in rendering the required performance services.

In a combined hardware/software system, the software elements of the system contribute in two major ways:

(a) operating software that runs continuously to support hardware elements during system operation;
(b) application software that runs as and when required upon user demands for provision of specific functions or services.

Dependability analysis of software sub-systems has to consider the timing requirements of the software application in the system operational profile as well as those software elements required for full-time system operation. Software modeling is needed for reliability allocation and dependability assessment of software-based systems.

1.3.6 Human Aspects

The human aspects of dependability consist of two distinct technical disciplines: human factors engineering/ergonomics and human reliability. Dependability management should exploit both disciplines for dependability value creation.

Human interactions during system operation can be viewed as a part of system functions or as it relates to an end user of the system. The role of the human in system performance can be beneficial with the human's ability to mitigate or control ongoing situations. However, most industrial incidents reported and major accidents studied can be traced back to human errors as the primary cause of system malfunction or disruption. Systems designed for human operation or use should incorporate human factors in the system design to minimize the risk of critical system failures, loss of property, security violations or safety threats. Dependability can be achieved by application of human factors engineering in design rules and simplification of tasks for human operation. The study of human factors involves a multi-disciplinary effort on gathering information about human capabilities and limitations for applications affecting human-system performance. The engineering aspects consist of the application of human factors information to the design of tools, machines, systems, tasks and jobs to provide an environment for safe, comfortable and effective human use. Training and education are important prerequisites for any system operation requiring human interaction. Human factors standardization facilitates system integration, enhances interoperability of system elements and improves serviceability and overall dependability performance.

The study of human reliability and application aspects associated with dependability technology should be exploited for better understanding and further development to achieve application benefits.

1.4 DEPENDABILITY MANAGEMENT FROM A BUSINESS PERSPECTIVE

1.4.1 Business Life Cycle and Market Relevance

Businesses are generally concerned with investments and operation to achieve profitable returns. A typical business life cycle profile is shown in Figure 1.7.

To explain the business life cycle profile, initial investments are injected for new product development and introduction, sales promotion and business growth to reach market maturity. Revenues are generated from product sales volume until market saturation. Market competition and product obsolescence affect product sales as volumes decline. The dependability application in practice closely follows the business life cycle profile for timing of life cycle process implementation. The cost of operation is minimized by mitigation of product malfunctions, product warranty and support services. The break-even point for net profit is very sensitive to delays, many of which are caused by dependability and quality problems.

This is a simple explanation of the business "growth-saturation-decline" cycle and market relevance for timing of product life cycle and renewal strategy. It also instigates

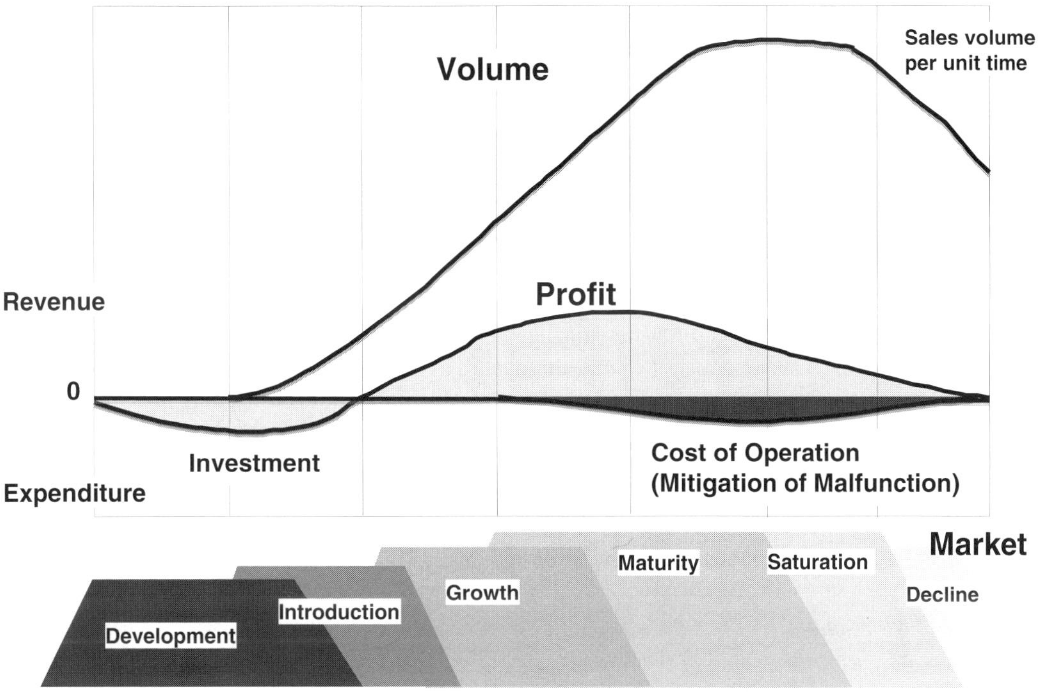

Figure 1.7 Business life cycle profile

the need for product/market diversification to sustain viable business operation. Market relevance from a generic business viewpoint is to provide the needed products and services that are timely and affordable by customers in the market place. Market relevance from a dependability management perspective is the development of relevant policy and efficient processes to support cost-effective achievements in product performance success that benefit industry applications.

1.4.2 Dependability Management Goals and Objectives

Organizations that embrace dependability principles and engage in proper dependability engineering practices are in the business of *dependability value creation*.

Dependability management is an organized approach for addressing dependability and related issues from a business-oriented perspective. Dependability is technology driven and greatly influenced by market dynamics, global economics and resource distributions, changing customer needs and a competitive business environment. Business strategies often need to adapt to anticipated changes in order to sustain viability in business operations. Dependability management should focus on stakeholders' interests in maximizing dependability values to enhance business objectives and increase return on investment. Management issues pertaining to achieving dependability should be considered in establishing a framework for managing dependability and developing a strategic business plan.

Value creation, both tangible and intangible, should be the primary goal of dependability management. Management objectives should focus on retention of dependability knowledge, cultivate and nurture available dependability expertise and resources, and establish realistic targets for achievements of dependability performance in products and services to gain user trust and customer satisfaction.

1.4.3 Changing Market Needs

The dependability management framework should maintain its agility and flexibility to accommodate business changes to reflect market relevance and needs. Dependability management strategies should advocate dependability value for successful performance and trustworthiness in the delivery of dependability of service. Change management should be addressed where appropriate in the formulation of dependability management directions and technical thrusts. Change management in the broadest sense reflects the transition from existing management policy and approaches to strengthening development of new initiatives.

Current market trends relevant to dependability applications may affect the way business is conducted.

(a) Economic trends worldwide have forced industries to make their operations more efficient and more flexible in order to stay competitive and sustainable. This has precipitated the growth of manufacturing and service industries in developing countries due to a disparity in material and labor costs. Current quality control and assurance methodologies have matured and become easily transportable from one region to another to set up the needed facilities and train staff for operation. Dependability has a limited role to play once the product has completed its design and testing for manufacturing and the essential service plans are well established for implementation. Business investments have diversified product manufacturing to selective worldwide locations due to scarcity of expertise, and there is a need to expedite product distribution and promote regional sales where the specific markets exist. This also reduces material transportation cost and service cost by instituting locally trained maintenance support personnel to attend to customer service needs.

(b) Fast-moving technologies and short product life cycles are affecting dependability design efforts based on traditional development process and product line manufacturing methods. Time to market often dictates the target schedule for competitive new product introduction. The ability of some companies to develop and introduce multiple products to meet market demands is accomplished by focusing on technology platform development. A technology platform permits realization of multiple products by using a platform supported by one technology to launch multiple products and manifest their product differentiation and deployment, hence satisfying the diverse market needs. Dependability design efforts should be focused on a technology platform to maximize return on investment. This adaptation in business strategies has resulted in the emergence of design-focused centers to consolidate scarce technical resources. Joint ventures and technology licensing are common practices in today's business arrangements and contract agreements to accelerate target market entries.

(c) The enormous growth of Information and Communication Technologies (ICT) over the past decades has triggered numerous innovations and fostered product evolution and enhancement of new product applications. The convergence of communications, computer and multimedia technologies involving a common architecture in system design is intended to ensure dependability performance and integrity in data transfer. Common assessment tools can be realized for assurance of safety, security, quality and dependability of network products and services. The integrity concept can be exploited to develop or utilize the needed technologies for realization of the integrity characteristic suitable for adaptive risk mitigation and containment of ICT product applications. Integrity is the assurance and credibility of information system data throughput to reflect trustworthiness. Dependability management should set guidelines for the extent of dependability involvement. A typical example is in the development of network dependability methodology for end-to-end communications in the provision of network services. To sustain a viable business in network services, it is prudent in the communications industry to provide the needed network service functions, adequate network capacity and performance capability, security of service, quality of service and dependability of service. Dependability has a primary role in the delivery of the needed dependability of service in communication network services.

(d) Growing concerns with security, health and safety, and environmental preservation necessitate dependability value creation to sustain societal confidence and timeliness in delivering relevant dependability of service through implantation techniques. Dependability may play a pivotal role to enhance service applications. For example, RFID (radio-frequency identification) as a technology is used for e-commerce and personal identification for control purposes. The RFID chip is small and non-intrusive, and it might fit the bill for an environmentally friendly application. With advancements in mobile computing and Web technologies, RFID is extensively used in enterprise supply-chain management to improve the efficiency of inventory tracking and asset management. It is also deployed for animal implantation to locate and track movements. The RFID fits the profile of rapid technology adaptation to worldwide industry applications beyond simply replacing the existing optical bar-code identification scheme. However, RFID is vulnerable to security hacking and identity theft in some applications due to the inadequacy or lack of protection mechanisms when used. This affects the integrity of user identification, as this information might be misused or stolen. This is one opportunity for dependability value creation to devise some means of protection with wide-ranging benefits and contributions.

1.4.4 Dependability Standardization for Evolving Systems

Evolving technologies are new and emerging technologies that are rapidly adapted for application in a broad range of form, fit and function in industrial and commercial products and services. They enhance product performance and interoperability, reduce cost of implementation, and greatly improve serviceability and updating of features to meet market demands. Examples of evolving technologies include system architecture, wireless technology, software engineering and life cycle processes. Typical products include cell phones, Internet services and hardware/software functions for industrial controls, safety and security. The dependability management framework and business strategies are greatly influenced by evolving technologies incorporated in systems and products due to global market competition and the diversity of ICT product features to meet increasing industry demands and user needs.

Standardization of evolving technologies is needed for common processes for design and development, new product introduction and system upgrade and service applications. This is to facilitate different product integration, enhance interoperability and provide interface or "handshake" for legacy products in backward/forward compatibility. The dependability management framework should advocate its role as an enabling agent to serve other product-specific standards with the needed framework references to deal with generic dependability processes and unique protocols for product development and applications. Standardization of evolving technologies has taken on a new dimension in knowledge capture and information dissemination. Hence the traditional approach in standards preparation and documentation is no longer suitable or cost-effective. New approaches are emerging and are being rapidly adopted in practice.

Standardization of evolving technologies could be realized. This involves the collaborative effort of industry, government and academic institutions to achieve a common standards objective. The standardization format is guided by industry forums, technology partnerships, business dialogues and information sharing. The role of standards is presented in the form of knowledge base, websites and electronic archives for accessibility and ready references. The information can be gathered through existing technology focus groups and other technology forums and consortia sponsored by government, industry and universities. The collaboration requires stewardship and guidance to harness the results for standardization. Dependability strategy could play a pivotal role in bringing global standardization effort to a common forefront.

1.4.5 Environmental Sustainability

Environmental sustainability is essential for consideration in today's market place. The environmental impact of product replacement and disposal presents a challenging situation to the developers and manufacturers of products. A typical user or customer is now starting to demand a take-back contract, whereby the provider or supplier of the product, such as a battery, is required to take back the used product prior to the installation of a replacement product. Buy-back contracts are also common in today's business, whereby the supplied quantity of spares retained or purchased by a user or customer that are not used at the end of an agreed time period must be bought back by the provider or supplier. In the European Union, where it is estimated that approximately 6 kg (13.2 lbs) of electrical consumer goods are scrapped every year, the European WEEE Directive [17] now regulates the scrapping and reuse of materials from these products. Many of the consumer products are scrapped due to failure of minor parts. Designing for sustainability implies design strategies for improving reliability [18]. In the design and manufacturing of products, consideration should be given to the re-use of disposable parts. Recycling of by-products in the production process to minimize waste disposal is another factor to be considered in the environmental impact studies. Reduction of emissions and wastes from the product environmental life cycle process should also be taken into consideration.

Dependability engineering has extensive influence on the selection of material and components, architectural design and manufacturing processes, maintenance and support strategies at various stages of the system life cycle, especially in the early design and development stages, to minimize or mitigate potential environmental impact.

1.4.6 Dependability and Asset Management

An increasing emphasis is being placed on the concept and practice of asset management. Growing out of PAS-55 [19, 20], there is a movement to bring asset management into the core of how an asset-intensive business is managed. The stated focus is on physical assets and asset systems, which are intended to encompass both hardware and software. However, it links itself to other categories of assets, including human assets, information assets, financial assets and intangible assets. Life cycle management is the basis of asset management, including life cycle costing and financial aspects. Risk management is also considered to be a major focus of asset management. A set of ISO standards on asset management is now under development.

A key threat to successful management of assets is a lack of dependability as it relates to all of its characteristics as enunciated in section 1.1.1. Dependability shares most of the aspects of asset management, including risk management, the life cycle, information management and quality. The major difference is that asset management claims a wider scope, reaching into strategic planning and high-level financial considerations. Nevertheless, dependability serves as the primary basis for asset management. Without proper consideration of dependability, asset management objectives could not be achieved.

REFERENCES

[1] IEC 60050-191: *International Electrotechnical Vocabulary – Part 191: Dependability*

[2] IEC 61069-5: *Industrial-process measurement and control – Evaluation of system properties for the purpose of system assessment – Part 5: Assessment of system dependability*

[3] Saleh, J.H. and Marais, K. *Highlights from the early (and pre-) history of reliability engineering*, Reliability Engineering and System Safety 91 (2006), pp. 249–256

[4] Mil-HDBK-217F *Reliability Prediction of Electronic Equipment,* U.S. Department of Defense (1991)

[5] U.S. Commercial Telecommunication Standard TR-332/SR-332 *Electronic Reliability Prediction*, Telcordia (Bellcore)

[6] *Capability Maturity Model® (CMM®)*, Software Engineering Institute, Carnegie Mellon University, Pittsburgh, PA USA

[7] *Capability Maturity Model Integration® (CMMI®) for Development, Version 1.2*; Software Engineering Institute, Carnegie Mellon University, Pittsburgh, PA USA 2006

[8] Lyu, M. R. (Ed.) *The Handbook of Software Reliability Engineering*, IEEE Computer Society Press and McGraw-Hill Book Company (1996)

[9] FAA, *Guidelines for Human Factors Requirements Development, AAR-100* (2004)

[10] ISO 9241-20:2008, *Ergonomics of human-system interaction – Part 20: Accessibility guidelines for information/communication technology (ICT) equipment and services*

[11] Strandberg, Kjell, 1990. *IEC/TC56 – 25 years of International cooperation*, Ericsson Telecom, Sweden

[12] J.C. Laprie, editor. *Dependability: Basic Concepts and Terminology*, Springer-Verlag, 1992

[13] www.iec.ch/smartgrid

[14] www.ieee-smartgridcomm.org/

[15] IEC 61069-2: *Industrial-process measurement and control – Evaluation of system properties for the purpose of system assessment – Part 2: Assessment methodology*

[16] www.wordIQ.com/definition/ilities

[17] Directive 2002/96/EC of the European Parliament and of the Council of 27 January 2003 on waste electrical and electronic equipment (WEEE)

[18] Loll, V, 2011. "Design for Sustainability," Tutorial, RAMS, The Annual Reliability and Maintainability Symposium, Jan 24-27, 2011, Lake Buena Vista, Florida, USA

[19] PAS 55-1:2008, "Asset Management, Part 1: Specification for the optimized management of physical assets," The Institute of Asset Management, British Standards Institute

[20] PAS 55-2:2008, "Asset Management, Part 1: Guidelines for the application of PAS 55-1," The Institute of Asset Management, British Standards Institute

Chapter 2

LIFE CYCLE APPROACH FOR DEPENDABILITY

2.1 INTRODUCTION TO THE LIFE CYCLE APPROACH

2.1.1 Why a Life Cycle Approach

A life cycle consists of a series of identifiable stages through which a system evolves from conception to retirement. The life cycle stages vary with application. For technological systems, the life cycle follows a sequence of process stages from concept/definition, through design/development, realization/implementation, operation/maintenance and enhancement, to system retirement or termination of service.

The rationale and justification of life cycle approach for dependability achievement can be viewed from the following perspectives:

(a) business perspective – the close association of the system life cycle to the business life cycle indicates where and when dependability achievement is best suited for early investments to harness business benefits of later product sales growth and marketing leverage;
(b) project perspective – the stages of a project closely follow the life cycle stages and are relevant to the business model used for implementation of major assets such as a processing plant;
(c) technical perspective – the achievement of dependability can be realized by design incorporation of reliable and maintainable system functions, assurance of dependable system architecture and implementation of effective maintenance support strategies to sustaining successful dependability performance in operation;
(d) safety perspective – adequate levels of safety can only be achieved by ensuring that safety procedures and processes are followed, including dependability design and incorporation of safety functions and auxiliary support at the various life cycle stages;
(e) environmental sustainability perspective – the impact to the environment is greatly reduced by adoption of "green" concepts for dependability design and processes to support reuse, recycle and reduce waste and minimize harmful by-products;
(f) economic perspective – implementation of processes such as responsible "lean" manufacturing and life cycle costing.

"Green" represents the global environmental movement for conservation. It advocates sustainable management of resources for protection and restoration of the natural environment and human habitat. "Lean" is a production practice that considers the expenditure of resources for any goal other than the creation of value for the end customer to be wasteful and thus a target for elimination. Dependability value represents the end product; in this case, this value is the system dependability performance that a customer is willing to pay for.

The cliché "what goes around comes around" has a sense of truth. This may imply the recursive nature of the system life cycle that "new" evolving systems continuously replace "old" legacy ones. From a dependability application viewpoint, the life cycle approach sustains market relevance of technological systems. It reflects the application needs for timely adjustments to seek proper balance of acquisition and ownership costs. From a life cycle costing [1] perspective, early detection and removal of a design fault will likely save 100 times in production cost and 1000 times or more in field operation. It is not uncommon for a product recall program to run into millions of dollars for payout to remedy the situation.

2.1.2 System Life Cycle Models

There are various life cycle models for product or system descriptions depending on specific applications. From an industry perspective, the system life cycle model that addresses the technical aspects and activity sequence of the system life is the most referenced and used. Specific life cycle models are also used to describe technology and product evolution such as the software evolutionary model. Cost models are also employed for costing of work breakdown structure and estimation and projection of acquisition cost and ownership cost such as life cycle costing.

From a project management perspective, the system life cycle model is used to describe how the system is designed and utilized to meet its service application and business objectives. The system life cycle model provides general information to assist project definition and plan tasks and activities to meet project objectives. Two life cycle models are presented herein to describe technological systems and evolving systems.

(a) Technological systems

Most technological systems are complex and may consist of a combination of hardware and software elements with multiple functional features to meet system objectives. Technological systems are new designs and generally employ current technologies to meet market demands. Figure 2.1 presents a typical life cycle model of the technological system highlighting the dependability engineering process.

The following describes the stages of the system life cycle.

- The purpose of the *concept/definition* stage is to identify market needs, define and identify the operational use environment and timeline, define preliminary system requirements and confirm feasible design solutions by producing technical specifications for the system design. Selection of design options is based on risk analysis, impact evaluation and practical engineering approaches. The key process activities of this stage include:
  – *requirements definition* to identify the system requirements for combined hardware and software elements in response to the users' needs and constraints of system applications;
  – *requirements analysis* to determine feasible design options and to transform the system requirements for service applications into a technical view for hardware and software sub-system design and system development.
- The purpose of the *design/development* stage is to plan and execute selected engineering design solutions for the realization of system functions. Design

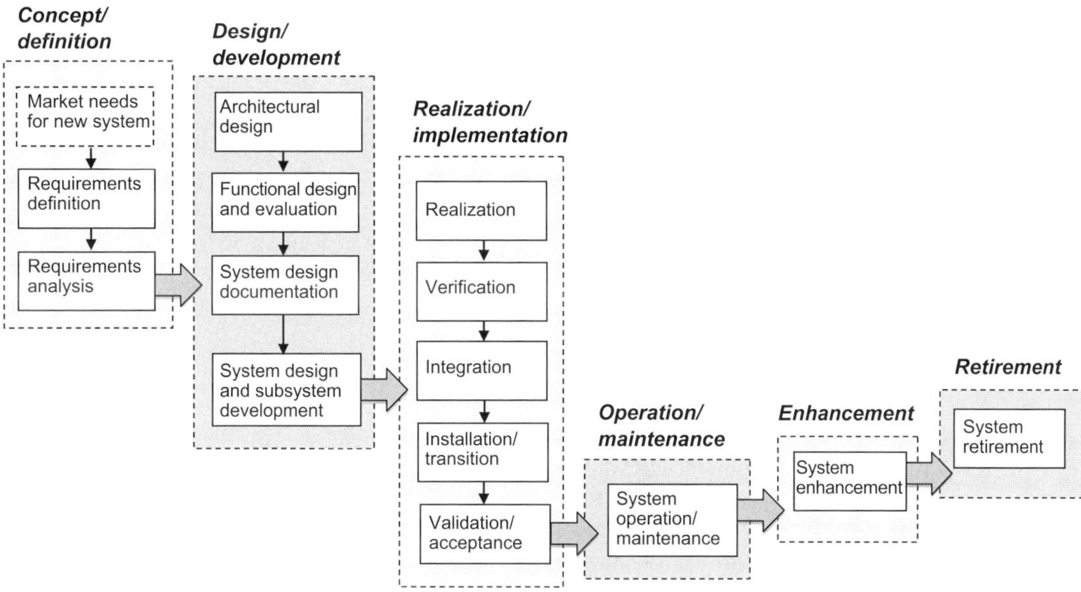

Figure 2.1 System life cycle model highlighting the dependability engineering process

activities include architectural design and functional design and evaluation to provide high-level system specifications. This is transcribed into an appropriate system development effort including engineering modelling, prototype construction, risk assessment and interface identification of system and sub-system elements. Systematic evaluation of the integrated system functions is conducted to verify interoperability of system performance and interactions with external environments to validate the final system configuration. Maintenance support planning, maintainability, operating procedures, and assurance and support processes should be well established prior to system realization. The key process activities of this stage include:

– *architectural design* to provide a solution for meeting system requirements by allocation of system elements into functional building blocks that establish a baseline structure for sub-system decomposition and identify relevant hardware and software functions to meet the specified requirements;
– *functional design and evaluation* to determine the practical means with hardware/software elements and human interactions for realizing the functions to facilitate design trade-off and optimization;
– *design specifications and documentation* to capture the detailed design data and functional interface information suitable for system and sub-system development;
– *system and sub-system development* to create the specified system and sub-system functions.

- The purpose of the *realization/implementation* stage is to execute make-buy decisions for acquisition and deployment of sub-system elements. The realization efforts deal with activities such as technology applications, manufacturing, packaging and supply sourcing to ensure the complete transformation from system design to the specified product or sub-system elements. The realized products or elements may comprise a combination of hardware and software

functions. Implementation includes such activities as integration of system functions, verification of sub-systems and installation of the system. For the construction of facilities such as process plants, on-site installation and commissioning is critical. System support and outsourced maintenance contracts should be ready with confirmed agreements. System acceptance procedures should be established with the customer for system trials in the actual operating environment prior to commissioning and service delivery. Validation should be a part of the trial to provide objective evidence of conformance to system specifications. The key process activities of this stage include:
- *realization* to produce the system and sub-system elements in hardware and software forms;
- *verification* to confirm that the specified design requirements are fulfilled by the system;
- *integration* to assemble the system and sub-systems consistent with the architectural design configuration;
- *installation/transition* to establish the system capability for providing the required performance service in a specified operational environment;
- *validation/acceptance* to provide objective evidence that the system fulfills the functional performance requirements.

- The purpose of the *operation/maintenance* stage is to deploy the system for delivery of the service and to support system operational capability by means of maintenance. The process activities include operating and maintaining the system for service in accordance with system performance requirements, operator and maintainer training to maintain skills competency, customer interface to establish service relationship, record keeping on system performance status and reporting failure incidents to initiate timely corrective and preventive actions. The system performance should be monitored and checked on a regular basis to ensure that dependability performance targets and quality of service objectives are met. The key process activities of this stage include:
  - *operation* to engage the system in delivering its capability for performing expected functional requirements and service;
  - *maintenance* to sustain performance and rectify failures and interruptions in service;
  - *monitoring* to verify system performance and effect improvements.
- The purpose of the *enhancement* stage is to improve system performance with added features to meet growing or changing user demands on the system. Routine process activities include software upgrades, hardware replacement, skills training, simplified procedures to improve operational efficiency and organizational procedural changes to increase expediency and customer value. No major capital asset investments to the technological system infrastructure are involved in the enhancement stage. The key process activities of this stage include:
  - *evaluation* to verify that proposed enhancements are worthwhile;
  - *enhancement* to improve system performance with added features or improvements such as efficiency.
- The purpose of the *retirement* stage is to end the existence of the system. Upon termination of system service to the customer, the system may be disassembled, redeployed for other use, or disposed of where possible without affecting the environment. For complex technological systems, a strategy for decommissioning or gradual withdrawal of services should be established. It is imperative in business for provision of customer service to formalize planning and

implementation of the decommissioning process to meet regulatory requirements. For consumer products, regulatory rules concerning return and reuse or disposal may be in existence. The key process activities of this stage include:
– *development of a strategy* to guide disposal of the system;
– *decommissioning of the system* to allow safe disposal;
– *disposal* to obtain residual value of the hardware and to support sustainability.

(b) Evolving systems
Evolving systems are systems that adapt to changes with time necessitated by technology advancements and incorporation of new features associated with ongoing market demands for service application. An evolving system may consist of a combination of new technological systems and legacy systems from previous generations during service implementation. There are also situations where a system grows in size and extends its scope of operation, such as a pipeline system or power generation plant that is expanded, leading to disparate system elements and equipment that have to operate together. The evolving system is designed with the objective of sustaining continuity of essential services, such as the provision of telecommunications and energy utility services.

Figure 2.2 presents a life cycle model of the evolving system.

While the early stages of the evolving system remain the same as the stages of the technological system, the difference is at the latter evolution/renewal stage, which replaces the enhancement stage. The retirement stage no longer exists in the evolving system, although individual legacy systems, including equipment and procedures, may be retired or terminated and replaced by new technological systems and processes in order to sustain continuity of essential but competitive service operation.

Evolution/renewal implies that progressive development and technology advancement has taken place to transform the service to a new level with well-defined goals and objectives. Examples include the transition to a next generation of technology, expansion of performance capabilities and improved provision of support services. The evolution/renewal process will likely entail infrastructure changes, major operational upgrades and addition of capital assets. New evolution/renewal project initiatives will activate processes for requirements analysis, risk assessment and management decisions. From a customer relations perspective, the evolution/renewal process will likely involve customer training and user familiarization to facilitate the implementation of new services.

Appendix B presents a typical system life cycle process for dependability application.

There are two types of organizations that engage selected stages of the life cycle model approach to run their business. They are generally known as system developers and service providers. System developers focus their business by engaging the concept/definition, design/development, and realization/implementation life cycle stages. This

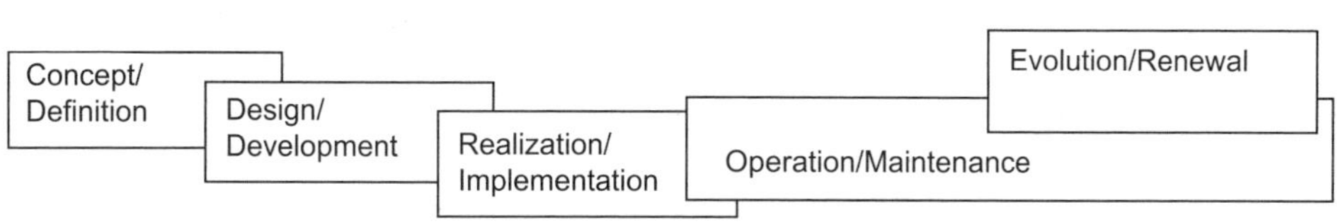

Figure 2.2 Life cycle model of the evolving system

allows the system development business to innovate, design, construct and produce new systems and products for clients. Examples include start-up companies with new product development and joint venture research organizations. Service providers on the other hand, focus their business by engaging the operation/maintenance and evolution/renewal life cycle stages. This enables the service provision business to invest their assets and resources in acquisition and ownership to sustain continual provision of specific services to customers. Examples include telecom operating companies and pipeline organizations.

2.2 DEPENDABILITY APPLICATION IN A BUSINESS ENVIRONMENT

2.2.1 Influences on a Business Environment

Business environment generally refers to the external influencing factors affecting the operation of an organization. The most important external factors include economic, legal, political, social and technological influences. The viability of a business can be affected positively or negatively by some of these influencing conditions that may impact new business ventures and the sustainability of business operation. Dependability application in system life cycle approach should focus on management support and technological solutions that are usually under project management control. Prudent application of dependability in the life cycle approach will likely achieve system performance values and succeed in a competitive business environment. The rules of engagement for dependability application under such circumstances include the necessity for value creation, boosting customer confidence and gaining the trust of all parties concerned in a dynamic business environment.

2.2.2 Dependability Focus on Management Support

The focus of management support in business operations is to incorporate dependability principles and implement relevant engineering practices in key functions of an organization. Management functions will benefit from incorporation of appropriate dependability value in management processes. Some examples include:

- *functional management* of different organizational functions or groups requiring coordination of dependability activities, usually by a project leader with assigned dependability responsibility for the coordination of management and technical effort;
- *resource management* requiring planning, acquisition of capital equipment, staff training and deployment, outsourcing and sub-contracting of technical dependability work;
- *development management* requiring dependability management involvement for assignment of specific expertise to assist design, development, realization, implementation and enhancement of project dependability needs;
- *process management* requiring planning, development, evaluation and application of management and technical processes involved in the execution of dependability disciplines;
- *configuration management* requiring dependability management inputs and confirmation to effect design changes of configured items;
- *operation and maintenance management* requiring dependability technical involvement to resolve dependability-related problems and assess impact consequences;

- *performance management* requiring assessment of dependability performance trends in field operation of systems and products and for the provision of support services;
- *support management* requiring dependability inputs on planning and implementation on system upgrades, product modification and service enhancement;
- *information management* requiring dependability contributions to establish and update dependability performance database for use by cognizant personnel;
- *knowledge management* requiring dependability involvement for dependability knowledge capture, copyright protection and patent registration, and dissemination of relevant dependability data and knowledge to cognizant personnel;
- *assurance management* requiring dependability management involved in planning, review, audit, verification and validation of ongoing project activities.

2.2.3 Dependability Application Focus on Technological Solutions

The dependability application focus on technological solutions in the business environment is to provide timely effective solutions on dependability issues in the life cycle approach. The technological solutions pertaining to application in business environment are presented from three interrelated perspectives: a) business life cycle profile, b) capability maturity of organizational processes, and c) dependability resource utilization.

(a) Business life cycle profile

Business exists because of ongoing market needs for the products and services provided by the business enterprise. The business life cycle profile represents the sequence of business events over its life cycle. The business event sequence includes development, introduction, growth, maturity, saturation and decline. For technological systems, the timing of business initiation, the leveraging of market advantages and associated risk implications may vary with the scope and intensity of applications and resource commitments in the business life cycle. The framework for technological solutions is based on technology platform development to speed up launching of new products to meet market needs. Each new generation of product launch is timed for market entry at the most advantageous and leveraging position from a business supply and demand perspective. A first-generation product establishes the foundation for market needs and product capability for the business venture. The second-generation product will greatly reduce the development time and resources to help speed up the product introduction and growth process to reach product maturity and profitability. The profit margin will increase significantly with long maturity duration and a period of stability before competition sets in. Market saturation and decline only exist when competitive products have an edge in business advantage. The third-generation product will further enhance the business proposition and increase in market share with the known product brand. The business life cycle profile for product advancements utilizing technology platform is shown in Figure 2.3.

The following shows typical examples of technological solutions associated with dependability issues addressed in these business events.

- Development – where the business enterprise contemplates innovative ideas and follows through with development and realization of a marketable new product or service. In most cases, the technological solutions are derived from an established technology platform on which the new product or service is prepared for the intended market launch. The extent of time and resource commitments involved in new product development is shared among similar

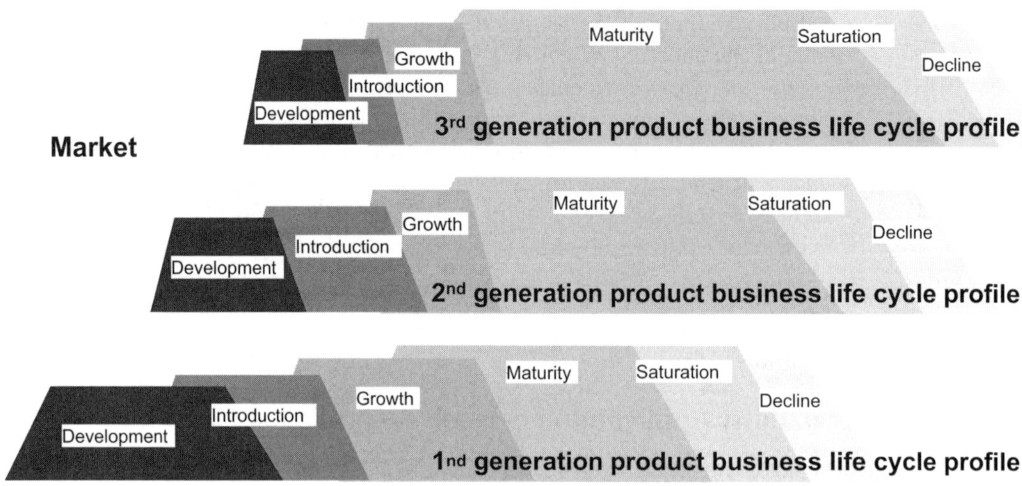

Figure 2.3 The business life cycle profile for product advancements

products generated from a common technology platform, hence reducing time and cost for realization of multiple products. The right timing and readiness for launching a new product are critical to suit the market window of opportunity. It is dependent on what the innovative technology will bring to meet the market needs. The technology market is highly volatile and competitive. Extensive research and development is essential to maintain the brand value of established product lines and sustainable supporting services. The dependability value of the new product being cultivated needs to be determined. Market relevancy, technology acceptance in the existing social context, financing and investment schemes, and product rollout plans are issues to be addressed in business development.

- Introduction – where the business enterprise introduces its new products and services to its potential customers. For technological systems, it is important that the capabilities and long-range plans of the business enterprise are made known to the intended customers. This is accomplished through marketing schemes, referral client contacts, trade shows and exhibitions, technical conferences and technology demonstrations. In some cases, test runs and customer trials are necessary to establish user confidence and ascertain fitness for use. Dependability is a key selling factor to technology users. As one technology industry leader has so insightfully noted: "Convince the geeks, the masses will follow."
- Growth – where the business enterprise continues to develop and deliver products and services with increasing customer demands. Business growth may be due to various reasons including, but not limited to, economic incentives for technology development, product novelty, unique technological solutions to specific problematic issues in process operation, user confidence in the after sales support and dependability of product performance. Business growth often leads to diversification of product line portfolios, acquisition of facilities for production expansion and outsourcing of support services. Regional profit centers are commonly established for managing business successes.

- Maturity – where the business enterprise has reached its capability to sustain ongoing business to cope with customer demands. Emphasis is placed on strategic and organizational operations to deliver dependability incentives and customer values. Organizational leadership is at its prime.
- Saturation – where the business enterprise encounters market saturation of similar products and services, with competitors offering alternate means or other incentives to meet customer demands for better value or at lower cost burdens. Customer retention is a key factor to sustain viable business operation. While the timing for renewal of products and services is essential, the relevant technological solutions have to be at hand to mitigate the situation to maintain business competitiveness. Technology is fast moving with relatively short advantageous durations for exploitation. Research and development needs are essential to sustain competitive business. New product mix with dependability brand value may reengage or regenerate customer confidence and user trust.
- Decline – where the business enterprise encounters market decline of the need for its products and services. Product obsolescence sets in as customers are more attracted to new innovations. Market decline will likely result in the inability of the business enterprise to sustain its operation and to meet its financial obligations. This will likely cause downsizing of the operation, leadership change and reorganization, streamlining the product line offerings and, in some extreme cases, restructuring the organization for refinancing. The business enterprise in a declined market situation is a likely target for takeover and a candidate for external acquisition.

(b) Capability maturity of organizational processes

Business organizations rely on institutionalized processes to manage their business operations in an orderly manner. Business vision and goals should be reflected in the organizational processes. The effectiveness of process development and implementation is dependent on how well these processes are being managed to achieve organizational goals and business objectives. The business capability maturity is a process-based concept and a structured approach to understand, appraise and describe how well the organizational processes, including practices and behaviors, can be relied upon to achieve consistent results or produce sustainable outcomes. While it is important for a business enterprise to realize its potential capabilities, the level of maturity achievable in practice in an organization should be ascertained.

The Capability Maturity Model Integration (CMMI) [2] describes the progressive maturity of the organization's management processes in doing business. CMMI was developed by a group of experts from industry, government and the Software Engineering Institute at Carnegie Mellon University. CMMI provides guidance for developing or improving processes that meet business goals of an organization. It is also used as a framework for appraising the process maturity of the organization. CMMI has been used by different organizations as an influential and credible tool for understanding and improving general business process performance. As with most conformance assessment tools, successful compliance with process requirements does not necessarily guarantee successful product outcomes.

The following are the key organizational processes focusing on product and service development that facilitate technological solutions in business environment.

The processes are grouped in five maturity levels to facilitate assessment. The purpose of each process is stated.

i) Maturity level 1: Initial – processes unpredictable, poorly controlled and reactive
   - *None specified*

ii) Maturity level 2: Managed – processes characterized for projects and often reactive
   - *Configuration management*: to establish and maintain the integrity of work products using configuration identification, configuration control, configuration status accounting and configuration audits
   - *Measurement and analysis*: to develop and sustain a measurement capability used to support management information needs
   - *Project monitoring and control*: to provide project progress status for appropriate corrective actions when project performance deviates significantly from plan
   - *Project planning*: to establish and maintain plans that define project activities
   - *Process and product quality assurance*: to provide resources and management with assurance objectives to drive product and process activities
   - *Requirements management*: to align and manage project requirements with project plans and work product
   - *Supplier agreement management*: to manage the acquisition of products and services from suppliers

iii) Maturity level 3: Defined – proactive processes characterized for the organization
   - *Decision analysis and resolution*: to analyze possible decisions using a formal evaluation process that evaluates identified alternatives against established criteria
   - *Integrated project management*: to establish and manage the project by involving relevant stakeholders according to an established and integrated defined process tailored for the organization's application
   - *Organizational process definition*: to establish and maintain a usable set of organizational process assets, work environment standards, rules and guidelines for effective team operation
   - *Organizational process focus*: to plan, implement, and deploy organizational process improvements based on thorough understanding of current strengths and weaknesses of the organization's processes and process assets
   - *Organizational training*: to develop skills and knowledge of people so that they can perform their roles effectively and efficiently
   - *Product integration*: to assemble the product from product components, ensure that the product, as integrated, performs properly for delivery
   - *Requirements development*: to elicit, analyze and establish customer, product and product component requirements
   - *Risk management:* to identify potential problems before they occur so that risk handling activities can be planned and invoked as needed across the life of the product or project to mitigate adverse impacts on achieving objectives
   - *Technical solution*: to select, design, develop and implement solutions to requirements
   - *Validation*: to demonstrate that a product or product component fulfills its intended use when placed in its intended environment
   - *Verification*: to ensure that selected work products meet their specified requirements

iv) Maturity level 4: Quantitatively managed – processes measured and controlled
  - *Organizational process performance*: to establish and maintain a quantitative understanding of the performance of selected processes in the organization's set of standard processes in support of achieving quality and process performance objectives, and to provide process performance data, baselines and models to quantitatively manage the organization's projects
  - *Quantitative project management*: to quantitatively manage the project to achieve the project's established quality and process performance objectives

v) Maturity level 5: Optimizing – focus on process improvement
  - *Causal analysis and resolution*: to identify causes of selected outcomes and take action to improve process performance
  - *Organizational performance management*: to proactively manage the organization's performance to meet its business objectives

(c) Dependability resource utilization

Dependability engineering is a technical discipline. The achievement of dependability in projects relies heavily on competent resources to support serious dependability endeavors and planned engineering efforts. Timing for dependability work implementation and project delivery targets are key success factors in doing business. For technological systems, the crucial business concerns include but are not limited to the following situations:

- *Meeting project delivery targets*: Scheduled contractual commitments should be fulfilled on time. Contract delays for any reason will lead to disputes and loss of customer confidence and user trust. Delays due to technical deficiencies and non-compliance will require appropriate technological solutions to promptly rectify the situation to stay on course and maintain scheduled commitments.
- *Dependability resource availability*: Knowledgeable dependability personnel are needed for assignment of dependability tasks. Technical resources are often scarce for project task allocation. Dependability experts that specialize in specific technical fields such as software development, system architecture, and failure diagnoses often have limited accessibility or are unavailable in times of need. External consultation with experts who have appropriate background knowledge may sometimes be utilized to facilitate timely problem resolution.
- *Customer satisfaction*: The customer needs to be involved in progress reviews to facilitate acceptance of final project delivery. Customer complaints should be reviewed regularly. Critical complaints and disputed issues should be taken seriously for prompt resolution to maintain good cooperative relationships and customer satisfaction.
- *Outsourcing and supplied products*: Outsourced work and supplied products used in the dependability project should conform to the same contractual requirements for system realization. Outsourcing and supplier requirements should stipulate qualification and acceptance conditions of the supplied products for system integration. Interoperability of products and equipment should be resolved prior to their integration into the system. Specific technical support services should be identified for probable maintenance and logistic support services when the system is deployed in field operation.
- *Dependability performance assurance*: Dependability performance for in-service system operation should be monitored, supported and assured. For technological systems, dependability assurance strategies should engage systematic processes to ensure achievement of dependability performance capability and delivery of

customer-focused dependability of service. An incident reporting system should be implemented for system outage control and mitigation in the event of system malfunction or degradation. End-user complaints and concerns with service provision should be addressed. System upgrade and scheduled maintenance procedures should be put in place for service implementation.

- *Regulatory compliance*: System should conform to all stipulated regulations and applicable standards. Internal and external audits should be planned and implemented to verify and validate compliance. Non-compliance with regulatory requirements will lead to extensive investigations and delays in project completion.

2.3 LIFE CYCLE APPROACH TO PROJECT MANAGEMENT

2.3.1 Framework for Managing Dependability Projects

The framework for managing dependability projects is highly dependent on the organizational infrastructure in which the project operates. The novelty of technology, the complexity of the system, the availability of resources, the capability of supporting facilities and services, the application environment and the cultural aspects of management operation will influence the efficiency and effectiveness of project management. The organizational infrastructure refers to the physical and functional structures and facilities needed for productivity in operation. An organization may include corporations of all types and sizes, public and private institutions, enterprises and non-profit associations where management is needed for infrastructure development, coordination and operation to pursue collective goals and common objectives.

Project management involves planning and control, resource allocation, coordination, decision execution, and review of project status and progress. Dependability engineering personnel should actively participate as members of the project team. The review of dependability activities should be done in conjunction with regular project reviews such that project issues, including dependability issues of concern, are brought to light to the project team for concerted efforts toward resolution. As with most management functions, the management of dependability requires control as well as vision, accountability and commitment. Technical knowledge is needed to address dependability issues, and prudent judgment is often necessary to confront the consequences of applied solution measures. Dependability contributions should be recognized for their specific achievements.

Dependability management should form part of the organization's management systems to achieve a clearly defined common objective. It is often impractical to set up an isolated or separate management system solely for managing dependability. This would likely create silos within the organization that can impede effective coordination of project management efforts. However, technical expertise and resources that are required to ensure ongoing project dependability issues and concerns are properly addressed on a timely basis are often scarce. Some organizations engage in matrix management schemes for assignment and rotation of dependability personnel to serve specific project needs while retaining core dependability expertise dedicated for research, training and consultation services. This would permit an organization to promote innovation, develop the dependability experience and cultivate technology infusion as fundamental value assets for knowledge enhancement. Dependability management does not require a complex organizational infrastructure and reporting hierarchy. The assignment of dependability responsibility along with adequate capability and resources within the overall management structure is generally sufficient.

A common database on dependability information should be established to serve as part of the organization's management information systems. The dependability information database would provide management insights on historical data and dependability-related performance records, enabling measurement of dependability status and improvements.

A training program should be institutionalized within the organization to update technological information, including dependability principles and practices for all personnel involved in the project assignments. This would facilitate a general understanding of the technical problems involved and an appreciation of the dependability issues at hand that need to be resolved.

2.3.2 Establishing Dependability Project Objectives and Task Requirements

Dependability project objectives should clearly define a customer's dependability requirements and expectations related to the system of interest. Where applicable, specific dependability characteristics should be established as quantitative targets of system performance measures for validation purposes. Project dependability thrusts and technical and management direction should reflect the principles and practices of the organization's vision and mission in doing business.

Dependability project tasks, where applicable from a life cycle perspective, should accomplish the following:

- identify the customer, the system of interest and the application environment;
- state the organization's understanding of the customer's needs and requirements;
- establish project responsibilities and identify the project leader and assigned project team personnel;
- develop a budget plan and resource requirements for the duration of the project;
- present a statement of work, schedules and target dates for completion;
- outline technical approaches for dependability solutions in the context of system requirements;
- provide specific resources for timely implementation of dependability activities;
- identify outsourcing requirements and short-listing of preferred suppliers;
- enable measurement of dependability progress status;
- perform risk assessment where appropriate to minimize exposures and mitigate potential risks;
- review dependability outcomes to assess the extent of project task accomplishments and deficiencies requiring further attention;
- recommend appropriate changes where needed for progressive improvements;
- institute customer interface and supply-chain collaboration;
- ensure system performance to achieve dependability value and customer satisfaction;
- incorporate assurance programs to sustain dependability performance in system operation.

Challenges in managing a dependability project over the system life cycle often involve more than one organization. Partnership alliances with other organizations and technical groups are usually engaged in the interaction and coordination of a dependability project. Over the life cycle, certain responsibilities are passed from one organization or technical group to another. Since organizational styles and procedures may vary, engineering practices and cultural differences in diverse operating regions should be appreciated. International system projects often collaborate and traverse over several countries and cross regional boundaries. The management of dependability should be adaptive to different situations and capable of being tailored to meet specific needs without losing sight of the project objectives. Where service functions such as maintenance and logistic support are outsourced, the responsibility for dependability aspects of outsourcing should be specified, monitored and controlled by the project. Team building, work collaboration and sharing of information are key to successful dependability achievement.

Prudent project management will accomplish one or more of the following objectives:

- achieved system dependability development targets;
- improved system interoperability in service operation;
- reduced risk exposures;
- enhanced security of service;
- improved safety operation;
- reduced environmental impact;
- increased integrity of information retention and data throughput;
- improved quality of service;
- improved system serviceability and support services;
- reduced life cycle costs;
- gained customer confidence and trust;
- sustained system dependability value.

2.4 TAILORING PROCESS

2.4.1 Tailoring of Dependability Projects

Tailoring is a process to adapt, adjust or alter an organization's set of established processes and activities to fulfill dependability and other objectives. The tailoring process applies to management processes, project activities and management reviews. Tailoring applies to any stage of the system life cycle. It is specifically important for tailoring during the initial design of the life cycle of a project. Tailoring is not necessary in all cases, such as when very similar products are being developed and produced.

Tailoring of dependability management needs to take into account the nature of the organization and the dependability tasks that need to be managed. The organization could vary from a technical consultancy to a multi-national conglomerate requiring appropriate dependability management of diverse disciplines, organization and specialization. Management approaches often seek technology transfer, knowledge infusion or expert consultancy to deal with critical short-term technical gaps. There is also a broad range of supporting disciplines and enabling systems to facilitate achievement of dependability management objectives. Examples include maintenance and logistic support management, customer care services, failure reporting, analysis and corrective action systems, and a dependability knowledge base. Project reviews should be readjusted to address specific dependability issues.

Tailoring of dependability should consider the organization's technical and administrative processes, including their constraints and influencing factors. These constraints and influencing factors include, but are not limited to, customer requirements, regulatory requirements, delivery targets, allowable budgets, available resources, technical capability, environmental impact and risk exposures, novelty of technology involvement and provision of sustainable services. They affect the outcome of successful dependability application to achieve the trustworthiness in system performance from a customer or user perspective.

The general tailoring of dependability activities, including those for projects, consist of the following:

(a) identification of the organizational policy and infrastructure;
(b) analysis of requirements and contract stipulations, characteristics and objectives that can be difficult to realize or to deliver;
(c) capability and resources needed and actually available for implementation;
(d) determination of the specific life cycle stages that are applicable;
(e) identification of the product- or system-related characteristics, such as the features and functions, past history of similar products or systems, their intended end use and anticipated application environments;
(f) selection of applicable dependability actions relevant to the specific life cycle stages identified;
(g) identification of the timing and duration of dependability program elements and activity applications for resource allocation;
(h) consideration of regulatory requirements and applicable standards;
(i) documentation of the rationale in formalizing the tailoring decisions as part of the organizational or project plan.

Tailoring for dependability management is usually applicable for short-term projects. The tailoring process is subject to the constraints and environment of the project. However, for long-term projects it may be more efficient to create new processes specifically suited to the project's objectives.

2.4.2 Tailoring for Project-Specific Applications

The system of interest in project application may span multiple technological domains. The process for tailoring multiple domains in project-specific applications is to consider possible domain relationships and timing for implementation to facilitate cross-domain collaboration. The following key influencing factors should be considered in the light of dependability achievement in tailoring for project-specific applications.

(a) The health and safety concerns of the system operators, maintainers and users should be investigated to determine potential health risks and unsafe usage conditions to be stipulated for further assessment. Specific health concerns and recommendations on design requirements and implementation procedures should be identified to prevent health and safety hazard occurrences. It should be noted that health and safety issues may persist even after the system or product is retired or terminated from service.
(b) Issues related to interoperability of system operational functions and human-machine interfaces should be considered in design and implementation processes to minimize service interruptions and mitigate graceful system degradation in fail-safe condition. System restoration and maintenance support services should be managed for cost-effective operation.
(c) Issues related to security risks should be thoroughly investigated to minimize risk exposures. Security risks include:
    - access and damage to personnel, properties and information;
    - corruption, theft or compromise of sensitive information;
    - denial of approved access to property and information;

- unauthorized system access;
- loss of life or property.

(d) Issues related to environmental impact as a result of disposal of waste products or obsolete products should be identified and investigated to avoid or minimize the potential damage to the environment.

2.5 PROJECT RISK MANAGEMENT

2.5.1 Risk Management for Dependability Application

Risk management [3] is an essential part of an organization's strategic management system. Risk is the uncertainty of meeting objectives and can be defined as the combination of the probability of an event and its consequences [4]. Dependability is a risk-based approach, which embraces the risk management principles in value creation. Dependability engineering activities address risk by following the generic risk management process. From a dependability perspective, the achievement of dependability in system performance relies on dedicated technical efforts and prudent judgments to attain dependability value in successful project outcomes. The dependability process engages in risk avoidance and risk control methods to manage dependability risk issues of the system life cycle.

Different organizations may have different objectives in managing risks. It is essential for the organization engaged in the risk management process to define the scope and strategies for organizational implementation. The responsibility for risk management and the capability and resource requirements should be defined. The methodology and procedures for risk assessment and risk treatment should be established to facilitate process evaluation and decision-making concerning the disposition of risk issues.

A set of risk criteria suitable to evaluate the significance of risks that are the primary concern of the organization should be developed. Dependability risk issues should be defined in terms of products and services in which the organization is engaged. Since businesses are generally involved with customers and suppliers, it is important to identify the risk factors for consideration when confronted with risk problems. The following factors relating to dependability risks and exposures should be considered:

- the nature and types of risk consequences that can occur;
- the likelihood of risk occurrences;
- the methodology for risk assessment and evaluation;
- the level of risk to be determined;
- the level of risk considered to be acceptable or tolerable;
- the extent of risk treatment;
- the relevant risk data required for information capture.

2.5.2 Risk Management Process

The risk management process should form part of the organizational processes and be tailored to the principles and practices of the business operation. A typical risk management process for dependability application is shown in Figure 2.4 and depicts the flow of activities.

For dependability application, relevant information and technical expert advice on problems associated with the dependability risk are needed at times to reinforce and support project management to reach an informed decision. The dependability risk problem, whether known or unknown, could exist in any stage of the system life cycle that demands project management attention. The discovery of probable risk exposures is generally

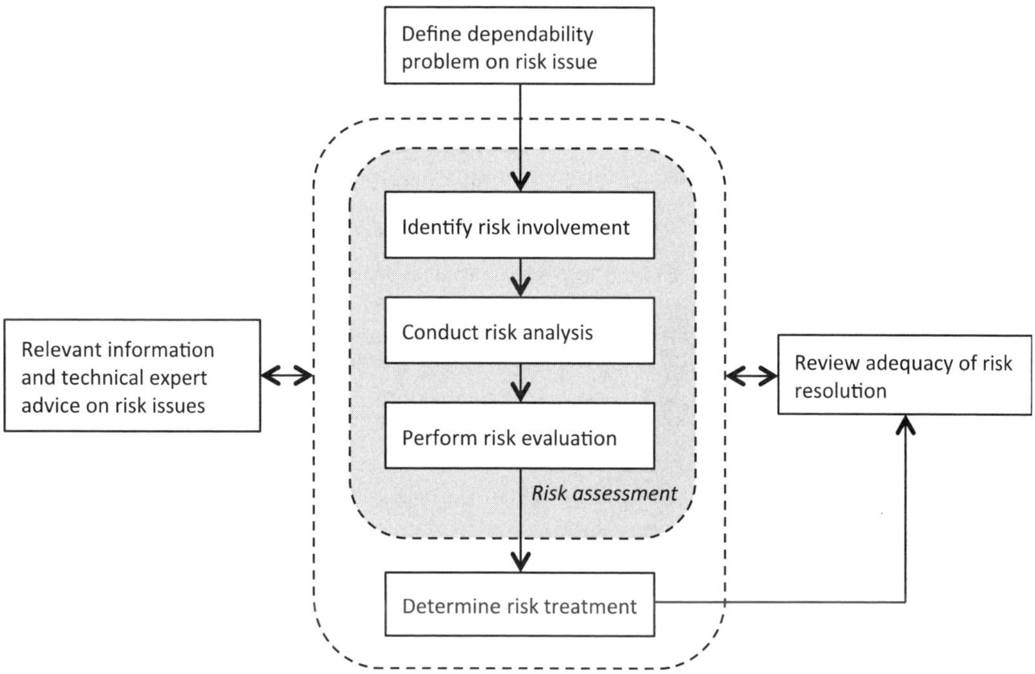

Figure 2.4 Risk management process

triggered by an incident and identified as potential risk with significant impact that warrants serious project consideration. Some risk management activities are planned in normal business operation where risk resolution is required. Examples include bid/no-bid decision, capital acquisition, outsourcing, and management reviews of the system life cycle process where decision on resource commitment is needed to move the project forward.

2.5.3 Scope of Dependability Risk Problems

In the context of achieving technological system dependability, the objective of an organization is to attain system performance value to enhance business operation. Dependability value is realized by incorporation of dependability characteristics in system development and in service assurance implementation. Figure 2.5 shows examples of the achievement of dependability value, where there could be vulnerable risk exposures if not appropriately addressed. This is due to increased business competition and challenges in operational problems with ongoing technology enhancement. Risk must be managed in all aspects of business operation to sustain dependability of service.

Dependability risk factors are potential problems associated with dependability risk exposures that may have a negative impact on dependability activities in any stage of the system life cycle process. The risk factor is characterized by the probability of the risk occurrence and a potential loss to the organization should the risk event occur. The dependability risk factors are driven by strategic risks and operational risks.

The specific dependability risk should be noted for possible problem investigation. The rationale for recommending risk investigation should be documented. Background information and relevant technical data are gathered to support the need for initiation of risk assessment. A project case on the dependability risk problem is raised for appropriate

Figure 2.5 Achievement of dependability value vulnerable to risk exposures

management action. The disposition of the dependability case is subject to review and decision at the discretion of the project management.

In practice, not every dependability risk problem identified needs to go through the formal risk management process. The risk problem is usually entertained in terms of risk exposures that would result in the needs to mitigate the technical and/or financial consequences. If not treated properly, the risk exposures might have further impact on the environment or cause safety, security or other concerns for the organization. Some dependability risk problems, such as disposal of used equipment, product safety recalls, or disclosure of sensitive information, may be under regulatory control or governed by specific jurisdiction of the operating regions. These dependability risk problems would likely invoke some form of official risk management process.

2.5.4 Dependability Risk Problem and Resolution – An Example

The following presents a dependability risk problem and resolution as a case study example. The risk management process activities are described.

(a) Define the dependability problem concerning risks
An organization has developed and produced a product line of control systems for utility service application. An estimated ten thousand control systems have been sold and deployed for service over the past 10 years. They are still reporting successful system operation in most utility service stations, some installed in remote rural areas. Current sales forecast of the control system is sustaining, which is attributed to the dependability of service commitments and product brand value of the organization. The organization has maintained a major market share on this product line. However, there are recent new start-up companies developing novel control systems for possible entry into this market. The novel control system design provides added features for remote monitoring and control capability of the service site operation. This added feature will facilitate the long-term plan of the utility provider for optimization of the overall utility service provision.

A critical device is used in the control system, which has good dependability performance records. A supplier had specifically designed this critical device a decade ago as a sole source to the organization for the control system. A limited spare device inventory is maintained on-site and at the organization's depots to sustain replacement returns. The device failure returns are about 100 per year. The dependability program has so far run well with the operational and maintenance service procedures. Authorized maintainers visit all installed sites annually to carry out routine preventive maintenance service. Special maintenance service calls are made in the event of system failures.

The original supplier has announced the discontinuation of the production of the critical device due to technology obsolescence, escalating production cost and limited sales projections. The device supply will likely cease within 6 months. No next-generation or technology upgrade of the device is planned for the foreseeable future. Several functionally compatible new devices from offshore suppliers are now available at a much lower cost and sold as COTS products. Direct substitution of the new device to replace the old device will require modification of the existing control system assembly in field operation. The dependability performance of the new device is unknown. This will likely require test verification and further substantiation of field performance records for similar applications. The organization also has the option to develop the next-generation control system for its replacement, thus providing the opportunity to design out the old device in the new control system. However, this will take at least one year of development effort and resource commitment to pursue a saturated market.

The project objective of this dependability case is to resolve the device replacement problem to sustain ongoing business of the organization's product line on control systems.

(b) Identify the risk involvement
The risk identification process involves a systematic approach to determine the risks associated with the dependability case to meet the project objective. The risk scenarios of the dependability case can be summarized as:
(1) Risk of device replacement to sustain control system operation;
(2) Risk of new device dependability performance;
(3) Risk associated with the modification of the control system assembly;
(4) Risk associated with development of the next-generation control system.

(c) Conduct risk analysis
The risk analysis process examines the identified dependability risk factors for probability of occurrence and potential impact to the project objective in terms of risk levels by qualitative or quantitative means. For the purpose of this case study, the risk levels can be classified as low, medium, and high. Risk analysis forms the second part of the risk assessment process. The results of the risk analysis for the risk scenario are summarized in Table 2.1.

(d) Perform risk evaluation
The risk evaluation process compares the estimated risk against given risk criteria to determine the significance of the risk. Risk evaluation forms the third part of the risk assessment process. The results of risk evaluation for the risk scenarios are summarized in Table 2.1.

TABLE 2.1 Summary of risk assessment results and risk treatment options

Risk involvement	(1) Risk of device replacement to sustain control system operation	(2) Risk of new device dependability performance	(3) Risk associated with the modification of the control system assembly	(4) Risk associated with the development of the next-generation control system
Risk analysis	• The risk exposure is running out of replacement spares for a long duration, causing multiple control systems outages. • The risk of old device replacement is "low," with a failure rate about 1%.	• The risk is determined from the device dependability performance characteristics by testing and field application history. • Several compatible new devices are available in the market to compete on dependability performance.	• The risk associated with assembly modification is the ability for the maintainer to carry out the task on site. • The risk is "low" since the maintainers are very experienced.	• The risk is associated with development of the new control system to meet current and future market needs. • Technical and financial resource commitments are required to ensure new development project completion. • The risk is "high."
Risk evaluation	• The new device dependability has yet to be determined, but would likely not exceed 1% failure rate due to 10 years of technology advancement.	• The risk of new device dependability performance is not directly relevant to this dependability case since multi-supplier sourcing is available.	• The assembly modification is a routine maintenance task and should not pose any risk problem.	• With increased competition on the horizon, timing for introduction of the new control system is critical to lead and leverage market shares and protect the organization's product brand value.
Risk treatment	• There are sufficient old device spares to last for 1 year that could be supplemented by replacement with new devices when needed. • No immediate action is required.	• No immediate action is required.	• A modification kit with instruction and training of the maintainer would likely not involve additional risk exposures. • Initiate the modification process.	• A progressive new product introduction plan is essential to phase in the new product while maintaining the existing products in operation to meet customer needs.

(e) Determine risk treatment
The risk treatment process selects and implements appropriate measures to modify unacceptable risk. Risk treatment may include risk avoidance, risk mitigation, risk transfer, risk retention or other measures. The risk treatment options for the risk scenarios are presented in Table 2.1.

(f) Review adequacy of risk resolution
(1) This dependability case is triggered by the organization's concerns with the risk of a shortage of devices for replacement needs. The immediate concern is to deal with finding a new compatible device for replacement due to the discontinued production of the old device. Since there are several compatible new devices commercially available in the market, the exposure to this risk is low.
(2) The risk of new device dependability performance is not directly relevant to this case. There is a 10-year time gap of technology advancement between the old device and the new device. Improvements must have been made for the technology-driven new device in the development and production process. There are several competing new devices in the market for selection. The best-in-class device would survive customer scrutiny to gain acceptance of dependability performance in the market place.
(3) The modification kit solution is a stopgap measure to provide a simple technical procedure for device replacement with routine maintenance service in case the organization runs out of spares replacement devices. With the 1% per year device return rate as the current acceptable risk level, the organization could sustain continued operation for at least one year. The risk exposure for implementing the modification kit solution is low. A plan for scheduling the modification solution activities should be initiated to address priority in terms of customer service needs and criticality of the control system failures requiring immediate maintenance action.
(4) The organization has 10 years of successful operation of the control system product line with good performance records. The success in operation reflects revenue generation over the past decade without major renewal or enhancement of the control systems. The current market competition with new start-up companies has prompted the need to review the existing marketing plan to stay competitive. It is important to recognize that good times do not last forever. The investment to develop the next-generation product is long overdue. A transition plan is needed to ensure progressive phasing in of the new system and phasing out of the old one. Provision of dependability of service and maintaining customer loyalty and satisfaction are key to protect the hard-earned product brand value to sustain a viable business operation. The risk is high if the new product is not successfully introduced on time into a saturated market. The existing control systems will soon be obsolete for lack of added features to cope with the utility service industry trends.

The lessons learned from this case study have demonstrated the close relationships of the risk management process with dependability application. The case study results illustrate the risk assessment approach to verify the facts and rationalize the assumptions in decision-making to engage risk treatment for dependability problem resolution. The project objective of this dependability case is not only to solve the short-term device replacement problem but also to sustain the long-term business goals of the organization's product line on control systems. Dependability value is a key driving factor in system life cycle process achievements.

2.6 REVIEW PROCESS

The objective of the review process from a system life cycle perspective is to ensure that all planned activities are implemented to provide adequate confidence that the established requirements are fulfilled. The review process ascertains that proper oversight with regular checks and balances are carried out in managing the business operation and reporting progress status of system activities. There are generally four levels of review activities from an organizational perspective:

(a) Management review – This is to ensure that organizational business objectives are being met by setting a course of action for policy direction and process implementation. Management reviews are conducted at the senior management level of the organization on a regular basis to address business policy, capability maturity in operation, intellectual capital, available resources and investment opportunities. Dependability contributions to management reviews engage in the formulation of technical strategies, facility planning and technology directions to guide the organization's business.

(b) Project review – This is to ensure that planned activities of a project are completed on schedule to meet stipulated project requirements and satisfy project milestone commitments at a designated life cycle stage. Project reviews are conducted at the project management level involving the project team and technical experts contributing to the project activities. In some situations, the customer or supplier associated with the project may be invited to participate in the reviews. The outcome of project reviews reports the achievement status of project-specific issues. The review information is used to justify the transition from one life cycle stage to the next stage where resource commitments are needed. Project reviews are sometimes known as "gate" or "phase" reviews, reflecting the decision-making process at the specific life cycle stage for transition. Dependability contributions to project reviews engage in project planning, risk assessment and reporting on status of dependability activities and follow-up actions.

(c) Design review [5] – This is to ensure that system and product designs are implemented and assessed for conformance to design requirements. Design reviews can be an informal technical discussion among design team members to resolve a specific technical problem. Formal design reviews involve independent and documented examination of a design and its requirements to assess the ability of the design to meet specified and implied requirements of the system or product concerned. Dependability contributions to design reviews engage in requirements review, preliminary design review and detailed design review.

(d) Technical review – This is to ensure that system performance and support services are adequate to sustain continual operation and provision of user services. Technical reviews are conducted as required during system operation from a service provider perspective. Dependability contributions to technical reviews engage in configuration management review, supply management review, product acceptance review, system performance review and maintenance and logistic support review.

All reviews are part of the assurance processes. Review records should be maintained as objective evidence for quality audits. The review information can be used to facilitate follow-up actions where needed. An action register is normally used as a control mechanism for monitoring closure of open action items.

REFERENCES

[1] IEC 60300-3-3 *Dependability management – Part 3-3: Application guide – Life cycle costing*
[2] *CMMI-DEV (Version 1.3, November 2010)*, Carnegie Mellon University Software Engineering Institute. 2010
[3] ISO 31000, *Risk management – Principles and guidelines on implementation*
[4] ISO/IEC Guide 73, *Risk management – Vocabulary*
[5] IEC 61160, *Design review*

Chapter 3

SPECIFICATION OF DEPENDABILITY REQUIREMENTS

3.1 STARTING A DEPENDABILITY PROJECT

3.1.1 Where to Begin with Technological Systems

Dependability engineering projects involve strategic planning and technology applications over the system life cycle. The activities include but are not limited to the following:

- system design to achieve dependability in functional performance;
- project management and coordination to ensure dependability process implementation;
- knowledge base development and advancement to sustain dependability of service;
- value creation to support enabling systems in technology transition and evolution of new product introduction;
- dependability assurance to ascertain successful system performance in operation.

The scope of dependability extends beyond mere utilization of engineering tools and techniques for problem resolution and embraces the challenges of technology transfer and infusion. Dependability activities act as enabling mechanisms to bridge the technological gaps in design trade-offs, process innovation, support service optimization, and sustainment of legacy systems. The objective outcome is to create dependability value to enhance system performance.

Dependability activities should be organized and managed as part of engineering or other programs or projects for effective operation. A project has a definite duration with specific start and end dates and a project plan. A program is usually longer term and ongoing, and can consist of several related projects to achieve common organizational objectives such as technology platform development, dependability process improvement and logistic support systems integration.

The scope of dependability involvement in projects is highly dependent on the intent and application of the system of interest. The objectives of dependability can be defined or quantitatively expressed, such as a 25-year service life of a solar energy system or an expected 95% availability for operational readiness of a fleet of transport trucks. The dependability objectives can be specified by the customer or user, or determined by means of requirements analysis, or both, where the latter approach is used to verify the adequacy of the former for assurance purposes. The stated objectives of system application in operation define the scope and requirements of a specific dependability project.

The process for defining the scope of dependability projects is described in the dependability engineering process outlined in Chapter 2, Figure 2.1. The steps for the concept/definition stage of the system life cycle include requirements definition, where the dependability needs associated with the system application is identified, and requirements analysis, where the system operational scenario related to the expected dependability performance is determined. These activities ensure that relevant dependability characteristics are specified in the system as part of its performance functions. The prerequisite for specifying system dependability is to fully understand the purpose and objectives of the system.

3.1.2 Understanding the System

A system is a physical/virtual entity: physical because it has mass and occupies space, and virtual because it may contain information and data. Technological systems are man-made and created to serve specific purposes. A system has its inherent properties that show the desirable system features created by design and construction. For a dependable system that can inspire confidence in fitness for use, the system properties should exhibit user trust in the features of its operation and demonstrate capability in providing the desired performance. The main groupings of system properties from a dependability perspective include the following:

(a) functionality: the extent to which the processing, monitoring, control and other functions are provided;
(b) performance: the extent to which the provided functions can be executed under defined operational and environmental conditions;
(c) operability: the extent to which information can be effectively communicated via the human interfaces and established protocols;
(d) supportability: the extent to which the system can be supported and maintained for continual and trouble-free operation;
(e) dependability: the extent to which the system can be relied upon to perform its intended functions as and when required under given conditions;
(f) application specific: the extent to which the system can be designed for risk avoidance and risk containment, such as safety precautions, security operational measures and immunity to specific interference.

Specific performance characteristics relevant to the system design functions can be selected and assigned under each main grouping of system properties [1]. This is to identify the specific performance characteristics relevant to the design of system functions. Figure 3.1 shows an example of system properties and related performance characteristics. The performance characteristics represent the system output as perceived or experienced by the end users. The specific system performance characteristics reflect the selected functional and non-functional requirements designed and incorporated into the system.

A system during operation is exposed to external influencing factors that interact with and affect system performance and are generally known as conditions of use. The areas of influence or domains affecting the condition of use include the following:

(1) task requirements imposed on the system;
(2) human interface with the system;
(3) processes involved with system operation;
(4) environment to which the system is exposed;
(5) support services available to the system;

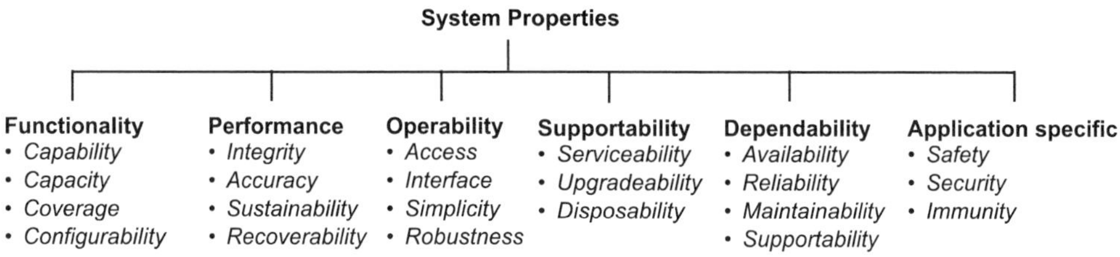

Figure 3.1 System properties and related performance characteristics

(6) utilities needed to operate the system;
(7) external interacting systems.

Relevant influencing factors identified in these domains can be selected and used to determine the risk or impact of influences external to the system for specific application. The evaluation of the system performance characteristics associated with their respective influencing factors can provide a means for identification of constraints and limitations of the condition of use in system operation. Figure 3.2 shows an example of influencing factors grouped under each domain of influencing conditions.

The information presented in Figure 3.1 and Figure 3.2 permits capability assessment of the system operating under given conditions and constraints. The profile of a specific assessment provides the system's condition of use. The assessment results are used to determine the system's capability to withstand or meet the demands and duration of its assigned performance objectives.

A process model can be used to represent a system in operation consisting of:

- system inputs showing user action, application demand or project operating profile initiation;
- system outputs showing response to user action, successful completion of performance outcome or achieved objectives;
- system properties showing inherent design capability to transform input demands into desirable output performance of the required system functions;

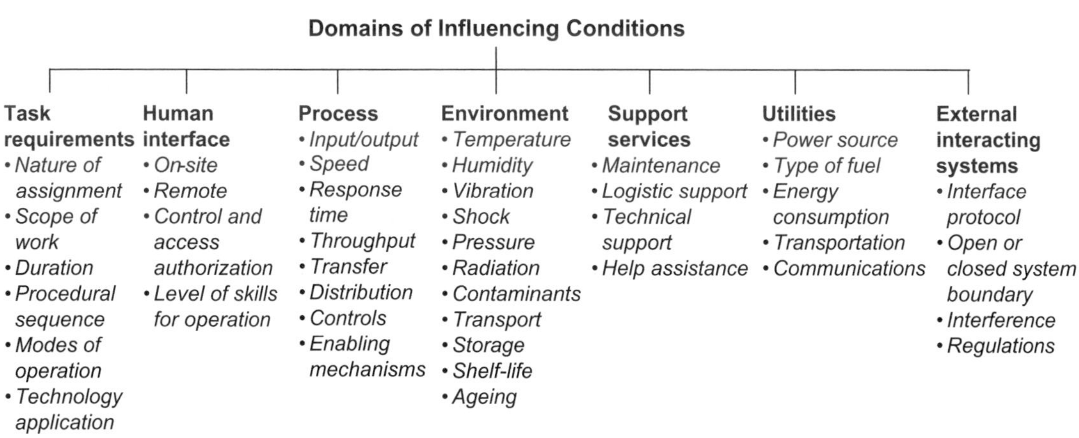

Figure 3.2 Domains of influencing conditions and related influencing factors

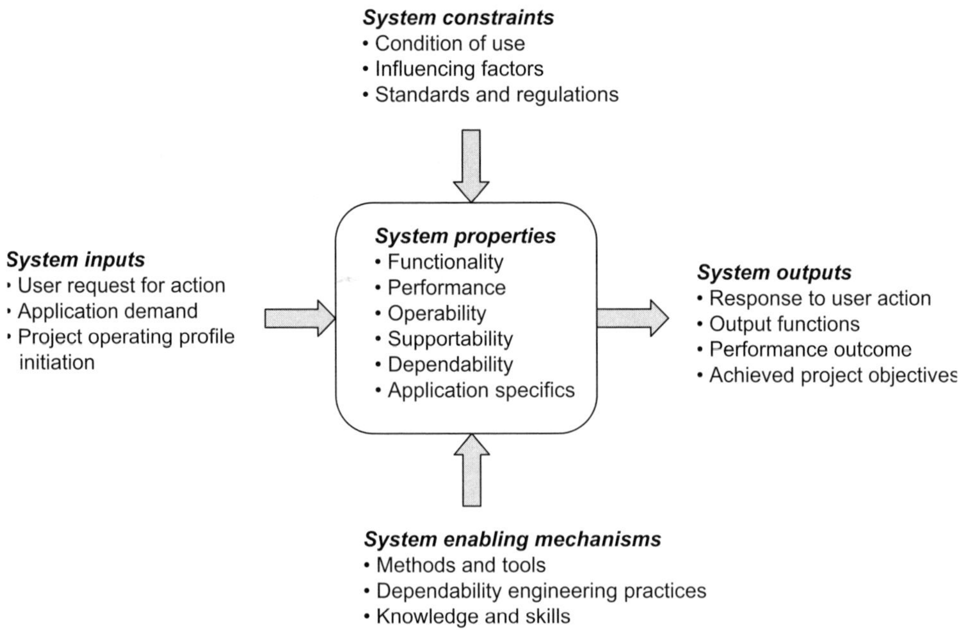

Figure 3.3 Relationships of system inputs and outputs subject to system constraints and enabling mechanisms

- system constraints reflecting the applicable conditions of use due to influencing factors affecting the inherent system properties;
- system enabling mechanisms acting as catalysts or agents to provide appropriate methodology to complement and assist in system development without directly becoming part of the system output functions during operation.

System dependability performance in system operation is achieved by ascertaining the appropriate transformation process of system inputs into desirable outputs. The relationships of system inputs and outputs subject to system constraints and enabling mechanisms are illustrated in Figure 3.3.

A system dependability specification resulting from the input-output transformation process cannot be completed in isolation. It requires the input of detailed information at the system planning stage to determine how the system is intended to perform for the entire duration of its operation as defined in the life cycle. This effort is essential to permit identification and selection of dependability value attributes, including other application-specific performance characteristics, and justification for design trade-offs and system architecture optimization.

3.1.3 Defining System Objectives

A system is designed for a purpose. A system must have a defined set of objectives to achieve its purpose. For example, the purpose of a remote monitoring system is to provide video and communication services for connecting remote sites. The objectives may include accessibility and connectivity of the remote sites for communications, high-definition picture display and sound quality, reliability and safety in service operation, and ease of installation and facility upgrade. A system may have a specific objective to perform a dedicated task, such as an aircraft carrying cargo to reach a delivery target. The objectives of a system may include the completion of a sequence of tasks, such as delivering different payloads to different destinations. Defining the system's generic or specific objectives to meet its intent and purpose is an important precondition in specifying system requirements.

A system with multiple functions and complex operational scenarios, such as a power generation facility or petrochemical plant, often involves external interacting systems to achieve its objectives. A system may also evolve with time, resulting from enhancements of its performance capability, to sustain service demands in operation and to leverage market competition. Examples include upgrading the performance efficiency of a communications network, termination of obsolete utility services and introduction of new product features.

System objectives are defined to establish relevant system requirements by means of requirements definition, which describes the user needs of the system functions or services. The information elicitation process includes the following activities:

(a) identifying market needs and customer or user expectations of the system for specific application and operating environment;
(b) identifying the key system functions and performance characteristics;
(c) identifying the constraints and influencing factors related to system operation;
(d) evaluating feasible design approaches and possible system solutions;
(e) developing system requirements for design specifications.

Figure 3.4 shows the flow of social, economic and business environment information that influences the requirements definition of system application in project operation.

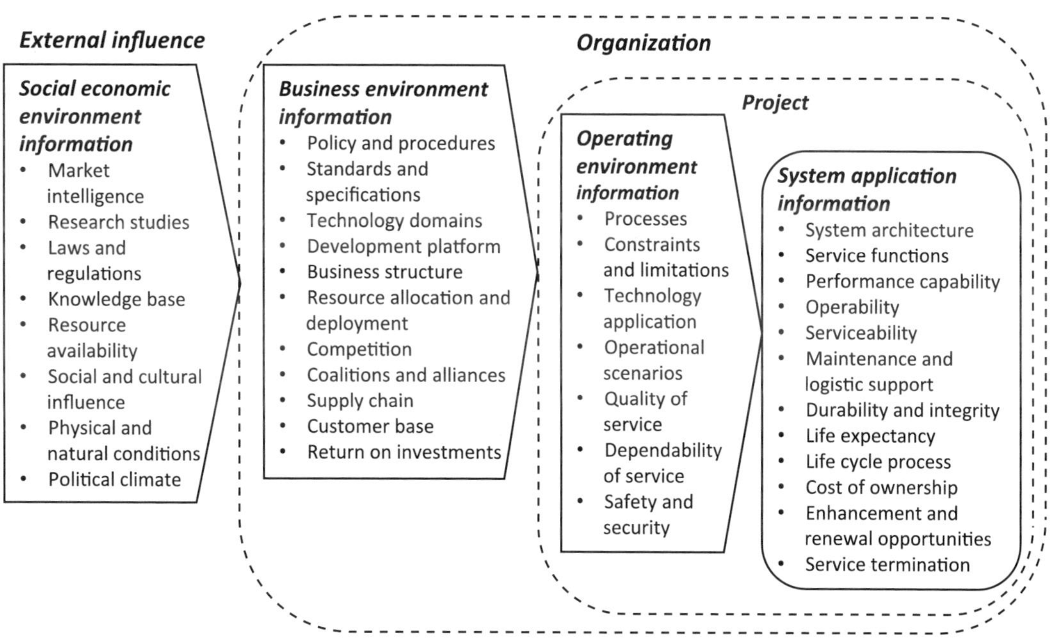

Figure 3.4 An overview of information flow for requirements definition

Organizations involved with technological system development and application are generally sensitive to information derived from external sources concerning the latest technology trends, innovative processes and industry best practices. Such information sources are highly sought after by industries. The crucial information has extensive influences on decision-making processes in most organizations dealing with technology. The respective information is prudently scrutinized and wisely utilized and appreciated in strategic planning, new business development, fine-tuning of product line portfolio management and facilitation in capital investments. Sometimes such information can be elicited from stakeholders or deduced from engagements in joint business ventures, industry- and government-sponsored projects, academic research studies and competitive insights in response to a customer request for proposal. An organization establishes its policy and management structure according to the knowledge of its business environments and customer base to permit pursuance of strategic goals and return-on-investment objectives. Organizational procedures and technical directions permeate project assignments and resource allocations to support system development and enhancement. In this respect, requirements definition facilitates relevant information capture to initiate a meaningful system design framework based on well-defined system objectives.

3.1.4 Identifying System Performance Functions

The defined set of system objectives represents the desirable outcome of system performance during operation. System functions describe the major system capabilities, application conditions and constraints. The term "function" is interpreted as a specific process, action or task that a system is able to perform. A system usually consists of a selected set of functions in order to accomplish the necessary tasks as an outcome of system performance. System performance reflects the collaboration of selected functions to provide the intended services to satisfy customer or user needs, or to deliver and complete the defined tasks to meet system requirements. Functional requirements describe the functionality of the system or what the system can do. Performance requirements describe the characteristics of performance functions as to how well the system performs its functions under given conditions.

The process of identification of system functions is aimed at developing specific system performance to meet project objectives. It establishes the hierarchical structure and relationships of the system and its constituent functions. The process provides a plausible architectural design representing a feasible system configuration. Identifying relevant system functions based on information derived from requirements definition is part of the transformation process from customer or user needs to system requirements prior to starting requirements analysis.

For example, a remote monitoring system may be represented by a system configuration consisting of three major sub-systems to meet its performance objectives. The interrelated sub-system functions are designed to provide remote video and communication services: 1) communication link, 2) audiovisual display and sound, and 3) configuration of hardware components. The interaction of these three functions is required to achieve the functional requirements of remote monitoring and deliver the dependability of service to meet performance requirements of the remote monitoring objectives. Table 3.1 illustrates the relationships of system objectives, functions, functional requirements and performance requirements for this example.

Functional requirements refer to the physical/virtual system functions needed to deliver operational performance and complete the designated tasks. They are expressed in the context of an architectural design configuration that reflects the capabilities of the

TABLE 3.1 Relationships of system objectives, functions and requirements

System and Sub-Systems	Objectives	Functions	Functional Requirements	Performance Requirements
Remote monitoring system	Provide video and communication services for remote monitoring	Capable of end-to-end audiovisual communications	Video monitoring and control of communication session	Dependability of service
Sub-system 1: Communication link	Achieve communication connection	Establish and maintain communication links of remote sites	Broadband network connection of remote sites	Accessibility, connectivity and availability
Sub-system 2: Audiovisual display and sound	Provide high-definition picture and sound quality	Control and deliver audio/video transmission and reception	Real-time audio/video display and reception, control of communication session	Reliability, quality and safety
Sub-system 3: Configuration of components	Ease of operation and facility upgrade	Simple to set up and disassemble components	Plug-in components for installation and replacement	Maintainability and operability

collaborating system functions in operation. Functional requirements are statements that can be verified to meet system objectives and are qualified by measurable conditions and bounded by constraints.

The capabilities of system functions are derived from technology applications and the incorporation of appropriate hardware and software elements. The interacting elements are selected and designed to provide specific characteristics capable of delivering the system functions and to carry out the tasks once these functions can be realized.

Performance requirements define the extent to which and how well the system functions and their designated tasks are to be performed. They also reflect non-functional requirements, which specify for the relevant functions the incorporation of associated performance characteristics, such as dependability. System performance characteristics can be expressed qualitatively or measured quantitatively for compliance to stated functional and performance requirements.

Figure 3.5 shows some typical performance characteristics applicable to technological systems.

The system is exposed to an external environment and various conditions of use. System functions are required to perform tasks according to planned activities or following an established event sequence within the operational scenario. Maintenance actions are to be performed as required to sustain system performance.

Dependability characteristics are time-dependent characteristics that interact and collaborate within the system by designated system functions to achieve the intended performance characteristics. The resultant activation of performance characteristics, responding to the variation of the dependability time function, exhibits the special functional features or demonstrates the specific task completion in performance operation. Dependability characteristics in this respect represent the specific performance values as inherent properties of the system by design and construction.

The following are examples of performance characteristics typical of technological systems. They are explained in the context of system application of performance functions.

- *Capability*: the extent to which the system is designed to provide the required performance functions to meet service demand;

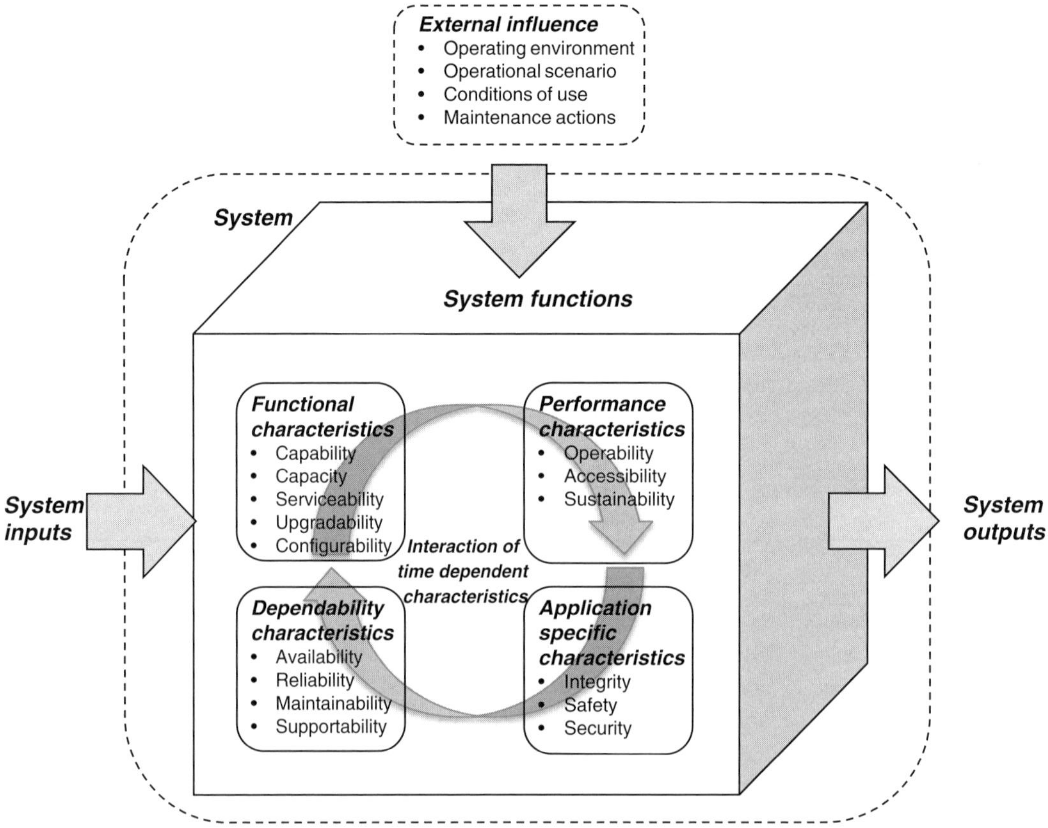

Figure 3.5 Typical performance characteristics applicable to technological systems

- *Operability*: the extent to which the system functions can be easily controlled and operated;
- *Serviceability*: the extent to which the system is designed to permit user access to the system service functions, and retention of the service functions for the duration requested once obtained;
- *Accessibility*: the extent to which the service provision of the system functions can be obtained and used;
- *Supportability*: the extent to which the system can be supported and maintained for continual successful operation;
- *Sustainability*: the extent to which the system or system function is designed to endure over time and maintain a level of service performance under given conditions;
- *Durability*: the extent to which the system or system function is designed to perform and sustain service operation over time before reaching its degradation limit;
- *Integrity*: the extent to which the system is designed for stable and robust operation, and consistency in service performance and use;
- *Safety*: the extent to which the system is designed to prevent harm in operation to people and the environment;
- *Security*: the extent to which the system is designed for risk avoidance and risk containment.

It should be noted that performance characteristics are used to describe a system or its functions for specific applications. A performance characteristic such as integrity may be interpreted differently according to the specific application function of the system. For example, the integrity of a pipeline system refers to the reliability and safety of the physical pipeline structure for transporting gas or a liquid, whereas the integrity of an information system refers to the security of the data throughput and storage. In this example, integrity has different interpretations in the context of different technology applications of the system function. The hierarchy and relationships of the characteristics cannot be determined until the system objectives and functions that characterize system performance are defined.

3.2 INCORPORATION OF DEPENDABILITY INTO SYSTEMS

3.2.1 Requirements Definition

The purpose of a requirements definition is to formulate a viable framework for the development of system requirements. Information and relevant data are used to gauge the market environment for business opportunities, identify resource availability for joint venture needs, benchmark the capabilities for technological thrusts, and plan strategy for project initiation. These are high-level strategic plans and crucial information needed to identify potential risks and rationalize project management decisions. Each project has its own specific mandate. Emphasis is placed on gathering information on customer or user expectations and concerns associated with dependability of service. The following presents a generic process for identifying risk impacts and business influences associated with requirements definition on technological systems for project consideration:

(a) identify system objectives and intended application in line with the corporate vision and business goals;
(b) elicit information from stakeholders and special interests on desired dependability performance of system features and new service offerings;
(c) identify potential customer base, distribution channels and service targets for system implementation;
(d) identify potential market share and sales volume, competitive advantages and projection for business growth;
(e) identify the scope and boundary of the system operation for provision of customer services;
(f) identify target budget estimates, project application conditions, technical constraints and process limitations in system development, realization and implementation;
(g) identify potential environmental impact resulting from activities generated by the project;
(h) identify applicable standards for system requirements and regulatory compliance;
(i) identify capital investments for new or dedicated assets required for project initiation;
(j) identify resource requirements and availability for project assignment on start date;
(k) identify support requirements for outsourcing and third party contracts;
(l) identify training requirements for technical update and development collaborations;
(m) justify benefits and advantages for timing of project initiation;

(n) develop project plan and risk management schemes;
(o) provide work statements and identify project deliverables and target completion schedules;
(p) establish project review schedules for monitoring progress;
(q) develop contingency plan for alternate plan execution and methods of introduction;
(r) incorporate information feedback and reporting systems.

3.2.2 Requirements Analysis

Requirements analysis is intended to examine and verify the transformation of customer or user needs into technical requirements of the system functions. The analysis process provides the essential technical information needed for development of system requirements and design specifications. The purpose of requirements analysis focusing on dependability engineering application is to foster the determination of:

(a) relevant characteristics of functional and performance requirements of the system functions;
(b) constraints affecting the design and realization from a dependability perspective of the system functions;
(c) traceability of significant data sources for potential risk identification;
(d) information database for development of system requirements and system dependability specifications.

The basis for requirements analysis starts with establishing a feasible configuration for the system of interest. The analysis process is both iterative and recursive to arrive at a plausible solution in determining the capability of the system to meet performance objectives.

A practical approach to perform requirements analysis is to construct a matrix displaying rows of major system properties (Figure 3.1) against columns listing the domains of influencing conditions (Figure 3.2). A matrix template as presented in Table 3.2 can serve as a guide to facilitate analysis and information capture for the requirements analysis.

Each matrix cell contains a set of relevant influencing factors that affect the performance characteristics of a specific system property. The relationships of these two sets of requirements information are examined for analysis of possible impacts contributing to the outcome of system performance. Technology applications and project resource conditions are often the key driving factors in the process for requirements analysis of technological systems. The matrix template helps to identify and delineate the specific influencing factors affecting the respective system performance characteristics.

The first matrix cell relates 1.1 Task requirements – Functionality. In the remote monitoring system example, the task requires end-to-end connection of remote sites. The Internet service function could provide the practical communications link. The 1.5 Task requirements – Dependability cell relates the time dependency of dependability association. System performance would demand high availability, such as >99.99%, to achieve dependability of service for the remote monitoring system.

It should be noted that not all matrix cells are relevant in the analysis process. This will depend on specific dependability project requirements and the nature of the system of interest. The relevance of the applicable cells can be determined by assigning numerical ranking of importance related to the desirable performance outcome of the system objectives. The requirements analysis should only consider relevant cells in the matrix.

TABLE 3.2 Requirements analysis matrix template

System Properties / Domains of Influencing Conditions	Functionality	Performance	Operability	Supportability	Dependability	Application Specific
Task requirements	1.1 Identify relevant functions to meet task assignment and technology and economic constraints imposed on the functions	1.2 Identify the adequacy of the functions to perform tasks and complexity in applications affecting performance	1.3 Identify operational needs and interfaces affecting the normal and specific operation of functions	1.4 Identify logistic support services to maintain operation of functions	1.5 Identify the technology impact on availability, reliability, maintainability, and maintenance support performance	1.6 Identify safety, security, regulatory constraints and legacy issues for specific application
Human interface	2.1 Identify the involvement of human interface with the functions	2.2 Identify skills needed for human performance of functions	2.3 Identify the degree of difficulty in human operations with the functions	2.4 Identify human resources needed to support the functions	2.5 Identify the reliance of human intervention affecting dependability performance	2.6 Identify human aspects in specific application of functions
Process	3.1 Identify procedures and methods for operating the functions	3.2 Identify the accuracy, consistency and repeatability in operating the functions	3.3 Identify ease of use and access to operate the functions	3.4 Identify the extent of maintenance and logistics support efforts to sustain operation of the functions	3.5 Identify availability and reliability in operating the functions	3.6 Identify any special processes needed for operating the functions
Environment	4.1 Identify the dominating environment affecting the design and operation of functions	4.2 Identify operating limits or restrictions in the performance functions exposed to the environment	4.3 Identify limitations to access and use of the functions exposed to the environment	4.4 Identify limitations to support or service the functions exposed to the environment	4.5 Identify the effects on dependability performance of functions exposed to the environment	4.6 Identify any special precautions when operating in extreme or hostile environments

Continued

TABLE 3.2 Requirements analysis matrix template (Continued)

System Properties / Domains of Influencing Conditions	Functionality	Performance	Operability	Supportability	Dependability	Application Specifics
Support services	5.1 Identify the need for support services to maintain operation of functions	5.2 Identify the effects of support services needed to maintain accuracy of performance functions	5.3 Identify the effects of support services to enhance system operability	5.4 Identify any special support services needed for system modification, upgrade or disposal	5.5 Identify the effects of support services to sustain dependable performance of functions	5.6 Identify any special support services needed for specific application
Utilities	6.1 Identify infrastructure for operation of functions	6.2 Identify the adequacy of utilities to sustain performance functions	6.3 Identify the effects of utilities to enhance system operability	6.4 Identify any special power, energy, fuel, etc., to support system operation	6.5 Identify the needs and effects of utilities to sustain dependability performance	6.6 Identify any specific tools or enabling systems needed for specific application
External interacting systems	7.1 Identify the influence of interacting systems on the functions	7.2 Identify the effects of interacting systems on performance functions	7.3 Identify the dependency of interacting systems on system operability	7.4 Identify the system boundary and jurisdiction for supportability of interacting systems	7.5 Identify the effects of interacting systems on dependability performance	7.6 Identify the interacting systems for specific application

Completion of the requirements analysis for a specific project provides the essential information to construct a baseline configuration to meet system objectives. The system configuration identifies the necessary system functions and their hierarchy and relationships as a starting point for further evaluation. Each system function identified represents a functional requirement of the system. The technical approach to realization of the system function has yet to be determined. The realization process should consider various options relating to the use of hardware and software elements, as well as the need for human intervention. Performance requirements of system functions cannot be validated until the realization process is complete.

3.2.3 Establishing the Operational Scenario

The purpose of establishing the operational scenario is:

- to identify the sequence of events or tasks involved by the system in operation;
- to describe specific task requirements of system functions and determine performance capabilities of the system's operational usage;
- to set criteria for continuation of tasks under normal operation or emergency condition.

A system operational scenario describes the task requirements of operational usage and event sequence in system application. The following aircraft performance scenario shows the operations of takeoff, in flight and landing. An operating profile describes the specific sequence of task performance by the system to achieve project objectives. The operating profile represents a specific operational scenario for the system.

The example of aircraft performance identifies takeoff (Task 1), in flight (Task 2) and landing (Task 3) as representing the performance tasks in normal operation. The aircraft may also operate under abnormal situations, such as changing the flight altitude due to air turbulence (Task 4), landing at an alternate airport due to bad weather conditions (Task 5), or an emergency landing due to a mechanical malfunction (Task 6). In an abnormal situation, a specific operation to mitigate the task situation is required. Figure 3.6 shows the relationships of the operating profile and system operational scenarios from a system life cycle perspective.

The essential key system performance functions to complete each designated task are determined from the technical knowledge and configuration of the system functions capable of delivering the task performance. In the scenario example for aircraft takeoff, the key system functions include the capabilities of a) engine propulsion, b) landing gears to taxi on runway, c) navigation and instrumentation, and d) piloting the aircraft under the air traffic controller's instructions.

Associated with each specific task, there are specific activities and conditions set forth for the aircraft to perform according to the flight plan to complete the mission. Criteria are established at the completion or start of each task to determine its success before continuation or to abort further operation.

3.2.4 Determining Dependability Requirements

Dependability requirements form part of the system requirements. Project activities advocate the need for early determination of dependability requirements pertinent to the system objectives in operation. The system consists of a set of system functions to perform the intended tasks. It is essential to incorporate the appropriate time-dependent dependability characteristics relevant to the system functions. Determining dependability requirements of

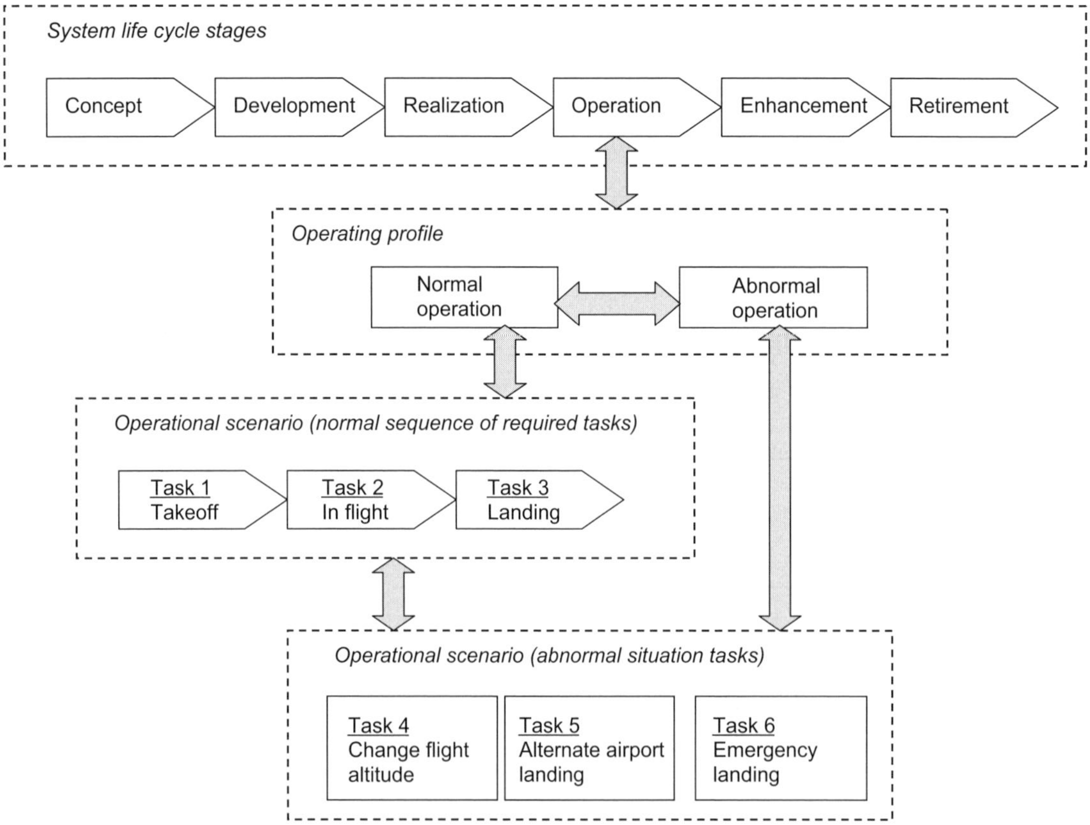

Figure 3.6 Relationships of project operating profile and system operational scenario

the system function requires specific technical information and knowledge of its intended application [2]. The design and implementation of a plausible system function need to consider the functional application and influencing factors that may affect the realization of their outcome. This includes the incorporation of relevant dependability characteristics pertinent to the functional performance in system operation.

The classic definition of dependability is the ability to perform as and when required. From a system perspective, dependability can be interpreted as the system's capability to perform upon demand under given conditions. It also implies that the system is able to complete its objectives or the assigned tasks. Dependability is a collective term expressing the time-dependent characteristics that can be incorporated into the system performance functions.

For example, the dependability of an aircraft can be interpreted as its ability to perform the operations of takeoff, in flight and landing, with the implied safety needs for the flight mission. For the aircraft takeoff operation, the key system performance functions require a) engine propulsion, b) landing gears to taxi on runway, c) navigation and instrumentation, and d) piloting the aircraft under air traffic controller's instructions. Each of these key functions may consist of sub-functions with contributing specific performance characteristics, such as the durability of brake assemblies as part of the landing gears. The aircraft safety needs, where applicable, may be extended down to the lowest level of sub-functions, designed or selected to perform with specific characteristics, such as the durability of nuts and bolts to withstand vibration and shock when used in the brake assemblies.

The following summarizes the dependability requirements of the aircraft takeoff operation. The descriptions are greatly simplified to convey the concept of dependability application rather than to expound on technical details of aircraft design and operation.

(a) Engine propulsion: the availability and reliability of engine capability with power to propel the aircraft from a stationary ground location to airborne cruising speed; maintainability to facilitate engine overhaul and maintenance services.
(b) Landing gears: the reliability and robustness of the landing mechanisms and wheels to sustain the taxiing duration on a runway for takeoff; the durability of landing gear component assemblies such as the brakes.
(c) Navigation and instrumentation: the reliability of instruments and cockpit displays to facilitate navigation and performance of flight procedures; the integrity of flight data and information for communications.
(d) Piloting: the reliability and credibility of the pilot's ability and experience to fly the designated aircraft with instructions and information communicated with the air traffic controller.

A system with multiple functions and complex operational scenarios often involves external interacting systems to achieve its objectives. In the aircraft takeoff example, the success of the pilot's task completion in takeoff is dependent on the integrity of information provided by the air traffic controller. The control tower facility is where the air traffic controller's operating station may be considered as an external interacting system to the aircraft system in mobile operation. The control tower facility has a different operational mandate, such as directing multiple aircrafts for takeoff and landing with rigorous safety operation rules and air traffic regulations. The takeoff operation has to take into account various environmental factors and runway conditions such as visibility, aircraft size and weight, runway length and obstacles.

The aircraft takeoff and landing operations share the same key system performance functions and sub-functions. The task difference between takeoff and landing is the differentiation in the pilot's operating procedures. The landing operation may encounter additional conditions such as runway approach, traffic congestion, fuel limitations and a flight path holding pattern. For these and other prevailing reasons, the aircraft performance scenario identifies separate task operations for normal landing, alternate airport landing and emergency landing as shown in Figure 3.6. The implied safety needs for the original flight mission will change accordingly as necessitated by the task change for an abnormal landing operation. Specific dependability requirements and performance characteristics have to be incorporated into the respective system functions to reflect the changed situation, such as an emergency task operation.

The dependability performance characteristics for a takeoff operation are presented by the set of dependability requirements relevant to the specific sub-functions according to the system hierarchy and event sequence of aircraft operational usage demands of functions and sub-functions.

Figure 3.7 presents a simplified view of a specific set of dependability performance requirements for illustration purposes. It should be noted that many of the functions and sub-functions in aircraft operations are shared and used among designated tasks, such as the engine propulsion sub-function, which is needed for aircraft takeoff, in flight and landing. There are other aircraft functions, such as the fuselage, aircraft stability control and on-board communications, that are not included in Figure 3.7. Additional dependability requirements are required to form the basis for development of a complete set of aircraft system dependability specifications.

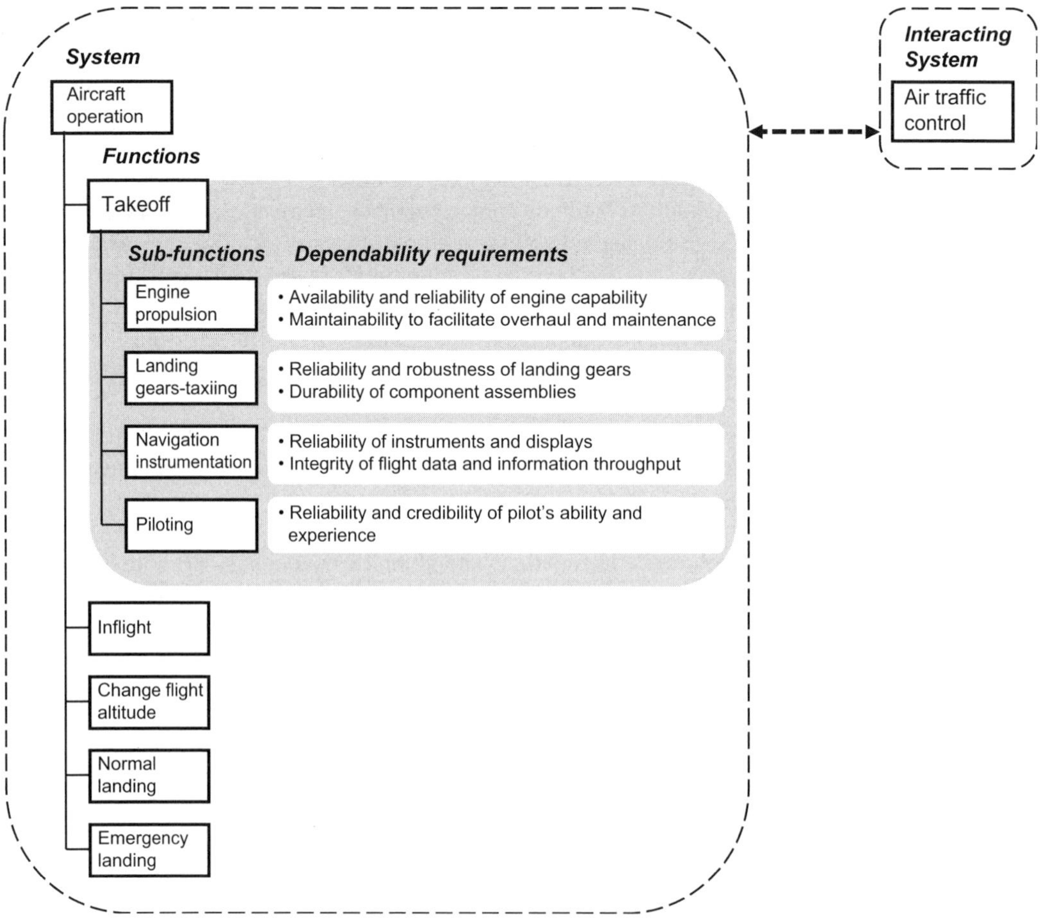

Figure 3.7 Dependability performance characteristics for aircraft operation

Each system function has its own unique dependability requirements consisting of specific performance characteristics. Selective system functions collaborate and contribute to the successful completion of designated tasks to achieve system objectives. Each contributing system function or sub-function has its own specific performance characteristics to designate its respective dependability value or trustworthiness. Dependability measures are not necessarily quantitative because the measurement values are achieved by a combination of diverse performance characteristics. The main dependability characteristics, such as reliability performance, can be quantified and measured in terms of probability of success. Other application-specific dependability characteristics, such as data integrity, can be expressed in qualitative terms of credibility of information throughput. There is no single, unified, measurable unit that can cover the measurement for all dependability characteristics based on established dependability concepts. In this respect, dependability value can only be expressed collectively in terms of trustworthiness reflecting its specific designated characteristics in value contribution. The dependability concept is not the same as the safety concept. The safety concept is based on a tolerable risk level for safety assessment.

The dependability assessment is based on a subjective value judgment as perceived by the experience of observers and users from their diverse viewpoints. This interpretation of dependability is based on the definition of dependability and its collaborating characteristics [3] with added practical experience.

It should be noted that dependability characteristics considered for a system depend upon its application. The influencing factors vary with application; hence, different dependability characteristics are emphasized. Some applications may include additional characteristics such as integrity, safety and security; other applications could be satisfied with a single main dependability characteristic, such as reliability. While some applications use availability as a singular performance measure for dependability, others do not, thus precluding generic standardization. For dependability engineering application in technological system projects, the IEC-recommended standards on dependability management and technical processes [4] should be referenced. This will facilitate international trade and global standards harmonization. The value of dependability is measured by its contributing performance characteristics to reflect specific standards conformance in system application. There have been some recent suggestions among IEC experts that standardization of dependability levels would simplify the specification of standards on dependability requirements. So far, no serious studies on this subject have been noted in the published literature or presented in propositions for technical discussion. A review process is necessary to justify the rationale for any proposed change of the current criteria in dependability concept and methodology for acceptance. The impact of dependability levels in contract specifications and business agreements has yet to be determined and evaluated.

3.3 FRAMEWORK FOR DEVELOPMENT OF DEPENDABILITY SPECIFICATIONS

3.3.1 Rationale for Dependability Specifications

A specification is an explicit set of requirements to provide a detailed description of the system of interest for the purpose of development or validation. The specification describes specific system functional and performance requirements, constraints and design characteristics, and may include qualification conditions and procedures for each stipulated requirement.

A dependability specification is used to characterize system functions or sub-functions in terms of dependability performance characteristics and provides specific quantifiable units or limits for verification or measurement. A service level agreement [5] is an example where dependability specification limits, such as availability and downtime, are used to define acceptable standards for compliance to quality of service [5] contracts in provision of public utility services. Numerical units or limits typical of specifying dependability characteristics as verifiable system performance parameters are expressed as:

- availability: percentage uptime of the system in operation, 99.99%;
- failure rate: one failure per 1,000,000,000 hours or $1/10^9$;
- expected life: 25 years;
- downtime frequency: 1/year;
- outage duration: < 1 minute;
- restoration time: < 30 minutes.

Dependability specifications are commonly quoted in product information statements, used in development contracts, and stipulated in acceptance agreements for complex system projects. The conditions set forth in specifying dependability requirements for critical performance parameters should be rationalized and evaluated to minimize risk or consequences of non-compliance. Prudent engineering judgment is needed to ascertain the reality of the guaranteed acceptance conditions in specification statements. Typical examples that need rationalization and justification include:

- lifetime warranty;
- demonstration of 99.99999% reliability;
- guaranteed system performance of 4 minutes downtime in 40 years.

In this respect, one of the key success factors in meeting dependability requirements in projects involving technological systems is to be prudent in the validation and assurance of critical dependability characteristics stipulated in the contract agreements. This may include the requirement of a contract specification for a system reliability demonstration.

A technological system development project case is used as an example. The development contract specified a system reliability demonstration during the first year of field operation under warranty prior to final acceptance of the system. In this case, the inherent reliability failure rates from outdated proposal estimates were used inadvertently as criteria for acceptance. This seemingly minor oversight created a serious problem in proving successful achievement of the specified reliability demonstration target. No credible results were observed for the whole year of testing. The reliability demonstration in this case incurred extensive cost in the acceptance process and created a negative perception of system reliability. It took months to resolve the problem that should have been properly addressed up front. Common practice in business arrangements is to withhold a percentage of the full contract payments until final acceptance when warranty conditions have been cleared. The case in point is that commitments to an unrealistic requirement–in this case, the inherent reliability failure rates used as basis for reliability demonstration–would likely have difficulties in reaching contract acceptance by all concerned parties. It should be noted that inherent reliability calculated from handbook component failure rates generally represents an ideal design target. It often reflects an optimistic expectation in prediction forecast rather than what would be actually experienced in field operation. There are discrepancies between predicted reliability and actual field performance reliability that need serious consideration when used in practice.

3.3.2 Evaluation of Dependability Characteristics

The evaluation of dependability characteristics is an essential process to determine the significant impact of influencing conditions affecting the outcome of system performance. The guidance for the requirements analysis in the matrix template (see Table 3.2) is intended to cover general applications for all systems. Selective matrix cells applicable to technological systems are addressed in the context of dependability application and possible implications. The following describes some examples of relevant influencing factors for evaluation of technological systems based on practical experience and the significance of their impact. References to matrix cells are identified in brackets.

(a) Economic constraints (matrix cell: 1.1 Task requirements – Functionality)
Economic constraints are often due to budget restrictions, spending limitations or market timing and competition issues for new system development and legacy

system enhancement. Economic constraints influence project management decisions and often affect dependability achievements. Early planning regarding design-to-cost, supply chain management and life cycle cost analysis will help to identify potential problem areas, provide insights to address cost avoidance and identify potential benefits and opportunities for improvement. Project risk analyses are often used to determine the criticality of the situation permitting the project management to chart an appropriate course of action.

(b) Regulatory constraints (matrix cells: 1.6 Task requirements – Application specific; 1.2 Task requirements – Performance)
Regulatory constraints may be imposed on systems such as public utility services and pipelines, where considerations for safety and environmental impact are important and of public concern. The applicable legislation generally defines the restrictions, limitations and recommended approaches for problem resolution or possible mitigation. Impact on dependability may limit the choice of technology applications and affect the efficiency of system performance.

(c) Type of system application (matrix cells: 1.2 Task – Performance; 3.5 Process – Dependability)
The type of system application determines the technology and configuration of the system design and its operation under given conditions. Dependability design needs to consider usage scenarios that reflect the incorporation of technologies employed for automation to facilitate operation. For example, an automated production line is designed to minimize human intervention in the manufacturing process, hence, reducing the needs and utilization of assembly personnel. The automation process can also incorporate sensing equipment in the production line to detect non-conformances and test for acceptance to enhance quality control and product assurance. This system performance outcome would also improve dependability. It should be noted that any introduction of new products or technologies would require a learning period for customization.

(d) Criticality of operation (matrix cell: 3.5 Process – Dependability)
Criticality is the extent to which the events of system failures may have severe consequences. Criticality of operation implies that specific system functions responsible for critical operation should be designed for risk avoidance and the mitigation of consequences in the event of failure occurrence. Backup support, redundancy configuration and timely human intervention should be considered in system operation and instituted in operating procedures. Dependability design and operation of critical systems should consider implementation of appropriate methods and tools for fault prevention, fault tolerance and fault forecasting to facilitate failure resolution. Effective restoration is another dependability characteristic used to address the criticality of system operation.

(e) Dependency of interacting systems (matrix cell: 7.3 External interacting systems – Operability)
Dependency of interacting systems is essential for most applications of technological systems. The type and nature of external systems needed for interaction to complete the system mission should be determined in the requirements definition and requirements analysis process. Dependability performance to achieve system interoperability should consider proper protocols, compatibility of

interconnections, sequencing and synchronization for access, retention and disengagement of the interactions. Authorization and authentication of interacting system connections are often required.

(f) System configuration and structure (matrix cells: 3.5 Process – Dependability; 2.5 Human interface - Dependability)
System configuration is based on the complexity of system functions and the physical elements in the structure for implementation. Dependability characteristics such as reliability can be apportioned and assigned to system functions based on an established system configuration. The system physical structure will depend on the partitioning of hardware and software elements for incorporation where dependability can be assessed and evaluated. Human factor engineering plays a significant role in the design and implementation of system interfaces. The overall system dependability performance can be assessed by users' experience.

(g) Maintenance and logistic support (matrix cell: 3.4 Process – Supportability)
System maintenance and logistic support are essential to sustaining a successful system during operation. The respective system functions need to be designed for maintainability. The level of maintenance should reflect the maintenance strategy for on-site replacement of replaceable units, depot workshop repairs, or return to factory for major overhaul and refurbishment. Logistic support entails the ability of the support system to manage the expediency of replacement items and spares provisioning for the system during operation. Third party maintenance contracts involve a dependency and assurance effort of supply chain management. The efficiency and effectiveness of maintenance and logistic support performance is an important criterion for dependability evaluation.

(h) Application environment (matrix cell: 4.6 Environment – Application specific)
Systems are exposed to application environments. These environments may include climatic, mechanical and electromagnetic exposures. System design should incorporate the appropriate requirements for environmental conditions. Dependability performance of system operation is influenced by the applicable environmental conditions under various classifications of environmental standards [6]. For example, climatic conditions include a controlled climate for a clean room, indoor or outdoor use, and storage; mechanical conditions will vary depending on whether the system is stationary, portable, mobile, being transported, or vulnerable to earthquakes; electromagnetic conditions include exposure in customer premises, outdoor locations, laboratories, and portable and mobile applications. A cell phone handset would likely be exposed to all climatic conditions, influenced by mechanical vibrations and shock for portable and mobile applications, and affected by electromagnetic interference in operation. Industrial equipment may be operated in arctic, offshore or dusty environments.

(i) Utility services (matrix cells: 6.5 Utilities – Dependability; 5.6 Support services – Application specific)
Technological systems are not always self-contained in every aspect when in service. Some systems are designed to use local utility services such as electricity, fuel, water, and third-party maintenance support services. The quality and dependability of local utility services should be thoroughly assessed prior to reaching long-term agreements with the utility service providers. The ownership of the

system and on-going long-term operation and maintenance responsibilities should be clearly established. Some systems dedicated for special applications in remote locations where no utility supplies are available should consider alternate means for generating self-sustaining utility supplies. Examples include the use of renewable energy generation such as solar, wind and hydraulic power generators with back-up battery or other power sources. Routine on-site maintenance visits are required to ensure sustainment of system operation.

(j) Legacy issues (matrix cell: 1.6 Task requirements – Application specific)
New systems may have to work or interact with existing old systems that use technologies from previous generations. The business goal is to maintain continuity and integrity of the service infrastructure without serious impact or interruption of customer services. Dependability performance should be transparent to the customers or end-users when integration of the old and new systems takes effect. The transition from old to new takes time for familiarization of new features and services replacing the old ones as they are retired. Maintaining legacy assets is due mainly to economic constraints or other resource limitations to develop new systems to completely replace the old ones. A new project takes time for development and implementation. Progressive introduction of new services replacing old ones over a period of time makes business sense. A long-term improvement plan is an appropriate approach to justify stepwise improvements.

3.3.3 Procedures for Specifying System Dependability

The following procedures serve as guidelines for specifying system dependability [7].

Step 1: Identify the system.
The system under consideration should be identified. System identification should include name, intended application and conditions of use or operation.

Step 2: Describe system objectives.
The purpose of the system in terms of its primary application should be stated. The system performance objectives to be achieved or the assigned tasks to be completed should be described.

Step 3: Identify the functions to meet system objectives.
The key functions needed to achieve system performance should be identified. The purpose of each function should be stated from a system requirements perspective. Where appropriate, the relationships of these functions should be clarified.

Step 4: Describe the functions.
Each identified function needed for the completion of system task operation should be described to provide a scope and objective for assessing the feasible realization of the function.

Step 5: Identify influencing conditions.
The influencing factors affecting each function should be identified to assess their impact on system performance. The matrix template in Table 3.2 can be used as a guide for evaluation of dependability characteristics relevant to the completion of system tasks.

Step 6: Evaluate the technical approach to realize the functions.

The technical approach to realize functions should be evaluated. This is to assess the feasibility and practical approaches of achieving functions within established technical application limits. Maintenance and logistic support needs associated with functions to sustain system operation should be addressed.

Step 7: Describe the system elements involved in the functions.

System elements consist of hardware, software and human interactions in the realization process of system functions. Evaluation of feasible designs and design options of system functions is conducted to determine practical approaches and rationalize cost-effective means to realize the required functions.

Step 8: Determine the system operational scenarios.

The system operational scenarios should be determined from a dependability performance perspective. An operational scenario describes the specific task requirements of system functions and determines performance capabilities of the system's operational usage and the sequence of tasks needed for system performance under stated conditions.

Step 9: Describe the system configuration.

System configuration is based on feasible system architectural designs. The selected system configuration associated with the operational scenario for the project operating profile should be described. The relationships of each function involved in a specific configuration should be established to facilitate system design trade-off and assessment. Where interacting systems are involved in specific operations, they should be identified and their boundaries and interfaces established for system evaluation.

Step 10: Determine the dependability requirements.

The dependability requirements of the system functions should be determined by establishing the system operational scenario and the requirements of the specific functions identified in the evaluation process. Each dependability characteristic associated with the key functions should be translated into a qualitative evaluation statement in the specific event sequence of the project operating profile. Where possible and appropriate, a quantitative value should be assigned. This characterization process forms the basis for capturing the dependability specification of system functions to satisfy the dependability objectives.

Step 11: Documentation of the system dependability specification.

The system dependability specification and relevant data captured should be documented. The dependability information should be recorded for future reference to provide design specifications of applicable functions to meet overall system requirements. The documentation can also serve as supporting data to facilitate design review for system verification and modification.

Documentation of the system dependability specification should contain the following data as part of the system specification:

1. System identification;
2. System objectives;
3. System functions;

4. System operating profile;
5. System configuration;
6. Dependability requirements;
7. A statement on system dependability.

Appendix C presents an example of a system dependability specification.

REFERENCES

[1] IEC 61069-1, Industrial-process measurement and control – Evaluation of system properties for the purpose of system assessment – Part 1: General considerations and methodology
[2] IEC 60300-3-4, Dependability management – Part 3-4: Application guide – Guide to the specification of dependability requirements
[3] IEC 60050-191, International electotechnical vocabulary – Part 191: Dependability
[4] IEC 60300-1, Dependability management – Part 1: Dependability management and application
[5] ITU-T Recommendation E.800, Definitions of terms related to quality of service
[6] IEC 60721 (all parts), Classification of environmental conditions
[7] IEC 62347, Guidance on system dependability specifications

Chapter 4

DEPENDABILITY ENGINEERING IN SYSTEM DESIGN AND REALIZATION

4.1 DEPENDABILITY ENGINEERING IN SYSTEM DESIGN AND DEVELOPMENT

4.1.1 Introduction

System design and development is based on well-defined system specifications so that detailed engineering can take place. Most systems consist of a structural framework of designed functions incorporating new design features with existing capabilities, add-on commercial-off-the-shelf (COTS) products and materials that form the basis of fabrication or construction. For example, a pump station will consist of equipment such as pumps, drivers, valves and instrumentation manufactured by an original equipment manufacturer (OEM), and supplemented by items such as a piping assembly and control system software that are incorporated and housed in a building to meet specific requirements.

The design and development activity is carried out at two levels:

- Product development and manufactured items to facilitate selection and incorporation of available OEM and COTS products in system configuration;
- Specific application design and construction to establish system architecture and service functions.

COTS items come with an inherent reliability and maintainability along with prescribed conditions of use and maintenance, which the final design has to take into account. Usually the most important dependability characteristic is availability since this is normally linked directly to the provision of service. Design options such as redundancy and choosing COTS items with the lowest downtime for maintenance can be considered to meet the final system objectives.

As suppliers, the OEM relies heavily on reliability and maintainability for their products. Their long-term success is dependent on delivery of dependable product performance. Their general reputation as preferred supplier factors in greatly with customers when selecting their products among competitors. The overall dependability of the system is quite reliant on the overall system design and specific selection of its components. In contrast to the OEM, the key success factors for the engineering/procurement/construction (EPC) contractors as contracted by an operating company are cost and schedule. Dependability is usually of secondary importance. It is quite important for the operating company to place emphasis on incorporating dependability into the design contract requirements. Appropriate balances for acquisition cost and ownership cost should be ascertained. This will optimize the total life cycle cost with improved performance and provide crucial benefits to the business operation in the long run.

4.1.2 Architectural Design

Architectural design is a process for defining system elements and their inter-relationships to establish a framework for system development. The system elements may be comprised of hardware, software and combinations of the two to derive the needed system functions in meeting system specifications. System architecture can be represented in various forms, such as system breakdown structure, functional block diagram, logic flow graph, physical mock-up and engineering prototype. These architectural design representations serve different purposes for different applications. The architectural design is aimed at rationalizing the system structure and functional contents to provide sufficient data and information for system-level analysis and evaluation. The process is iterative and recursive, hence the necessary flexibility for design trade-offs to achieve system objectives and to optimize performance expectations. Dependability as an enabling mechanism in value creation is most influential in ensuring the adequacy of structural integrity and the capability of functional configuration in satisfying system objectives. It is important to maintain mutual traceability between architectural design and system requirements.

Major dependability contributions to the system architectural design can be summarized.

(a) Dependability planning for system architectural design evaluation – This will enable the establishment of a viable architectural design baseline with dependability focus for system configuration and integration of system functions. The functional interface requirements are incorporated into the architectural design solution.
(b) Determination of the adequacy of technology choices and design options to minimize faults and risk exposures – This will permit selection of the needed system functions to set dependability priorities for system availability allocation.
(c) Establishing the failure criteria of system functions – This will permit selection of appropriate technologies associated with the hardware and software elements for development of system functions.
(d) Determining the adequacy of system availability performance – Limitations of the architectural design configuration and options are examined to provide quantitative values for assessment. Reliability Block Diagram (RDB) [1] and Markov analysis techniques [2] are generally applied at the functional level for system availability determination. Relevant data to support dependability calculations are based on similar application experiences and failure rate estimations of system functions.
(e) Identification and assessment of critical system functions – System availability performance is dependent on the constituent system functions in performance. Failure of a critical function would likely lead to total system outage requiring mitigation. Fault Tree Analysis (FTA) [3] is a top-down method for identifying the criticality of system function. Reliability prediction methods [4, 5, 6] are used for failure rate estimation of functions. Failure mitigation approaches can incorporate redundancy designs, backup functions and activation of graceful degradation procedures to minimize risk exposure due to abrupt changes or an outage of system performance during operation.
(f) Establishing maintainability criteria for design of system functions – This will facilitate development of maintenance support strategies for testability and diagnosis of system failures and malfunctions.

The architectural design process transforms the design concept for the intended system application to realizable solution that meets system specifications. The process identifies and explores implementation strategies consistent with the system's technical and business requirements and risks. The architectural design solution is defined in terms of the requirements for the set of system functions for system configuration. The specified requirements resulting from the architectural design process form the basis for functional design and evaluation.

4.1.3 Functional Design and Evaluation

4.1.3.1 Establishing Functional Design Criteria

A function is designed to perform a system task or a series of related tasks. In establishing the functional design criteria, the technical and business requirements should be considered. This is due to possible impact on practical realization and evaluation of the function. Dependability engineering activities involve analysis and evaluation as a general methodology for dependability assessment. Chapter 3 has extensively addressed the performance characteristics and influencing factors on system functions and their relationships. The functional design criteria should consider the following:

- functional requirements and application environments;
- timing and budget for development or acquisition of the function;
- capability for in-house design or suitable for outsourcing;
- technology selection for use in the function;
- reusability of the intended function to be designed;
- methods for realization of the function;
- functional interface, interoperability and dependency on other functions;
- human interface in functional applications;
- criticality of the function;
- testability of the function;
- maintenance support of the function.

4.1.3.2 Dependability Design Approaches

The approaches for dependability design are application-dependent and often unique to suit specific project implementation situations. Dependability designs may include a combination of the following approaches:

- analytical approach – this involves activities such as design analysis and evaluation, functional performance simulation, conformance assessment to established standards and design compliance analysis to specification requirements;
- experimental approach – this involves activities such as performance testing and technical evaluation of the design functions, engineering prototype modeling and physical assembly mock-ups, OEM product integration and demonstration, product warranty testing and evaluation prior to acceptance;
- consultative approach – this involves activities such as expert reviews, use of industry best practices, suppliers' consultation on product information, customer survey and user feedback, supply-chain participation and collaboration and infrastructure development and enhancement;
- negotiated approach – this involves activities such as establishing acceptable risk limits for system operating exposure to the environment, conditions for product

deployment in specific regions, recycling of by-products and waste disposals, economic incentives and social benefits in contract agreements, and compliance to changing regulations.

4.1.3.3 Dependability Design Strategies

The dependability design strategies should focus on two main aspects in engineering dependability into systems.

(a) The application focus is to meet project specific applications for compliance of contractual requirements. The essential assessment activities are focused on the analysis and evaluation of system dependability at major decision points of the applicable system life cycle. The methods and tools deployed for assessment are commonly used for product verification and system or sub-system validation.
(b) The technology focus is to evaluate the selected technology for system design and functional development for dependability performance achievement. The essential assessment activities are focused on evaluating the technology leverage that can be exploited to support system service operation. Issues concerning technology evolution and obsolescence should form part of the assessment strategies.

4.1.3.4 Designing for System Application Environments

The system and its constituent system elements are exposed to the application environments during operation. The system elements may be exposed to different environmental conditions. For example, an industrial control system consists of a central processing computer designed to operate in a standard office environment where no extra precautions are needed. The operating equipment controlled by the computer is housed in a plant exposed to a corrosive or dirty environment due to the fluids and gases being produced at the plant where extra precautions for equipment protection are required. A handheld instrument used in both the office and the plant environments should be designed to withstand the most severe environmental conditions for practical application. The computer and the equipment are designed for stationary application, while the handheld instrument is designed for portable application.

The environmental exposures include electromagnetic [7], climatic and mechanical [8] conditions. The system application environments include various application premises, such as climatically controlled offices, indoor and outdoor premises, and mobile, storage and transport environments. Figure 4.1 presents a matrix framework for mapping system application environments to exposures. Specific environmental characteristics are identified in this matrix framework. They are used to provide the classification of application environments and relevant exposures to facilitate determination of the design envelope for specific product applications. The linkages between exposures and the performance characteristics associated with the product functional and physical designs should be considered.

Electromagnetic emission and immunity are exposed to the following application environments:

- E1 Controlled environment premises
- E2 Outdoor locations (sheltered, unsheltered)
- E3 Customer premises (industrial, business office, residential)
- E4 Portable and mobile applications

EXPOSURES	Characteristics (parameters)	SYSTEM APPLICATION ENVIRONMENTS							
		Controlled environment premises	Customer premises: Industrials	Customer premises: Busines office	Customer premises: Residential	Outdoor	Mobile	Transportation	Storage
Electro-magnetic	**Conducted immunity** (electrostatic discharge, electrical fast transient) **Radiated immunity** (radiated E-field) **Radiated emissions** (radiated H-field, radiated E-field) **Conducted emissions** (AC/DC power)	E1	E3	E3	E3	E2	E4		
Climatic	**Air Temperature** (high/low,change) **Air** (pressure, velocity) **Air Purity** (sulphurous, chorinous, nitrous, ozone) **Humidity** (relative high/low) **Water** (precipitation, spraying) **Particles** (dust, sand, aerosols) **Biological** (mold, fungus) **Chemicals** (salts, acids, detergents, solvents) **Radiation** (sun, heat)	C1	C3	C2	C2	C4 C5	C4		C6
Mechanical	**Vibration** (displacement, acceleration) **Shock** (acceleration, duration) **Bump** (acceleration, duration, repetition frequency) **Drop/Topple** (height, angle) **Free fall** (weight, height)	M1 M2 M5	M1	M1 M2	M1 M2	M1	M3	M4	

Figure 4.1 Mapping system application environments to exposures

Climatic conditions such as temperature and humidity are exposed to the following application environments:

- C1 Controlled climate
- C2 Indoor temperature controlled
- C3 Indoor temperature not controlled
- C4 Outdoor protected
- C5 Outdoor unprotected
- C6 Storage

Mechanical conditions such as vibration and shock are exposed to the following application environments:

- M1 Stationary
- M2 Portable
- M3 Mobile
- M4 Transportation
- M5 Earthquake

The information presented in Figure 4.1 is limited to ground-based applications for systems, products and equipment. Specific application environments and exposure conditions should be investigated such as marine, underwater, underground, mining, airborne and space application environments where different sets of environmental conditions and exposures may exist for their applications.

4.1.3.5 Designing for Human Interaction

Human aspects of dependability [9] provide the process of incorporating human interactions into system designs for service operation. A human can act as a part of the system during operation such as an operator. The incorporation of human interactions in system operation has the advantage of the human's intuitive reaction, flexibility to adapt to situations and the capability to perform many functions and tasks. A person also has limitations in cognitive and physical capabilities for task performance. The inherent qualities of human involvement in system operation can be exploited for design trade-offs with the system hardware and software elements. The aim is to maximize overall system capabilities in operation.

Human factors engineering activities at early design stages have a profound influence in maximizing return on investment and optimizing total system performance. The critical impact areas for human factors engineering participation in system design and operation achieves dependability value in the following:

- early identification of critical system functions that are considered suitable and advantageous for human interaction by analysis of a system operating scenario;
- user-oriented task designs for timing and operating task sequence for ease and expediency in human-machine operation;
- understanding human capabilities and limitations when making function allocation decisions for cost-effective application and training needs;
- identifying human requirements with respect to decisions, information display requirements, access, special skills and any other physical, cognitive, organizational or social constraints;
- integrating human requirements into the system design process for optimizing the system performance compatibility of human-machine interface and interoperability;
- addressing human roles in diagnostics and maintenance;
- the process of engineering human into systems includes analyses of the systems engineering process and the human factors engineering process. The relevant tasks, decisions and information common between the two processes can be used as a basis to identify areas of interactions between the human and the system.

Human factors design criteria are established to enhance the following dependability values:

- Fitness for use
  - make system durable, reliable and applicable for its intended use;
  - allocate functions appropriately;
  - accommodate physical characteristics of the users;
  - test with users.
- Simplicity
  - design for simplicity;
  - incorporate self-correction/self-healing features where practicable;
  - minimize training;
  - make functions obvious.

- Consistency
  - make design consistent;
  - be consistent with user experience with real-life objects and similar systems.
- Standardization
  - standardize hardware and software where practicable;
  - maintain identical interfaces for identical functions;
  - make controls, displays, markings, coding, labeling and arrangement uniform;
  - make appearance distinctive;
  - standardize terminology, look and feel;
  - make functionally similar equipment interchangeable.
- Safety and security
  - incorporate safety and security features;
  - provide fail-safe and intrusion protection design;
  - make system error-resistant and tolerant;
  - warning of potentially unsafe actions;
  - provide emergency and recovery procedures.
- User-centered perspective
  - understand user roles, responsibilities, decisions and goals;
  - provide timely and informative feedback;
  - use familiar terms and images;
  - design within user abilities;
  - maximize human performance;
  - minimize training requirements;
  - facilitate transfer of skills;
  - accommodate physical diversity.
- Maintenance support
  - provide logistic support where needed;
  - design for common tools;
  - make system easy to maintain and accessible for service.

4.1.3.6 Evaluation of Design Functions

Evaluation of the functional design includes the following dependability activities:

- conduct reliability assessment – this is achieved by reliability prediction, analysis and evaluation of the function to determine the adequacy of the function to meet reliability requirements; where needed or warranted, the evaluation process may involve additional assessment effort such as simulation for sensitivity studies, testing to confirm durability and safety requirements;
- conduct maintainability assessment – this is achieved by maintainability analysis to determine the ease of maintenance required by the function, and by testability of the function for fault identification and isolation where built-in test or self-healing procedure may be contemplated for incorporation as enabling mechanism (see Chapter 8);
- conduct functional level failure modes, effects and criticality analysis (FMECA) [10] – this is to determine the causal effects and criticality of failures affecting system performance; the analysis also identifies possible failure mechanisms causing the failure of the functional design; the objective is to determine practical means for failure prevention and mitigation and to recommend an appropriate recovery process for implementation in realization of the function;

- conduct functional level design trade-off, fault tolerance and risk evaluation – this is to identify areas for design trade-off improvements, incorporation of fault tolerance design schemes and prevention of critical failure occurrence in system performance degradation; risk evaluation is performed on critical functions to determine appropriate risk treatments to avoid or reduce probable risk exposures;
- establish plan for maintenance support and logistics [11] – this is to define the maintenance support strategies and procedures appropriate to support the recovery of system functions;
- establish process for supplier evaluation for quality assurance and reliability conformance – this is part of the supply chain management process to ensure that outsourced design functions are evaluated according to quality assurance procedures;
- establish process for off-the-shelf product evaluation and acceptance [12] – this is part of the supply chain management process to ensure that purchased items are evaluated according to quality assurance procedures.

4.1.3.7 Assessment Value and Implications

The assessment effort should be rationalized in terms of value gained for practical application. The assessment results should be completed within reasonable time limits to realize the expected value and benefits to the project. This also establishes the needed confidence to support project decisions. Typical examples are shown to highlight their major implications to project outcomes.

(a) Timing of assessment is crucial to provide meaningful results. The assessment value greatly diminishes when the assessment results are not available at the time needed to support major decisions. Reliability prediction conducted during system design provides valuable insights into the proper technology selection, architectural design, partitioning configuration, and choice of OEM products and system components for realization of system functions. Reliability prediction conducted after design completion has limited value to influence design improvements when the system is configured for production.

(b) Project planning for effective management requires assessment to rationalize the technical approaches and cost benefits prior to project initiation. The continual improvement process in quality management systems (QMS) [13] is commonly used for planning assessment activities. Investment analysis related to specific project needs is critical to justify major capital expenditures and new facility acquisitions.

(c) The infrastructure support should be adequate to foster project implementation. This may involve technical procedure changes, training for new product development and adjustments in engineering practices that would consume both time and effort. The extent of infrastructure support in migrating from an existing development process to the new process may sometimes be a major endeavor for any corporation. Both technical resources and management culture would require adjustment to attain the best industry practices.

(d) Contingency planning is essential to avoid unexpected project outcomes or unscheduled delays. This may impact resource allocation and existing workforce redeployment. Contingency plans should be included as part of the assessment process, such as identifying alternate suppliers in case of supply disruption, the deployment of technical expertise to work on critical designs to meet stringent delivery targets and exploring the means of viable financing for capital investments.

4.1.4 System Design Documentation

Documentation is an essential process to capture and retain important data and relevant information to serve various application purposes. System design documentation should conform to the project requirements and be guided by business policies and procedures on distribution and information controls. Where applicable, the relevant documents for external distribution should comply with contract data requirements. The complete set of system specifications and application procedures should be established and documented for project implementation. The essential documents containing relevant dependability information and data to support system design specifications and technical applications are shown below.

(a) Dependability engineering project plan – provides critical information on project tasks and delivery time schedules to support system design needs.
(b) Dependability assurance plan – provides assurance strategies and planned dependability engineering activities to ensure design conformance to established specifications and standards to meet system requirements.
(c) Configuration management system [14] – provides continuous update of baseline system structure and configuration of constituent hardware and software system functions to facilitate document controls and maintain records of approved design changes.
(d) System reliability allocation – provides relevant reliability data apportioned to system functions reflecting the latest system configuration on system reliability estimates.
(e) System dependability assessment – provides analysis and evaluation data documenting the results of FTA, FMECA and other relevant risk assessment information affecting the system design and operational performance.
(f) System design dependability specifications – provide relevant dependability characteristics to facilitate quantitative assessment and measurement for verification that system design dependability requirements are met.
(g) Functional design dependability specifications – provide relevant dependability characteristics to facilitate quantitative assessment and measurement for verification that functional design dependability requirements are met.
(h) Functional test specifications – provides procedures for testing of system functions for performance verification.
(i) System test specifications – provides procedures for integrated testing of system for performance verification.
(j) Failure reporting, analysis and corrective action systems – provide systematic capture of relevant information and failure data from incident reports during functional testing and integration and acceptance tests to maintain a historic record of system dependability performance history (see Chapter 6).

The information and data contained in these documents form the basis to establish dependability requirements for incorporation in system design and sub-system development.

4.1.5 System Design and Sub-system Development

System design entails project planning to initiate development of the specified sub-systems. The development effort requires major commitments to both capital and resource investments. Business decisions are needed to ascertain the following:

(a) contract agreements on terms and conditions for the development project;
(b) business plan on capital and resource investments and strategy for implementation;

(c) project plan on work statements and target delivery schedules;
(d) resource planning for recruitment and training to support the development project;
(e) manufacturability and testability of the development product outcome;
(f) manufacturing and production capability available and established;
(g) test verification facilities established;
(h) configuration management process established;
(i) plan for maintenance support and logistics established;
(j) customer liaison and supplier collaboration agreements established;
(k) supply chain management process established;
(l) outsourcing and subcontracting procedures established;
(m) assurance procedures established.

The development process needs to ensure that product development outcomes, such as the system function realizable in hardware or software, can be produced and tested for acceptance within project product budget and time constraints. In development of technological systems, there are often unique or special products designed with novel technologies that are convincing on paper in concept but lack practicality for product realization. In such cases, special tooling and fabrication processes are often required to make the product work, resulting in additional resources and effort. These extracurricular activities will likely outweigh the benefits of the product novelty against common-sense engineering to provide a practical solution to the problem. Project management "buy-in" with full disclosure of the risk involvement is needed if novelty products are introduced into the technological system for development. A similar situation may exist if unknown by-products or excessive waste are generated during the development process. Project reviews should address the novelty technology and environmental impact issues. This would avoid misunderstanding by resolving the situation before it becomes a problem.

4.2 DEPENDABILITY ENGINEERING DESIGN ISSUES

4.2.1 Safety Design

Almost all engineering and design of systems are concerned with safety. The methods for ensuring safety are highly application-dependent and can be designed by following four major principles [15]:

- Inherently safe design by excluding potential hazards instead of controlling them;
- Safety reserves by over-engineering or strengthening the design through the use of safety factors;
- Applying the "safe fail" concept that the system fails without causing harm;
- Procedural safeguards such as general safety standards and training and control of staff behavior.

These safety principles are only effective if they are combined with risk and safety assessment using such methods as probabilistic risk analysis (PRA), probabilistic safety analysis (PSA) and other dependability analyses such as fault tree analysis and Markov techniques. A main contribution to safety from dependability is the identification of potential failure modes and the subsequent reduction of failure rate. Maintainability also contributes to safety of maintenance personnel.

The key concepts of functional safety and safety integrity for Safety Instrumented Systems (SIS) are described in IEC 61508 [16] for electronic systems and in IEC 61511

[17] for the process industry. These and other related standards distinguish the four Safety Integrity Levels (SIL) by defining a probability of failure on demand (PFD) and a risk reduction factor (RRF) for each SIL. The integration of dependability engineering and the safety life cycle is further enunciated in [18].

The concept of safety barriers is now gaining wide acceptance [19]. Safety barriers consist of functions that prevent, control or mitigate the propagation of an incident into an undesirable event or accident. These safety functions may be a physical or inactive barrier such as a firewall, or they may be an active mechanism that performs an action such as a control system. They are normally described in barrier diagrams. A similar method in use in the chemical industry is LOPA (layers of protection analysis) [20].

The methods for probabilistic safety assessment are well established, but methods for human reliability assessment (HRA) are less well known. This area is still receiving considerable attention [21], particularly in applications such as the nuclear industry [22].

Redundancy of components and sub-systems is part of the design of reliable complex engineering systems used in several industrial applications, including power generation facilities. Such engineering systems may be in one or more intermediate operating states at various times. For example, these systems could continue operating during the repair of failed components and during maintenance intervals. For this reason, they do not lend themselves well to traditional reliability modeling techniques that do not account for a system's progression through various intermediate states. The use of traditional reliability modeling techniques may lead to significantly different estimates of the performance of a complex system required to exhibit high reliability.

As an example of a system that requires continuous operation and can become significantly complex in large industrial applications, a hypothetical heating, ventilation, and air conditioning (HVAC) system was chosen for conceptualization and modeling [23]. In applications where the HVAC system is expected to filter air that can be potentially contaminated with hazardous particles and may need to maintain negative differential pressure for containment of hazardous releases within a facility (such as in nuclear facilities), the estimated failure rate and its related uncertainty become crucial aspects of the design. The problem with applying traditional FTA is that it does not lend itself well to analyzing the reliability and availability of systems with the possibility of repair, since they consist of several components that can exhibit multiple states (e.g., normal, failed, standby). Markov techniques can be used to estimate failure rates for systems such as these, which may also have different modes of operation, as well as to assess the level of vulnerability of the system and the effect of a specific maintenance approach. This approach was applied to the HVAC system described in Figure 4.2 to illustrate the benefits of such methodology.

In this system, for either section, one train (i.e., functional loop) is operating and the other one is in standby at any time. If a failure occurs in the operating train, then the controller automatically places the standby train online, if the standby train is available; otherwise, a system failure occurs. Failure of the system is defined as the inoperability of both trains in either section. A train can become inoperable due to maintenance or component failure. Subsequent system failure occurs if a component fails in the running train. In addition, for this system, a single controller is used for which failure of the controller results in failure of the system regardless of the system state prior to the controller failure.

To develop the Markov model, a set of discrete system states is defined, and a set of transition rates from one state to another is determined. The transition rates are developed from component failure rates and component repair/maintenance restore rates. Figure 4.3 shows the system states identified in the Markov model for the hypothetical HVAC exhaust

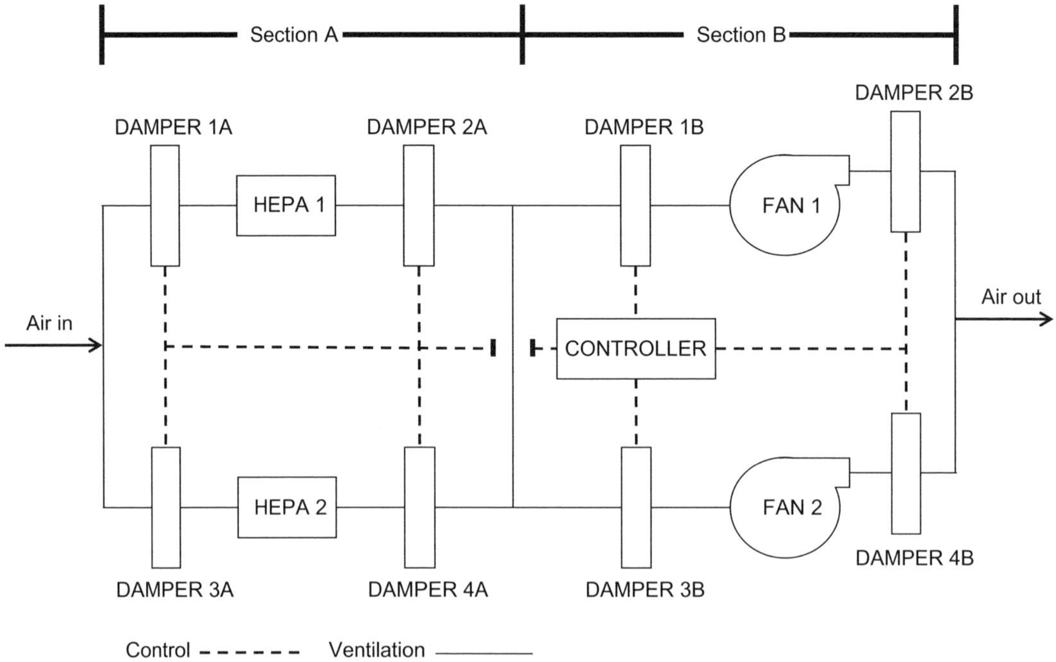

Figure 4.2 Schematic of a hypothetical HVAC exhaust system [23]

system. There is one normal operating state, identified as state "N," three intermediate operating states in which one or more trains are inoperable (i.e., I1, I2, and I3), and five states in which system failure occurs (i.e., F1, F2, F3, F4, and F5).

Component failure rates and repair/restore rates common to all approaches were estimated from data sources on the assumption that mean time to repair a failed component or restore a train from a maintenance activity is arbitrarily varied from 4 hours to 48 hours. Failure rates were estimated over this range using fault tree and analytical and simulation-based Markov analyses.

A fault tree model was also developed and the failure rate for the top event was calculated. The analytical and simulation-based Markov results were plotted along with the FTA results and are shown in Figure 4.4. An uncertainty analysis was performed for the analytical Markov technique, as described in the previous section. The uncertainty is represented by the mean value, the 15th percentile, and the 85th percentile of the estimated failure rate.

The failure rates for the hypothetical HVAC exhaust system compare well between the analytical Markov model and the stochastic simulation over all values for mean time to repair/restore, as expected. However, estimated failure rates calculated from the FTA diverge from the results obtained from the Markov approach as the mean times to repair/restore increase, making the FTA more conservative at these higher times.

The primary reason for this divergence is that the analytical and simulation-based Markov methods account for the operation of the system as it transitions from the normal state to possible intermediate states in which the system is still functional, whereas the fault tree model considers transitions from the normal operating state directly to one or more failed states. The fault tree technique generates an estimated failure rate of the hypothetical HVAC system that is essentially linear with mean time to repair/restore, while the Markov techniques generate an estimated failure rate that increases more slowly as mean time to repair/restore increases. These results show that

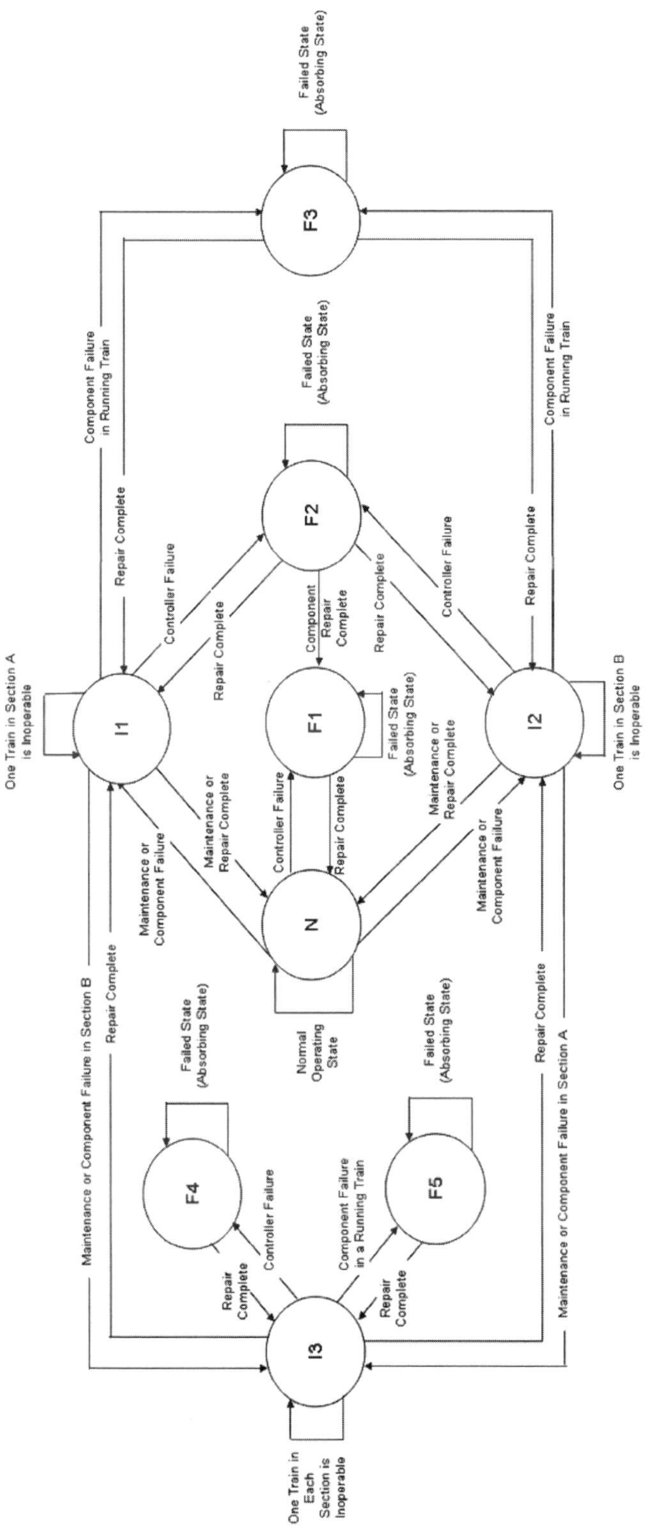

Figure 4.3 System states for the Markov model [23]

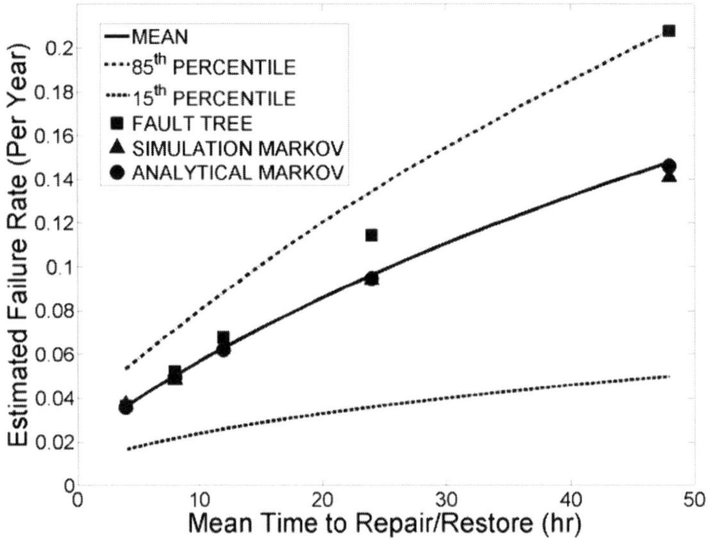

Figure 4.4 Comparison of failure rates for fault tree and Markov models and uncertainty analyses [23]

consideration of these intermediate states plays a significant role when estimating the failure rate of a system subjected to maintenance or repair actions that may last for several hours. In this sense, the Markov techniques may generate a more realistic estimate for the failure rate of complex systems, and the fault tree model may result in more conservative estimates for these systems. Furthermore, the inclusion of an uncertainty analysis in the Markov approach also allows for the quantification of the confidence in the estimated results, similar to FTA, given that published component failure rates can exhibit variability. However, since the system state is modeled explicitly as a stochastic process, the uncertainty analysis benefits directly from the Markov modeling capabilities in handling standby redundancy.

This result indicates that fault tree techniques may not adequately capture the performance of complex engineering systems that can be in several intermediate operating states, whereas Markov techniques can be used for these systems. The consideration of uncertainty is beneficial for quantifying the estimated failure rates to evaluate the confidence in the established maintenance and repair policy for a complex system. This can be particularly useful for cases where a high level of reliability is expected to fulfill specific requirements established by a client and/or regulations.

4.2.2 Dependability-Centered Design

Manufacturers of major equipment such as gas turbines pay great attention to the dependability of their products. An example is the design and validation of what is currently the largest gas turbine in operation, the Siemens SGT5-8000H, which is the prime driver of a combined-cycle power plant [24]. At 340 MW for the gas turbine and a total output of 530+ MW with an efficiency of 60%, reliability and availability take on an even more important meaning. Six-sigma methodology was used to effect a robust design in spite of the new technologies needed to achieve reliable, low cost and environmentally friendly power generation.

The design made use of extensive field experience as a base for an extensive R&D effort. Risk during the conceptual, basic and final design phases was mitigated by a series of failure mode and effects analyses (FMEA). Alternative designs were evaluated and compared for reliability.

Serviceability was enhanced by enabling rotor blades to be removed without lifting the rotor, also a significant enabler of higher availability. Robust design was achieved by blade designs that resist foreign object damage (FOD) and advanced 3D design tools.

Testing and validation efforts were intense, with full-scale testing at all operating and load conditions with an extensive amount of instrumentation to prove mechanical integrity and overall performance.

An example of how dependability is accounted for in design is found in Engelbert et al. [25]. For the development of new power generation equipment like gas turbines, the targets are laid out very early in the project. Since the development of gas turbine components can be a multi-year undertaking, it is paramount that the targets contain this element of future customer expectations. This ultimately drives the design team to challenge the technical boundaries – a gas turbine parameter that is satisfactory in today's market environment might not be pleasing in tomorrow's changing environment. This is not only valid for the engines thermal performance but also for its reliability and availability. Siemens industrial gas turbines and components are therefore being developed with "Reliability Centered Design" in mind. This report [25] describes the application of a predictive reliability engineering methodology in the development of new gas turbine components and how it influenced the design team's decision making. The reliability analysis and prognosis is based on a predictive fault tree model with supporting Markov models to address consecutive failures or failure probabilities of standby equipment. It has been validated with field data. While observation data retrieved from operational engines are being used for direct improvements of existing turbine designs, the model-based approach has its merits in supporting new designs by considering design or system alternatives. Furthermore, it provides the sensitivities of the reliability of the gas turbine with respect to the reliability of components. The model comprises not only the core engine but also the auxiliary systems within the package.

The overall reliability structure of a gas turbine and its package is a complex combination of series and parallel systems that are best described by fault trees and Markov chains. When using fault trees, it is advisable to use absolute numbers to express reliability like mean time between failures (MTBF) and mean down time (MDT). These figures are measurable and can be expressed in hours for how long on average equipment can survive between failures after repairs (MTBF), or how long it takes on average until the installation is available again (MDT).

Fault trees can be used to model independent failures and repairs of components. Sometimes, however, equipment is redundant but cannot be repaired without shutting down the system. A failure of standby equipment might also not be detected and hence not be repaired until it is needed. In this situation, fault tree models are not applicable, so Markov models are used to model and evaluate consecutive failures, which take into account dependent failure scenarios and repair strategies. A difficulty for the application of Markov models is the exponentially growing number of states that arise for even small sub-systems. For systems composed of more equipment, the number of states becomes too large for the manual creation of Markov models. There is the possibility to build a behavioral model of the system that includes the failure events and the repair times. Figure 4.5 shows such a model for the system of the three lube oil pumps and their associated valves.

TOMSPIN is a generalized stochastic Petri net tool developed by Siemens for performance and dependability analysis to model initially the behavior of computing systems [26]. Each failure or repair event and its effect on the state of a component are modeled. The strong advantage of this modeling is that it is possible to formulate prerequisites like a working component (e.g., "isolation valve") for the repair event of another component

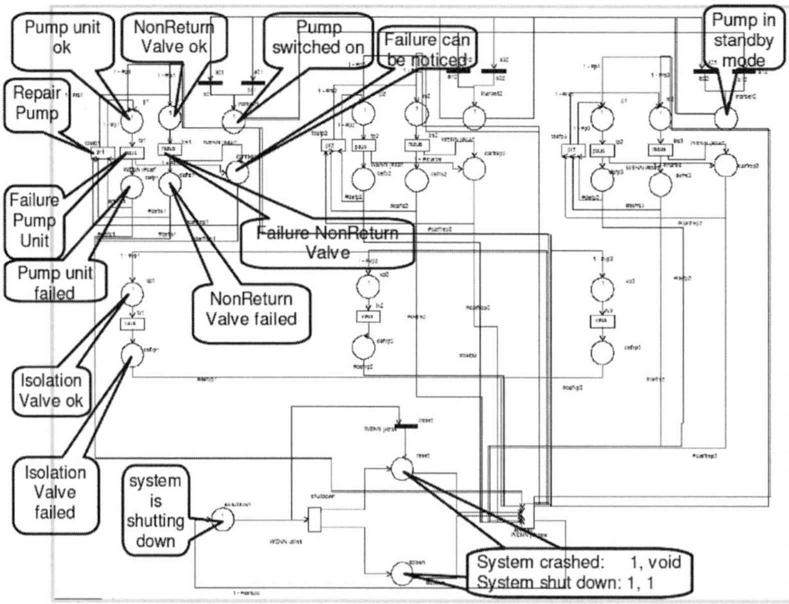

Figure 4.5 TOMSPIN model for three lube oil pumps and their associated valves [25]

(e.g., "pump"). According to the failure and repair rates, this model computes the probabilities for all states in the model, especially for the system states "shutdown" and "crash." A comparison for the mean times between shutdowns and between crashes shows the used Markov model to be conservative. In the Markov model, failures of non-return valves and isolation valves have been modeled to be unrecognized failures of a pump. It neglects the time that this pump could still keep on working while the valves are stuck. This leads to a smaller MTBF of 18.7 years for the Markov model instead of 19.3 years.

The reliability model states a prognosis for a future system while the system or component is still to be designed. To trust the computed results, it is valuable to compare model scenarios for existing engines with actual field data. If there is good correlation of the computed reliabilities of the sub-systems and the field data for the modeled component failures, it can be concluded with a high degree of certainty that a model that takes into account incremental changes for new designs will provide trustworthy estimations for their future reliability. Hence extrapolating and applying the model to new designs gives a fairly good prognosis for MTBF and MDT. The values show whether a new design is improved in its MTBF and MDT prediction relative to the baseline model. It also answers the question whether the design team stands a chance to meet its reliability target theoretically.

The computational model for reliability not only provides the MTBF and MDT of the gas turbine, but also the sensitivity of the system's reliability with respect to the components' reliabilities. These indicate which component's reliability should be increased as its improvement will have the strongest effect on the system level.

The application of this process to a major package redesign exercise has shown that after iterating the finally proposed design changes will improve the theoretical MTBF of the whole installation including core engine in excess of 10%. The real benefit of this approach shows when it is being applied to major design changes, re-designs or new developments.

Increasing system complexity and the advent of complicated networks of interacting systems are increasing the difficulty of performing accurate dependability analyses.

It is apparent that many applications constitute what is known as multi-state systems that have to be modeled by techniques such as Markov analysis and Petri nets [27] along with Monte Carlo simulation. For example, power generation has to deal with varying daily and seasonal loads. The demand for delivery from a natural gas pipeline similarly varies with time. Another example is the failure analysis of satellites and their sub-systems [28] using a statistical approach.

4.2.3 Structural Design

Reliability-Based Design and Assessment (RBDA) methodology has been developed in different applications in order to optimize the design of structures including ships, pipelines and bridges. Onshore pipelines have traditionally been designed with a deterministic stress-based methodology [29]. The changing operating environment has, however, imposed many challenges to the pipeline industry, including heightened public awareness of risk, more challenging natural hazards and increased economic competitiveness. To meet the societal expectation of pipeline safety and enhance the competitiveness of the pipeline industry, significant efforts have been spent for the development of RBDA methodology.

Since the origins of long-distance transportation of hydrocarbons by pipeline, design has been based on allowable stress methods. In such methods, for specific design checks, stresses are limited to some fraction of a conventionally defined reference stress that notionally represents the "strength" of the material. The difference between the maximum allowable stress and the reference stress is perceived as representing a margin of "safety" against structural failure. For structural elements that are expected to fail by overload, this approach is intuitive and is simple to apply. Nevertheless, this approach suffers from fundamental deficiencies that appear to be limiting in the pursuit of further advances. Most significantly, the design process is only marginally related to the failure mechanisms that have been observed historically and that will have to be addressed in future, long-distance pipelines. Certain failure causes simply cannot be addressed, in any useful way, using a stress-based approach. There is very little integration between the design and the operation and maintenance processes, even though pipeline integrity is critically dependent on both.

These deficiencies, which could be critically limiting for the development of future long-distance pipeline systems, can be addressed through the adoption of a reliability-based limit states approach, which can be applied to both the design of new pipelines and the evaluation of existing ones. The approach is thus referred to as RBDA. RBDA can be considered as a sub-set of limit states design, in which all the failure modes and mechanisms that can apply to a specific pipeline are addressed, and design decisions are based on ensuring an appropriate level of conservatism against them; the conservatism required will be dependent on the severity of the consequences of failure. RBDA makes use of reliability theory, which takes into account the statistical variability of all the parameters that influence a specific limit state (failure mode) and failure mechanism in determining the probability of failure ("reliability" in this sense is simply 1 minus the probability of failure). The related assessment then involves comparison between the calculated reliability, for all failure mechanisms, against a target (minimum) value that is calibrated to account for the severity of the consequences of exceeding that limit state.

Initial work was carried out to provide robust general guidelines for RBDA, together with some valuable analysis tools. The approaches proposed were further developed and refined in a project funded by PRCI [30, 31, 32]. This work resulted in the provision of much more detailed and comprehensive guidelines, aimed at facilitating the application of the approach by pipeline practitioners, as well as considerable refinement of the proposed target reliability levels. It also included a more detailed analysis of the economic implications of the RBDA approach, in particular illustrating the potential that it offers for finding an optimum trade-off between design and maintenance measures in optimizing lifetime cost.

The guideline document [31, 32] lays out the six technical steps in the application of RBDA as illustrated in Figure 4.6.

The process for developing target reliability levels was critical since it has been maintained, though with significant refinements and improvements, in subsequent work intended to lead to the adoption of RBDA in standards. The process for the establishment of target reliability levels was considerably developed and refined. For the calibration of ultimate limit state targets, a matrix of 240 design cases was analyzed, covering five diameters, three operating pressures, four strength grades and four class locations. Weighting factors were derived from a survey of approximately 90,000 km of North American gas transmission pipelines, and representative population densities from a survey of over 19,000 km of right of way. The primary reliability targets were based, as described previously, on the weighted average societal risk, expressed as the average expected number of fatalities, with an additional requirement based on individual risk that only takes effect for low safety consequence levels. The target reliability for ultimate limit states is presented in Figure 4.7 and the three risk criteria used for calibration are clearly reflected in the three straight segments in the target reliability curve.

The target reliability described above was defined in terms of per km-year. Recognizing the fact that integrity threats to a pipeline were often located at specific locations and reliability was not uniformly distributed, a reliability evaluation process was developed that was based on the average reliability over a characteristic evaluation length. The characteristic evaluation length was suggested to be 1600 km maximum to reflect the process for determination of the current location classes, based on which many data used in the target reliability calibration process were collected. Supplemental study was also conducted to rationally combine the distributed failure probability (e.g., contributed by mechanical damage) and the location-specific failure probability (e.g., contributed by ground movement) in order to determine the total failure probability. Assessment (RBDA) of natural gas pipelines has now been incorporated into CSA Standard Z662 [34] and ASME B31.8 [35].

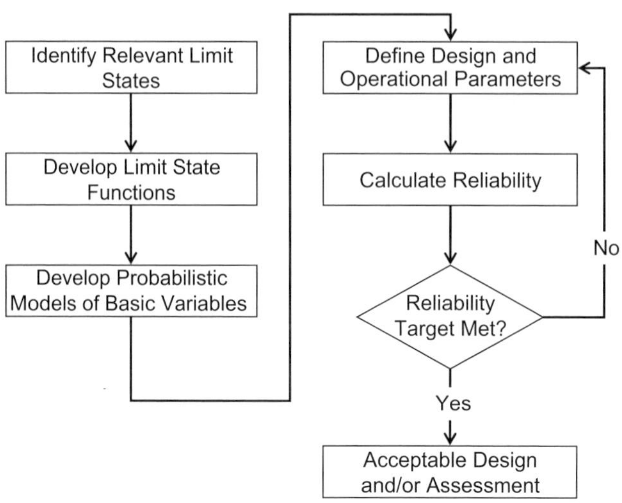

Figure 4.6 Overview of RDBA process [31, 32]

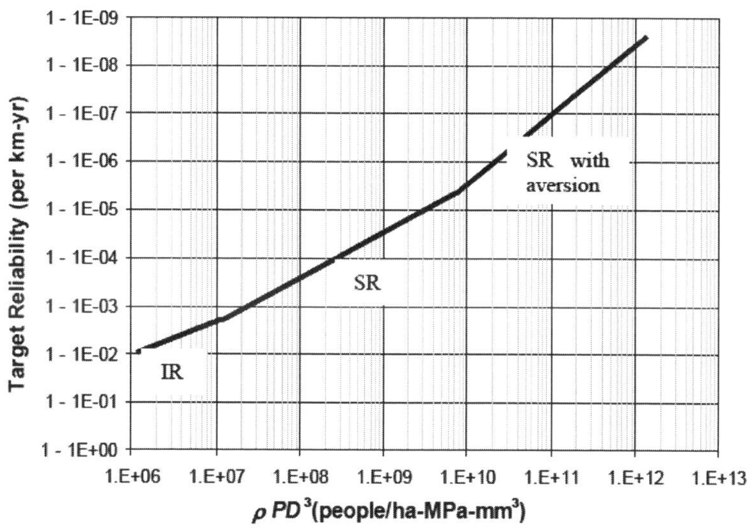

Figure 4.7 Proposed target reliability level for ultimate limit states of onshore natural gas transmission pipelines [33]

A similar reliability-based design process has been applied to offshore towers such as those used for drilling and supertanker terminals [36]. In this case, referred to as reliability-based design optimization (RDBO), its procedures include structural analysis, reliability analysis and sensitivity analysis for both optimization and reliability.

4.2.4 Life Cycle Cost

Basing equipment selection on acquisition cost only may well lead to the wrong decision from a life cycle point of view. Life cycle cost (LCC is also known as TCO or total cost of ownership) studies are best done during equipment acquisition and used to compare alternatives. The major steps in a LCC analysis consist of:

- preparing a breakdown structure for applicable costs;
- determining costs for each breakdown element;
- collecting failure and repair data (MTBF/mean time to repair or restoration (MTTR) or Weibull) from industry sources or actual experience;
- analyzing system availability and reliability;
- selecting an LCC model (e.g., LCC = Acquisition cost + Operating cost + Failure cost + Support cost – Net disposal value);
- estimating costs for each component of the LCC model;
- applying discounting over the time period of the study;
- determining the final LCC based on net present value (NPV);
- comparing alternatives.

An example of a LCC comparison for pumps [37] is shown in Figure 4.8. This challenges the common practice of basing equipment selection on acquisition cost. The primary cost is that of operation, with energy costs being the largest component. This is the main reason why the smart pumping system, which uses a variable speed drive and better controls to stay in the optimum operating range, has a lower LCC.

Pump Type	Cost Element	Cost	Percent of LCC
Conventional Pumping System	Initial Cost	$ 20,600	3.0%
	Operating Cost	$ 636,900	92.5%
	Maintenance Cost	$ 31,000	4.5%
	Life Cycle Cost	$ 688,500	100%
Smart Pumping System	Initial Cost	$ 19,800	4.40%
	Operating Cost	$ 410,700	91.44%
	Maintenance Cost	$ 18,600	4.16%
	Life Cycle Cost	$ 449,100	100%
Parallel ANSI Pump	Initial Cost	$ 13,500	8.5%
	Operating Cost	$ 110,715	70.19%
	Maintenance Cost	$ 33,503	21.31%
	Life Cycle Cost	$ 157,718	100%
Multistage Centrifugal Pump (Present Study)	Initial Cost	$ 31,000	3.5%
	Operating Cost	$ 838,750	94%
	Maintenance Cost	$ 21,346	2.5%
	Life Cycle Cost	$ 890,265	100%

Figure 4.8 Example of a cost profile for various pump scenarios [37]

4.3 DEPENDABILITY ENGINEERING IN SYSTEM REALIZATION AND IMPLEMENTATION

4.3.1 System Realization

The system realization process creates the system with its constituent system elements according to the practices of the selected technology implementation. The process starts with the system configuration, where each system element is identified for realization as a product for building the system. The realization process transforms the specified functional design characteristics and interfaces into fabrication actions to create the product. The realization process outcome is a product consisting of hardware and/or software that meet the design requirements of the specified system function. The resultant product is tested to verify its conformance to design specifications. The system comprises a set of system functions that could be realized into various specified products suitable for integration into a system. The realized products could be in any form, fit or function according to their respective design specifications. These products are known by various designations according to their specific applications. Examples include system, sub-system, functional module or unit. It should be noted that service provision could also be considered a marketable product, where the realization process creates the methodology, procedures and mechanisms for service provision to clients. Examples include maintenance and logistic support services, software engineering services and independent laboratory testing services.

There are several means of achieving the realization process depending on the type of business and the technology involved in system applications.

(a) The system is designed, developed, fabricated and constructed by a vertically integrated prime organization. An example is a software organization developing software programs for provision of software system application services. The deliverable software systems include a set of software programs to run on existing host computers to provide client specific application services. Another example is an OEM organization of specialized instrumentation and measurement products, which are designed, fabricated, assembled and constructed for delivery of the OEM product. The OEM organization also engages in OEM product repair and calibration service.
(b) The system is designed and developed by a prime organization with system functions outsourced for supply and construction provided by several support organizations. An example is a utility service organization providing services to multiple clients and relying on other support organizations to provide their specific operating sub-systems and functional services necessary to sustain continuity of the utility service operation.
(c) The system is designed and developed by a prime organization and is constructed mostly with outsourced system functions while retaining its own fabrication of proprietary products. Typical examples include most technology organizations and new start-ups engaged in novel technology design and technological systems development.
(d) The system is designed, developed and operated by several organizations in a joint collaborative service operation. Each organization contributes its share of specific functional services within its jurisdiction of operation. The system of systems is providing an integrated network service to multiple clients. An example is a telecommunications network consisting of multiple telecommunications organizations operating in various regions in cooperative agreements to establish global telecommunications for provision of customer services.
(e) The system is specified by an organization for contracting out to qualified contractors engaging in the engineering/procurement/construction (EPC) business. The realization process involves project award to a prime contractor, or to several individual contractors to develop separate sub-systems for progressive completion of the system project. An example is the contracting out for public utility services such as engineering, construction and installation of a pipeline distribution system. For contracting out of major development projects requiring multiple engineering and construction contractors, the contracting organization serves as the prime project manager for coordinating the activities of the multiple engineering contractors selected for the system project. The project may take several years to complete before the system is ready and accepted for service operation.

It should be noted that system realization is not merely a component assembly process. In the context of technological systems, the realization process extends beyond making the parts and putting them together to work as a system. The dependability process implementation in the realization process ascertains the compatibility and interoperability of the system functions within the system functional application framework. Connectivity and interface protocols need to be established and verified to ensure integrated system performance. From a business economics viewpoint, the following realization process issues should be considered.

- Make-buy decision of the specified system functions – The cost benefits, in-house design capability, availability of outsourced products in meeting specified requirements, timing and budget for project completion should be fully investigated and evaluated to support the make-buy decision.

- Outsourcing development of specified system functions – There are custom design houses catering to a variety of client product services, such as electronic product manufacturing, hardware and software product development and material fabrication foundries, to meet diverse customized product needs. The major advantages for outsourcing include deferring investments in capital equipment acquisition and unplanned capacity growth outside of the organization's core capability and resources, overcoming the time and cost constraints in meeting project schedule delivery targets, and short-term project solution to maximize available resources within the organization.
- Outsourcing of service functions for external service provisions – There is a broad range of external services such as testing and calibration services, on-site construction and installation services, and factory-authorized third-party maintenance support services that contribute to system realization.

4.3.2 Product Verification

The verification process is recursive at the functional level, the sub-system level and the system level. The verification objective is to confirm that the specified design requirements of the products at different levels have been met in accordance with the established assurance standards [38]. Verification and validation are based on established system configurations at the time of product measurements. Hence configuration management is necessary to maintain an updated system configuration for control purposes.

Project management of successive system configurations is essential for dependability engineering to provide up-to-date assessment results for system verification and planning for maintenance support and logistics. The configuration changes are due to the variety of interfaces generated through different change versions, the increasing demand for interchangeability of hardware and software components, and the assurance of interoperability of system performance functions. This is particularly relevant for evolving systems using shorter-life components for replacement, which may change frequently in technology over the longer life of the system. During system development, a sound configuration management system contributes significantly to effective dependability achievement. Configuration management is essential for system change controls and meaningful dependability assessment efforts.

It is important to maintain records of all incidents occurred during the verification, integration, installation/transition, validation and acceptance processes. This will permit follow-up incident analyses of problem areas for preventive and corrective actions.

4.3.3 System Integration

The system integration process assembles the system in line with the architectural design. The process combines the system elements to form complete or partial system configurations in order to create the system as specified according to the system requirements. For complex technological systems such as a telecommunications network, it is likely that the integration process may be carried out in project phases or planned sequence of progressive work targets, such as first to provide core essential services, then to add application features and service extensions. Hence an integration strategy is required to identify the sequences of partial progressive integration until full service completion of the network configuration over an established time period.

The integration strategy permits verification against a sequence of progressively more complete system element configurations over time. This is dependent on the system element availability and consistent with a fault isolation and diagnosis strategy. The system configuration at any point in time of verification includes the interaction of human operators where applicable, which represents the actual system operation. Successive applications of the integration process and the verification process, and the validation process as and when required, are repeated for systems configured at successive levels until the system has been completely realized. For evolving systems, the system configuration is updated with time due to its sustaining renewal process. Enabling mechanisms, such as integration facilities, jigs, conditioning facilities and assembly equipment, are used where applicable to facilitate the integration process.

4.3.4 System Installation/Transition

The system installation/transition process is to establish the system capability on-site to provide services specified by the customer requirements. The process installs a verified system in the customer-designated premises and operating environment. Where applicable, the system is supported by maintenance and logistic support services, enabling equipment and tools, and training devices to sustain system performance and service operation.

The installation process can be a simple download of an updated version of the software operating system onto the computer hard drive, or putting together a plug-and-play home entertainment system. In major system installation projects, such as constructing a large petrochemical plant, the transition process may involve deployment of equipment and inventory assets, transportation and logistics, construction and inspection, assembly and testing, verification and validation, training and support services. The capability transition to the local system operator would require collaborative effort of all parties concerned. The project would likely be carried out over a scheduled period of time. Dependability assurance practices would require an implementation plan and agreements on a common strategy to guide the transition process. This is to ensure successful completion of the technical knowledge capability transfer to the new operation.

A most important step is project commissioning. This entails the application of engineering techniques and procedures to check, inspect and test every component, sub-system and system for required functionality. Some of these functional and performance checks are first carried out by the OEM as part of their quality program, while verification of performance and mechanical testing may even require a full-load shop test such as those typically done for gas turbines and centrifugal compressors.

A different example is the commissioning of a leak detection system for a crude oil pipeline, which is traditionally carried out at the start of operation, but due to stringent environmental requirements was in this case performed as the pipeline was filled [39]. This ended up being feasible for the statistical leak detection system but not for the hydraulic model-based system, which could only be commissioned when the line was full.

In some jurisdictions, environmental studies and impact assessments are prerequisites for system installation that may have health and safety concerns in service operation.

4.3.5 System Validation/Acceptance

The system validation process provides relevant objective evidence to confirm successful system implementation that satisfies customer requirements. For technological systems, validation is achieved by progressive demonstrations of conformance results at various functional and sub-system levels, and when the final installed system is ready for customer acceptance.

The objective evidence needed to support system and product acceptance at applicable system life cycle stages should include the following information from a dependability perspective:

(a) Evidence to substantiate that relevant system dependability characteristics and operating environment reflect customer expectations in commercial specification or proposal information. This provides information to start project planning and develop system dependability specification.
(b) Evidence to substantiate that system performance characteristics are stated in the system dependability specification. This provides information for establishing dependability design objectives and system architecture.
(c) Evidence to substantiate that reliability and maintainability performance characteristics for each system function are stated in functional design specification. This provides information for technology selection, make-buy decisions and establishing procurement requirements.
(d) Evidence to substantiate that reliability and maintainability characteristics are requirements for system in-service operation and maintenance. This provides information for logistic support planning, contract maintenance and special training needs.
(e) Evidence to substantiate that relevant dependability characteristics are demonstrated for product acceptance, verification compliance and validation of system performance results. This forms the basis for fulfillment of contractual agreements for deliverable contract items.
(f) Evidence to substantiate that all dependability project reports contain dependability analysis data, test status and demonstration results. This provides information for project reviews, design changes, procedural updates, corrective and preventive actions for progressive improvement.

All objective evidence should be documented and authenticated for auditing and contracting purposes.

From a system dependability performance demonstration perspective, the system may be required to conduct product accelerated testing [40] and system reliability growth testing [41] as prescribed in the customer contract requirements. In such cases, the duration of the warranty period should be specified and the validation results should be assessed prior to formal customer approval and acceptance of the system.

4.4 DEPENDABILITY ENGINEERING CHECKLISTS

Dependability engineering checklists are designed to support engineering projects at major decision points to facilitate management reviews. These checklists identify the critical issues that need to be addressed to identify the status of key system dependability activities associated with the implementation of project tasks. Regular project reviews between major decision points are recommended for progressive reviews to determine the extent of dependability achievements. The project review objective is to ensure that all critical issues have been assessed and resolved within target time schedule. The review records can be used as objective evidence to support the dependability assurance and project auditing process. The relevant dependability engineering checklists [42] are provided in Appendix D for specific applications.

(a) System life cycle project applications – The checklists reflect the processes for transfer of project responsibilities and transition of design authorities and system ownerships during the entire system life cycle. A checklist for each stage of the system life cycle is provided to guide project reviews to support management decision-making.

(b) Technology design applications – The checklists for hardware, software and human factors engineering design applications can be used for engineering dependability into systems. They facilitate the selection of hardware and software for designing the required system functions and providing an opportunity for design trade-offs. Human factors play an important role in maximizing system dependability performance.

(c) Outsourced products used in system applications – The outsourced products include the COTS products that are widely used in system applications. COTS products are generally market-driven, and their fitness for use has been demonstrated by a broad spectrum of commercial applications. A COTS product provides a ready-made package for commercial purchase. The purchaser of a COTS product has no influence on the product features and its operational specifications. Selecting the right COTS products for system incorporation is of prime importance in engineering dependability into the system. There are certain risks involved in selecting a COTS product, and validating its suitability for a specific system application is essential. Using a COTS product for critical system application will require additional evaluation effort for assurance. The checklists are provided to facilitate requirements identification, performance evaluation and assurance of the COTS product for suitable incorporation into system application.

REFERENCES

[1] IEC 61078, Analysis techniques for dependability – Reliability block diagram and Boolean methods
[2] IEC 61165, Application of Markov techniques
[3] IEC 61025, Fault tree analysis
[4] IEC 61709, Electronic components – Reliability – Reference conditions for failure rates and stress models for conversion
[5] Mil-HDBK-217F, Military Handbook, Reliability prediction of electronic equipment
[6] Italtel: IRPH, Italtel Reliability Prediction Handbook
[7] IEC 61000 (all parts), Electromagnetic compatibility (EMC)
[8] IEC 60721 (all parts), Classification of environmental conditions
[9] IEC 62508, Guidance on human aspects of dependability
[10] IEC 60812, Analysis techniques for system reliability – Procedure for failure mode and effects analysis
[11] IEC 60300-3-12, Dependability management - Part 3-12: Application guide – Integrated logistic support
[12] IEEE Std 1062, IEEE Recommended practice for software acquisition
[13] ISO 9000, Quality management systems – Fundamentals and vocabulary
[14] ISO 10007, Quality management systems – Guidelines for configuration management
[15] Moller, N. and Hansson, S.O., 2007. "Principles of engineering safety: Risk and uncertainty reduction," Reliability Engineering and System Safety 93 (2008) pp. 776–783.
[16] IEC 61508, Functional safety of electrical/electronic/programmable electronic safety-related systems.
[17] IEC 61511, Functional safety—safety instrumented systems for the process industry
[18] Lundteigen, M.A., Rausand, M. and Utne, I.B., 2009. "Integrating RAMS engineering and management with the safety life cycle of IEC 61508," Reliability Engineering and System Safety 94 (2009) pp. 1894–1903.
[19] Duijm, N.J., 2009. "Safety-Barrier Diagrams as a Safety Management Tool," Reliability Engineering & System Safety, Vol. 94, No. 2, pp. 332-341, 2009.
[20] Center for Chemical Process Safety. "Layers of protection analysis — simplified process risk assessment," New York: American Institute of Chemical Engineers, 2001.

[21] Colombo, S. and Demichela, M. 2008. "The systematic integration of human factors into safety analyses- An integrated engineering approach," Reliability Engineering and System Safety 93 (2008) pp. 1911-1921.
[22] Kim, M.C. and Seong, P.H., 2006. "A computational method for probabilistic safety assessment of I&C systems and human operators in nuclear power plants," Reliability Engineering and System Safety 91 (2006) pp. 580-593.
[23] Adams, G. and Farrante, F., 2007. "Markov Modeling Application to a Redundant Safety System," Proceedings of POWER2007 ASME Power 2007 July 17-19, 2007, San Antonio, Texas.
[24] Eulitz et al, 2007. "Design and Validation of a Compressor for a New Generation of Heavy-duty Gas Turbines," Proceedings of POWER2007 ASME Power 2007 July 17-19, 2007, San Antonio, Texas.
[25] Engelbert, C., Nilsson, M., Sutor, A. and Montrone, F., 2008. "Application of a Reliability Model to Gas Turbine Design," Proceedings of POWER2008 ASME Power 2008 July 22-24, 2007, Orlando, Florida, USA.
[26] Klas, G. and Lepold, R., 1992. "TOMSPIN-a tool for modeling with stochastic Petri nets," CompEuro '92 . 'Computer Systems and Software Engineering', Proceedings, May 4-8, 1992, pp. 618-623.
[27] Sachdeva A, Kumar D, Kumar P., 2007. "Reliability modeling of an industrial system with Petri nets," Proceedings of ESREL 2007, Stavanger, Norway, vol. 2, 25–27 June 2007. p. 1087–94.
[28] Castet, J-F and Saleh, J.H., 2010. "Beyond reliability, multi-state failure analysis of satellite subsystems: A statistical approach," Reliability Engineering and System Safety 95 (2010) pp. 311-322.
[29] Zhou, J., Rothwell, B., Messim, M. and Zhou, W., 2006. "Development of Reliability-Based Design and Assessment Standards for Onshore Natural Gas Transmission Pipelines," Proceedings of IPC2006 6th International Pipeline Conference September 25-29, 2006, Calgary, Alberta, Canada.
[30] Nessim, M., Zhou, W., Zhou, J., Rothwell, B., and McLamb, M., 2004. "Target Reliability Levels for Design and Assessment of Onshore Natural Gas Pipelines," Proceeding of International Pipeline Conference, Calgary, Alberta, October 4-8, 2004.
[31] Nessim, M. and Zhou, W, 2005. "Guidelines for Reliability-Based Design and Assessment of Onshore Natural Gas Pipelines", A report prepared for Gas Research Institute (GRI), GRI-04/0229.
[32] Nessim, M. and Zhou, W, 2005. "Target Reliability Levels for Design and Assessment of Onshore Natural Gas Pipelines", A report prepared for Gas Research Institute (GRI), GRI-04/0230.
[33] Nessim, M., Zhou, W., Zhou, J. Rothwell, B., 2006, "Reliability Based Design and Assessment for Location-Specific Failure Threats", Proceeding of International Pipeline Conference, Calgary, Alberta, Sept. 25 – 29.
[34] CSA. 2007. "CSA-Z662 (2007), Oil and Gas Pipeline Systems." Canadian Standards Association, Mississauga, Ontario.
[35] ASME, 2010. "ASME B31.8-2010 – Gas Transmission and Distribution Systems". American Society of Mechanical Engineers, New York, New York.
[36] Karadeniz, H., Togan, V. and Trouwenvelder, T., 2009. "An integrated reliability-based design optimization of offshore towers," Reliability Engineering and System Safety 94 (2009) pp. 1510-1516.
[37] Waghmode, L.Y., Birajdar, R.S. and Joshi, S.G., 2006. "A Life Cycle Cost Analysis Approach for Selection of a Typical Heavy Usage Multistage Centrifugal Pump," Proceedings of ESDA2006, 8th Biennial ASME Conference on Engineering Systems Design and Analysis, July 4-7, 2006, Torino, Italy
[38] ISO/IEC 15026-4, Assurance in the life cycle.
[39] Mabe, J., Murphy, K., Williams, G. and Welsh, A., 2006. "Commissioning a real-time leak Detection System on a Large Scale Crude Oil Pipeline during Startup," Proceedings of IPC2006 6th International Pipeline Conference September 25-29, 2006, Calgary, Alberta, Canada.
[40] IEC 62506, Methods for product accelerated testing.
[41] IEC 61014, Programmes for reliability growth.
[42] IEC 60300-3-15, Dependability management, Part 3-15: Guidance to engineering of system dependability

Chapter 5

SOFTWARE DEPENDABILITY

5.1 THE SOFTWARE DEPENDABILITY CHALLENGE

5.1.1 Software Dependability Implications

Software has widespread applications, such as smartphone apps for user communications and deployment of supervisory control systems for monitoring pipeline transmission. Dependability from a user perspective implies the confidence and trust in the software apps for use in smartphones. Dependability from a service provider viewpoint infers the ability of the supervisory control system to perform as specified and the reliability of the software application monitoring the system. The rapid growth of commercial and industrial software applications, Internet services and Web development has affected dramatic changes in the socio-economic landscape. Digital technology has revolutionized the means of communications. Standardized interfaces and protocols have enabled the use of third-party software functionality to permit cross-platform, cross-provider and cross-domain applications. Software has become a driving mechanism to realize complex system operations. It enables the achievement of viable e-businesses for seamless integration and enterprise process management. Software design has assumed the primary function in data processing, safety monitoring, security protection and communication links in network services. This paradigm shift has put global business communities in a situation of heavy reliance on software systems to sustain business operations. Software plays a dominant role to influence successful system performance and assure the integrity of data. The software dependability focus in this chapter attempts to provide current industry best practices and present relevant methodology to facilitate the achievement of software dependability. It identifies the influence of management on software design and implementation and provides relevant technical processes to engineer software dependability into systems.

Fundamental differences between hardware and software dependability can be distinguished by recognizing and comparing their respective failure characteristics as shown in Table 5.1.

A system failure exhibits the loss of the system's ability to perform as required from a dependability perspective. Failure is an event experienced by the system user. A hardware failure in a system can lead to gradual system degradation. A software fault is the software's inability to perform its function as specified. A fault is the state of the software that prevents it from functioning as intended due to various reasons. A software fault exists in a system such that, when triggered, it can cause abrupt system failure without any advance warning. It happens often with no indication that the fault has occurred. Bugs are latent software faults due to software design errors. Software containing bugs can still

TABLE 5.1 Comparison of hardware and software failure characteristics

• Hardware failures are caused by material deterioration, lack of maintenance, operating outside of design limits or under extreme environmental stress exposure.	• Software failures result from incorrect logic, coding error and incomplete input data or procedural omission.
• Hardware does wear out when used.	• Software does not wear out.
• Hardware failure cause is apparent.	• Software failure cause is often difficult to trace.
• Hardware can be extensively tested.	• Software testing is never-ending.
• Hardware maintenance can be preventive or corrective.	• Software maintenance is always perfective with an update version.

accomplish its intended function unnoticed to the user. Bugs are software problems that can trigger system failures.

5.1.2 Understanding Software and Software Systems

Software refers to procedures, programs, codes, data and instructions for system control and information processing. A software system consists of an integrated collection of software programs, instruction procedures and executable codes, and it is incorporated into a physical computer host for execution and processing to realize system operation and deliver performance functions. The software system is a structure representing the system architecture and consisting of sub-system software programs and lower-level software units. A software unit is designed as a set of executable codes or a program to perform a specified function. In some applications, multiple software units are required to construct a complex software function. A system engages hardware and software interactions to generate the required functions for performing a service.

In a combined hardware/software system, the software elements of the system contribute in two major roles: operating software to run continuously to sustain hardware elements in system operation, and application software to run upon user demands for provision of specific user services. Dependability analysis of the software sub-systems has to consider the software application time factors in the system operational profile. It is necessary to identify those software units required for full-time system operation or for part-time application on demand. Software modeling is needed for dependability assessment of software-based systems.

Software is ubiquitous in controlling equipment such as gas turbines, generators and compressors, often by means of software-driven industrial PLC (programmable logic controller) devices. Control systems used at a higher plant level often utilize specialized software such as a distributed control system (DCS), which runs on standard PC-based network. To overcome some of the limitations of a standard control system where load and set-point variations can lead to instabilities, traditional software tools are being enhanced with soft computing techniques such as genetic algorithms, neural networks and fuzzy logic [1].

Human aspects of dependability [2] play an important role in guiding effective software design and implementation. The human-machine interfaces greatly influence the efficiency of system access and interaction in system operation affecting the outcome of system dependability performance. Software dependability design strategies and perfective maintenance efforts are needed for optimization in the software life cycle processes [3].

Software codes written by programmers are susceptible to human errors. The software design environment and organizational culture greatly influence the quality of software products created. The software engineering disciplines and practices affect the reliability of software in applications. Software upgrades can also unknowingly introduce new errors in the maintenance support process. The nature of software faults and their traceability of cause and effects are not easy to determine in system operation. In most cases, the software faults leading to system failures cannot be consistently duplicated. Corrective actions on traceable system failures due to software faults do not guarantee total elimination of the root causes of the software problem.

5.2 SOFTWARE DEPENDABILITY ENGINEERING

5.2.1 System Life Cycle Framework

From a system configuration perspective, software fits within the system life cycle framework as system element, sub-system, product or component. Software creation, along with hardware fabrication, and user training are subsets of the system realization/implementation process. This relationship is shown in Figure 5.1, which references the model established for the technological system (Chapter 3, Figure 3.1) as the system life cycle framework.

From a life cycle perspective, software has its own life cycle as a unique product for development and application. Software by itself can be a stand-alone software application product or an integrated software-intensive system such as dedicated systems for financial transaction, airline and hotel reservation, and data processing and dissemination, where the software contributes essential influences to the design, construction deployment and evolution of the system as a whole to encompass specific applications [4]. Computer-based systems are software-intensive and primarily software-driven utilizing standard computer hardware hosts.

5.2.2 Software Dependability Project Management

Project management should adapt the principles and practices of the software life cycle processes [3] for software engineering operation. The dependability activities should be integrated in the respective project plans and incorporated in the system engineering

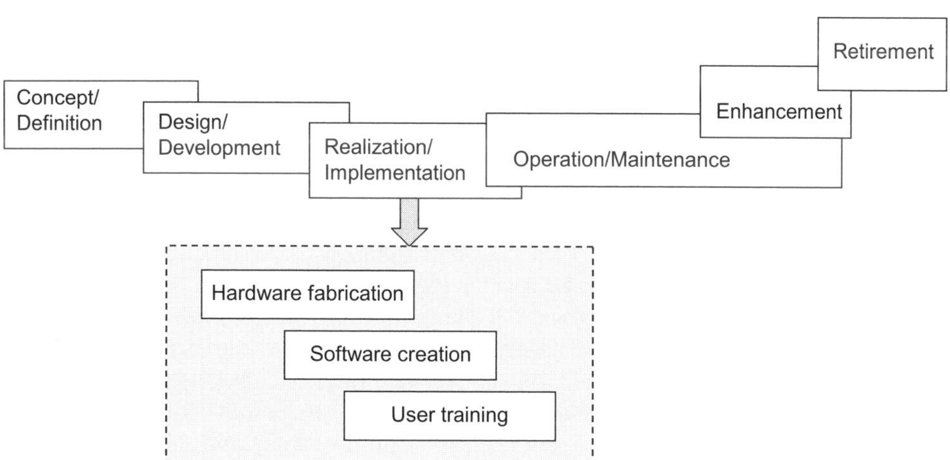

Figure 5.1 Software within the system life cycle framework

tasks for effective design and realization, implementation and application to meet project objectives. The guidance to engineering dependability into systems [5] influences the extent of dependability achievement for optimization of hardware design and software creation in development projects. Recommended processes and relevant methods are provided for general guidance in this chapter to address management concerns and technical issues relating to software aspects of dependability [6]. The following procedures are essential for software dependability achievement in project implementation:

(a) identify the software application objectives and requirements relevant to the specific software life cycle stage and application environment;
(b) identify the applicable software dependability characteristics relevant to the software project;
(c) review the adequacy of dependability management processes and available resources to support software project development and implementation;
(d) establish software requirements and dependability objectives;
(e) classify software faults and identify relevant software metrics for software dependability strategy implementation;
(f) apply relevant dependability methodology for software design and realization;
(g) initiate dependability improvement where practicable, taking into consideration various constraints and limitations for project tailoring;
(h) monitor development and implementation process for control and feedback to sustain software operability and assure dependability in system operation.

As software development has become more complex and diverse, the pressure to deliver high quality quickly has intensified. It is therefore even more important to manage and control the process of producing the software by measuring statistically project progress against quality, cost and delivery [7].

5.2.3 Software Life Cycle Activities

The software life cycle encompasses the following activities [6]:

(a) *requirements definition* identifies the system requirements for combined hardware and software elements in response to the users' needs and constraints of system applications;
(b) *requirements* analysis determines the feasible design options and transforms the system requirements for service applications into a technical view for hardware and software sub-system design and system development;
(c) *architectural design* provides a solution to meet system requirements by allocation of system elements into sub-system building blocks to establish a baseline structure for software sub-system decomposition and identify relevant software functions to meet the specified requirements;
(d) *detailed design* provides a design for each identified function in the system architecture and creates the needed software units and interfaces for the function, which can be apportioned to software, hardware, or both. The functions apportioned to software are defined with sufficient details to permit coding and testing. The software function can be labeled as a software sub-system and identified as a software configuration item for design control;
(e) *realization* produces the executable software units that meet verification criteria and design requirements, including lower level-activities in:

i) *coding* of the software units;
ii) *unit test* for verification of software unit to meet design requirements;
iii) *sub-system test* for verification of software program functions to meet design requirements;

(f) *integration* assembles the software units and sub-systems consistent with the architectural design configuration and installs the complete software system in the host hardware system for testing;

(g) *acceptance* establishes the system capability and validates the software applications to provide the required performance for specified system operations in the target environment; software acceptance tests include lower-level activities in:

i) *reliability growth testing* to increase the reliability of the software system, which can be conducted after the software system is fully integrated and executed in simulated field operational conditions representing the target environment;
ii) *qualification testing* to validate acceptance of the software system for customer release;

(h) *software operation and maintenance* engages the software in system operation, sustains the system operational capability and responds to application service demands to deliver specific operational services;

(i) *software update/enhancement* improves the software performance with added features;

(j) *software retirement* terminates the support of specific software service.

Figure 5.2 shows the key dependability activities important to the software life cycle identified for project implementation.

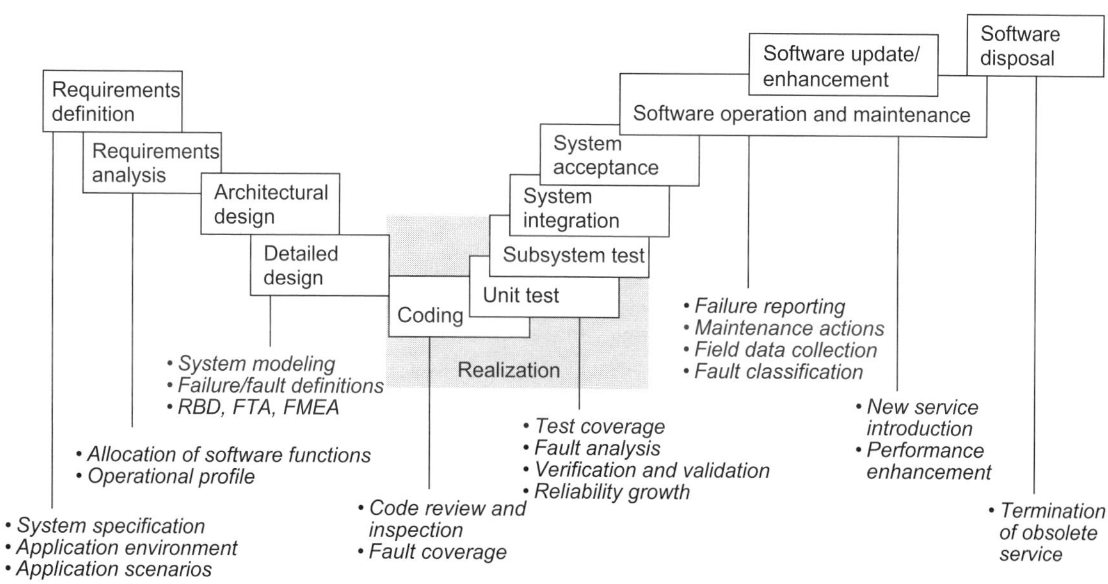

Figure 5.2 Software life cycle activities[1] (Courtesy of IEC)

[1] IEC 62628 ed.1.0 "Copyright © 2012 IEC Geneva, Switzerland.www.iec.ch"

5.2.4 Software Dependability Characteristics

Software dependability characteristics [6] are the specific dependability-related features and performance characteristics inherent in the software system by design and construction. They are selected for system incorporation to achieve combined hardware/software system dependability objectives.

The main software dependability characteristics include:

- *availability* for readiness of software operation;
- *reliability* for continuity of software service;
- *maintainability* for ease of software modification, upgrade and enhancement;
- *recoverability* for software restoration, following a failure, with or without external actions;
- *integrity* for correctness of software data information.

Other software dependability characteristics are application-specific. They include,, but are not limited to the following:

- *security* for protection from intrusion in software application and use;
- *safety* for prevention of harm in software application and use;
- *operability* for robust, fault-tolerant and non-disruptive operation;
- *reusability* of an existing software for other applications;
- *supportability* for provision and management of resources to perform software maintenance tasks;
- *portability* for cross-platform applications.

These software dependability characteristics form the basis for software system design and application.

5.2.5 Software Design Environment

The software design environment relies on an organized process to promote good design practices. This is to ensure error-free code generation, minimize mistakes in defining requirements, avoid ambiguity or misinterpretation of software verification results, and assure correctness in test validation prior to software release. The cultural aspects in the software engineering approach often adopt a Capability Maturity Model [8] concept for infrastructure development of the software management process. The software design environment and practice principles should be included in the organization's policy to establish procedural guidelines for dependability achievement.

The dependability management objective should provide a well-balanced design environment for creativity within project budget resources, time schedule and delivery targets. Organizations associated with software development and provision of software services are user application-oriented. The selection and adoption of applicable processes for engineering dependability into a specific software system is accomplished through the project tailoring process for effective dependability management. Opportunities for outsourcing design construction, software reuse and application of commercial-off-the-shelf (COTS) software products for system integration should be explored.

Software design often engages the application of computer-aided software engineering (CASE) tools [9]. An effective automated system provides computational accuracy, traceability of data, configuration management and a means for collecting the required measurements or metric inputs to the models automated. The data collection systems for field failure reporting, analysis and corrective actions are generally automated for capturing and processing performance history and experience data on software products and services. They are indispensable and provide valuable information to enhance operation.

5.2.6 Software Requirements and Influencing Factors

Software requirements and dependability objectives should form part of the overall software product specifications. Dependability activities associated with the software requirements are application-specific. They reflect the software design criteria and service application features needed to deliver the required system functions in meeting specific software performance requirements. Specific dependability objectives are derived from the selection of key dependability characteristics and their relevant quantitative metrics for assessment and test verification. Timing for implementation of relevant dependability activities is important. Dependability applications are time-dependent and have extensive impact on system life cycle cost [10]. Project tailoring is essential for design trade-offs and constraints resolution.

The influencing conditions on combined hardware/software system dependability specifications are provided in the system dependability specifications [11]. The general process for specification of dependability requirements is described in Chapter 3. The following influencing factors affecting dependability achievement in software development [6] are recommended for consideration:

(a) the organization's design culture and experience in software design and implementation;
(b) understanding the application environments and changing market dynamics for new platform or feature development for practical implementation;
(c) documentation processes such as failure reporting, data collection, software configuration management for control of software versions and maintenance of experience data records;
(d) application of software design rules for fault avoidance by controlling the design processes to optimize software performance in software complexity, program complexity and functional complexity;
(e) effective use of applicable software methods and tools such as structured design, fault tolerance, design review and software fault management to enhance reliability growth;
(f) selection of appropriate higher-order programming languages more suitable for specific software-structured development;
(g) established requirements for qualification and measurement of software dependability characteristics.

5.2.7 Classification of Software Faults

Software faults can be classified as specification faults, design faults, programming faults, compiler-inserted faults or faults introduced during software maintenance. Classification of software faults provides a means for capturing and grouping relevant software fault information. The classification process helps software designers discover unusual fault patterns for corrective actions. The objective is to eliminate the recurrence of the class of similar faults.

Orthogonal Defect Classification (ODC) [12] is a method used in software engineering for analysis and feedback of software fault (defect) data, targeting quality issues in software design and code in a procedural language environment. The term "fault" is used to designate the internal state of the software's inability to perform. The software faults noted herein are interpreted as software defects in the ODC process. A defect is a non-fulfillment of a requirement related to an intended or specified use of the software. The defect concept as interpreted in ODC may have legal connotations associated with product liability issues. In this context, a fault due to the inability of the software to perform its required functions exhibits the characteristics of defect attributes in the ODC scheme. Defect attributes are the signature of a defect, and they contain information related to the software fault. The ODC method captures the software fault information of the defect attributes for analysis and modeling. Analysis of ODC data provides a valuable diagnostic method for evaluating the maturity of the software product at various stages of the software life cycle. The ODC can also be used to evaluate the process by analyzing the types of triggers to identify specific technical needs to stimulate the missing triggers. The causal analysis of fault (defect) data presents a means for software fault reduction and reliability improvement.

5.3 SOFTWARE DEPENDABILITY STRATEGY

5.3.1 Software Fault Avoidance

Software codes are generated to produce a software product. A mistake made during coding design can manifest itself as a software fault leading to a system failure. Preventing faults from being introduced during design and maintenance are common approaches to lessen the existence of fault problems in software applications. The recommended software fault-avoidance strategy [6] includes fault prevention and fault removal:

(a) Fault prevention
- Establish fault prevention objectives in software engineering disciplines.
- Conduct early user interaction and refinement of the software requirements.
- Establish software standards and specifications.
- Introduce formal methods for software development where applicable and practicable.
- Implement systematic techniques for software reuse and assurance for application.

(b) Fault removal
- Initiate software code review.
- Detect and eliminate the existence of software faults by testing.
- Conduct formal inspection on finding faults, correcting faults and verifying the corrections.
- Perform corrective and perfective maintenance actions during software in-service operation.

5.3.2 Software Fault Control

Software faults are difficult to detect. Fault identification and removal can be achieved by various means, including rigorous software testing and inspection. Testing of software can also be deemed exhaustive in project management and are often cost-prohibitive with no guarantee of complete fault elimination. Software fault control employs fault tolerance and forecasting methods to minimize the manifestation of latent software faults or bugs that can still exist after the software is released for use. The recommended software fault-control strategy [6] includes fault tolerance and fault/failure forecasting:

(a) Fault tolerance
- Develop methodology for fault confinement, fault detection and fault recovery.
- Implement software design diversity and fallback schemes.
- Introduce multi-version programming techniques.
- Implement self-checking programming techniques.

(b) Fault/failure forecasting
- Determine fault/failure relationships in the operational environment.
- Establish data collection system to capture relevant data.
- Conduct reliability growth testing where applicable.
- Develop and implement relevant reliability models for fault/failure estimation.
- Refine forecasting techniques for time projection of software version release.

5.4 SOFTWARE DEPENDABILITY APPLICATION

5.4.1 Software Development Practices for Dependability Achievement

An organization's ability to develop software with consistency and its capability to deliver dependable products for intended applications reflect maturity in running the software business. The recommended technical approaches and management practices for dependability achievement [6] are provided for incorporation where applicable in software development:

(a) Standardize the methods for high-level architectural design, detailed design, coding and construction, and documentation to facilitate communications and fault avoidance.
(b) Use a modular design approach for software units and sub-systems with well-defined software functions and interfaces.
(c) Build simple, separate and independent software units to facilitate design interaction, maintenance, error traceability, fault mitigation and bug removal.
(d) Use design patterns of proven, reusable solutions from well-tested software as templates for solving software design problems to speed up the development process.
(e) Use formal methods where appropriate for control and documentation of software design and development process.
(f) Use software reliability engineering [13] techniques for software reliability assessment and enhancement [14].
(g) Consider reusing software available from a software library of well-tested software units and sub-systems for similar application and operational profile to reduce development cost and time and to minimize new design fault introduction.
(h) Develop regression testing methods to ensure functionality of existing software as new functionality is introduced or fault removal is performed.
(i) Test the software units and sub-systems to verify low-level design functions and validate integrated high-level design architectural system performance for progressive bug removal to prevent fault propagation.

(j) Inspect and review the software requirements, design specifications, software codes, user manuals, training materials and test documents to detect and eliminate as much as possible unintentional mistakes; utilize different review teams for comparison of results where practicable.
(k) Establish software configuration management for software process change control and software version release control to reduce fault occurrences.
(l) Identify and analyze the root causes of problems for continuous software improvement.
(m) Establish data collection system for knowledge base capture of software faults and performance data history.

5.4.2 Software Dependability Metrics and Data Collection

Software dependability metrics are measures of the dependability characteristics of a software system. The measurement of metrics provides a quantitative scale and method to determine the value of a specific characteristic associated with the software system performance. These metrics are obtained either by direct measurement or by deduction to serve different purposes in determining the software dependability characteristics. The following software metrics are de facto industry standards [6] for application:

(a) *Availability*: provides a measure of up time over the duration of system operation.
(b) *Failure frequency*: provides a measure of the number of failures over the duration of system operation.
(c) *Time-to-failure*: provides a measure of the failure-free time period.
(d) *Restoration time*: provides a measure of the time for restoration of a system from a failed condition (down state) back to normal operation (up state).
(e) *Fault density*: provides a measure of the number of faults contained per kilo source lines of code (KSLOC) and is used for software reliability assessment.
(f) *Function point*: provides a measure of the functional size of application software for software project planning by means of the *function point analysis* method [15].
(g) *Code coverage*: provides a measure of the degree to which the source code and the logical branches of a software program has been systematically tested; *code coverage* is an indicator of the thoroughness of software testing, and it is used to represent *fault coverage*, which indicates the percentage of faults detected during the test in code execution.
(h) *Fault removal rate*: provides a measure of the number of faults detected and corrected in a software product for a defined period of time or software execution duration; *fault removal rate* is used in reliability growth to establish reliability improvement trends.
(i) *Residual faults in software*: provides a measure of the estimated number of bugs still remaining in the software product after testing for bug removal.
(j) *Time for software release*: provides a measure of the estimated time for software release based on established criteria of an acceptable level of bugs still remaining in the software.
(k) *Software complexity*: provides a measure of the degree of difficulty for design and implementation of a software function or a software system.
(l) *Functional complexity*: provides a complexity rating value of low, average or high on specific software data or transactional function types according to the functional size measurement method [15].

Software dependability data are collected for use in software system performance assessments. Software metrics can be grouped into three general categories to facilitate data collection:

- *Fault data metrics*: intended to capture the software problem reporting data for measuring the impact of faults and the efficiency of the reporting process to improve software maintenance;
- *Product data metrics*: intended to capture software product information by categorizing the size, functionality, complexity, location of use and other characteristics to facilitate the use of user experience data as inputs to benefit new product development. The metrics provide performance history and data information of various software product groups;
- *Process data metrics*: intended to capture the software restoration process information and conditions at the time of fault detection and removal for reliability model inputs in reliability prediction.

The data collection process is critical for measuring software performance and relevant dependability characteristics. An effective data collection system should be practical for implementation. The amount and types of data should be relatively simple to collect, easy to interpret for data analysis and useful for dependability assessment and enhancement. The data collected is used to determine system reliability performance trends, frequency and time duration needed for software maintenance, response time for service calls and maintenance support strategy for restoration of degraded system performance.

5.4.3 Software Dependability Assessment

The purpose of dependability assessment is to appraise the extent of dependability achievement at any stage of the software life cycle. One-time assessment is often used to determine project status concerning specific dependability issues. The long-term assessment objective is aimed at ensuring software engineering process maturity through progressive dependability improvement. The dependability assessment process [6] includes the following tasks.

(a) Identify system performance objectives to develop relevant dependability specifications.
The system performance objective is provided in the customer requirements that are translated into project-specific technical specifications. The relevant technical information includes:
  - the system performance scenario and application environment;
  - the relevant performance influencing factors;
  - the system boundary and interfaces with external interacting systems;
  - the relevant system performance attributes;
  - the system architecture, hardware/software configuration;
  - the interoperating hardware and software functions of the system configuration;
  - the dependability characteristics of the relevant hardware and software functions, such as availability, reliability and recoverability associated with maintenance support criteria.

(b) Establish software operational profile.
An operational profile represents the sequence of required tasks to be performed by the system to achieve its mission or service objective. The system performance

is highly dependent on the environment in which the system operates. The intended application environment can affect the physical system structure and influence the system functions in task performance. The development of a software operational profile is a quantitative characterization on how the software is being used from the system application perspective. The operational profile is used as inputs for development of test cases to simulate actual operations and specific usage of the software functional features. The following are recommended processes for development of a software operational profile:

- determine the application profile by establishing the user needs and types of service intended for the software application;
- establish the user profile on the different types of users on interaction with the software for specific applications;
- define the system-mode profile on how the system is being operated and in what sequence or order expressed in terms of modes of operation in executing the software application;
- determine the functional profile by evaluating the system performance functions and service features in meeting their respective functional requirements;
- determine the operational profile based on the functional profiles established for system performance functions;
- determine the information profile by populating software application data at the software development life cycle.

(c) Allocate applicable dependability characteristics.
Allocation of dependability characteristics and metric measures for a software system is based on modeling of the system architectural functions to reflect the requirements of the system dependability objective. The apportionment of reliability and availability values to the various software sub-systems and functional units are assigned according to their complexity, criticality, estimated achievable reliability or availability performance targets and other influencing factors relevant to the allocation process.

The construction of a system model for software differs significantly from hardware due to its inherent operating characteristics. For each mode in system operation that involves the software program functions as configuration items, different sets of constituent software units of relevant configuration items are being executed. Each mode has a unique time of application associated with the software unit execution duration on demand in system operation. This indicates the time duration of each system mode. The software system modeling includes the number of lines of source code in each software unit, the code complexity and other information pertaining to software development resources, such as programming language and design environment. They are used to establish the initial failure rate for reliability or availability prediction of the software configuration items. Actual experience data should be used when system performance has sufficient data collected on failure rates associated with the identified software configuration items.

(d) Perform dependability analysis and evaluation.
The analysis and evaluation process is iterative for optimization of dependability design requirements to meet system performance objectives. Functional block modeling techniques such as reliability block diagram (RBD) and fault tree analysis (FTA) are commonly used to perform dependability analysis and evaluation of time-dependent performance functions.

The software sub-system functional model is constructed using software units as building blocks to deliver software program functions. A software unit is the lowest level of a configurable software item. Software units do not fail independently as with hardware components. Software codes are virtual entities and are not subject to physical changes. Software units fail in association with the system operational profile, which affects the configuration scheme of the software reliability model structure. The software system is influenced by the application time factors reflecting the system operational profile. Modeling software reliability must incorporate the operational profile information in developing the software configuration structure. The software sub-system program can consist of one or more constituent software units to deliver the required functions. A software sub-system program residing in a host hardware sub-system is configured to form a software configuration item. The interoperation and mutual dependency of the software sub-system and its designated hardware host are needed to deliver specific sub-system software functions for system operation. There can be one or more combined software and hardware sub-systems servicing different functions in the entire system configuration.

Figure 5.3 illustrates the functional relationships of a combined hardware/software system, which includes a configuration item consisting of hardware and software functions. The application software functions of the software sub-system are influenced by the application time factors associated with the user service demands reflecting the system operational profile while the operating sub-system is running continuously.

Software functions are often used for control of a critical system where a failure can cause catastrophic effects that result in severe consequences. The criticality of software functions should be identified early in system concept definition and evaluated during software architectural design of the system. The criticality of functional failures should be classified in the system specifications, such as critical, major or minor based on established criteria, and verified by analysis in system reliability performance.

The level of risk associated with the critical software function can be determined and evaluated by means of risk assessment techniques. Project risk management [16] should focus on fault prevention and fault tolerance where the severity of failure consequences can be mitigated. A risk-driven approach can also be used to test software reliability, where both consumer and producer risks are considered. Consumer and producer will have different perspectives on what would be considered as tolerable risk of software failure [17].

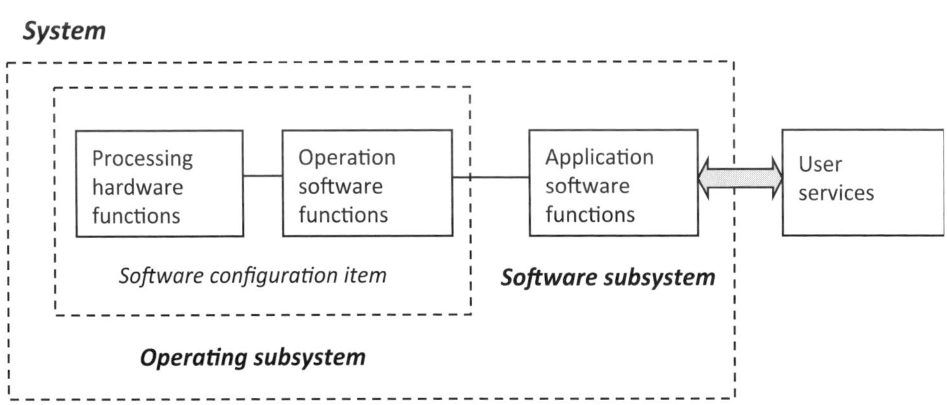

Figure 5.3 Functional relationships of a software configuration item

5.4.4 Software Testing and Measurement

(a) Software testing consideration.
Software testing is the process of executing a program or a set of coded instructions with the intent of verifying software functions and finding errors. The objectives of software testing vary with project needs, software product availability, software maturity status and scheduling for testing during the software life cycle. Planning for software testing should consider the following software-specific conditions:

- Software testing is an art that requires good testing practice. Good testing techniques demand the skills, experience, intuition and creativity of the tester to achieve dependable results. Maintaining test records is important in providing accuracy and traceability of test data.
- Testing is more than just debugging the software program to locate faults and correct errors. Testing is also used in software verification and validation, as well as reliability measurement.
- Test efficiency and process effectiveness are criteria for coverage-based testing techniques. Test automation can expedite software test time and reduce project cost. The selection of appropriate test tools for software testing should be evaluated for acceptability prior to using them for software test application.
- Testing is not necessarily the most effective means to improve software quality unless appropriate follow-up actions are taken. Alternative methods, such as code inspection and code review, should be considered.
- Software testing is only part of the software reliability growth and improvement process. Other assurance efforts should be used in conjunction with software testing to achieve dependability goals.
- Software complexity influences the extent of test completeness. The complexity problem often limits the tester's ability to detect and remove bugs by the testing process due to time and cost constraints.
- Latent software faults or bugs do exist in software after its release for service operation. Software reliability prediction provides a means to estimate the test time required to reduce the residual software bugs to an acceptable number before the next software version release.
- Testing beyond unit testing is recommended to be performed by separate testing teams that are independent of the teams developing the software.

(b) Types of software tests.
The following presents the types of software tests [6] performed during the software life cycle.

- *Unit test*: testing of one software unit that can be compiled before it is integrated into the software program or sub-system. The software unit is tested to verify that the detailed design for the unit has been correctly implemented.
- *Sub-system test*: testing of a sub-system software program consisting of one or more software units as a software configuration item to verify functional performance requirements.
- *Integration test*: testing of a software system in a hardware host as a whole consisting of integrated sub-systems to verify functional operation, expose problems in software interfaces, hardware interfaces and interactions between the hardware and software, and validate reliability performance.
- *Reliability growth test*: testing of software in an iterative process to improve reliability through testing until failure, analyzing failures, implementing corrective

action on the existing software version for upgrade and continuing the test with the newly upgraded software version. Termination of reliability growth test is based on when the established software reliability target is met.

- *Qualification test*: testing to demonstrate that the software meets its specifications when integrated in its host hardware system and is ready for use in its target environment. Before release of the final version for software distribution, alpha and beta testing are often conducted for quality assurance purposes. *Alpha testing* is an in-house trial carried out by software developers before release for external users. *Beta testing* is a field trial carried out by a limited number of users in its intended application to seek user experience information.
- *Acceptance test*: testing of a software system to validate that the customer's requirements are met. For acceptance testing of complex hardware/software systems where no prior information exists on similar systems, reliability growth and stress testing [18] should be considered as part of the acceptance test requirements.
- *Regression test*: testing of software that has been previously tested in an effort to uncover any maintenance errors introduced, new code being developed, improper configuration or inadequate source control.

(c) Testability of software.

Testability of software is the degree to which the software facilitates the establishment of test criteria and the performance of tests. Design for testability is a design/development methodology that structures development to enable testing. The process analyzes the software characteristics and predicts the likelihood of any bugs in the software that can be revealed through testing. The analysis is used to optimize the testing process to determine the adequacy of testing. It provides a means to manage test resources and determine the value or benefits of a specific testing approach.

(d) Test cases.

Test cases are developed to simulate actual software field operating conditions in which specific interest areas or potential problems could be encountered. A test case specification is developed to specify inputs, identify expected test results, and establish execution conditions for the test item. An effective testing process includes both manually and automatically produced test cases. Manual tests cover the depth of finding software faults, reflecting the developer's understanding of the problem domain and data structure. Automatic tests cover the breadth of fault investigations by executing the entire range of test values, including those extremes that human testers might miss. The automatic test process engages the use of a test case generator to accept source code, test criteria, specifications or data structure definitions as inputs to generate test data and determine expected results. Fault insertion test can be considered as a test case in which a deliberate fault is introduced in one part of the software system to verify that another part reacts appropriately. The test results are used to determine probable fault conditions and facilitate fault-tolerant software design. The fault insertion technique is also used to test the coverage of the test program by counting the fraction of the inserted faults found.

The execution of test cases for software testing provides valuable information and data capture for estimation of software reliability in field operation. The test case results can be used to validate the effective provision of the maintenance functions and confirm the efficiency of the maintenance support performance.

(e) Software verification and validation.

The software verification process determines that the requirements for the software are complete and correct as applicable to the software life cycle stages. The validation process determines that the system performance and services conform to the customer/user requirements. Appropriate enabling systems, such as test tools and equipment, facilities and supplementary resources, are required to support the implementation of the verification and validation processes. The enabling system does not contribute directly to the performance functions of the software or the system under test during operation.

The software verification process is intended to confirm that the specified design requirements are fulfilled by the software system. Verification process activities include:

- defining a strategy for software verification;
- developing a verification plan based on software system requirements;
- identifying the constraints and limitations associated with the design decisions;
- ensuring that the enabling system for verification is available and associated facilities and testing resources are prepared;
- conducting the verification to demonstrate compliance to the specified design requirements;
- documenting the verification results and data;
- analyzing the verification data for initiation of corrective action.

The validation process provides objective evidence that the software system performance meets customer/user requirements. The validation process activities include:

- defining a strategy for validation of the services in the operational environment and achieving customer/user satisfaction;
- preparing a validation plan;
- ensuring that the enabling system for validation is available and associated facilities and testing resources are prepared;
- conducting validation to demonstrate conformance of services to the customer/user requirements;
- documenting the validation results and data;
- analyzing, recording and reporting validation data according to the criteria defined in the validation plan;
- reviewing the results of validation data with the customer to determine software system acceptance.

5.5 SOFTWARE DEPENDABILITY IMPROVEMENT

5.5.1 Software Dependability Improvement Methodology

The objective of software dependability improvement is to sustain dependability performance in system operation. The software dependability improvement process establishes the criteria and methodology for implementing improvements in software projects. The improvement process advocates the improvement in software design, improvement through reliability growth testing, improvement in software perfective maintenance and software enhancement upgrades, and improvement in provision of customer support services. The improvement effort is not a short-term software patch-up for a temporary fix, which often creates long-term problems in software operation. The software dependability improvement process should be considered as a strategic approach for software design and implementation of industry best practices. Depending on the magnitude of the software problem encountered in a software project, some improvement activities can be relatively simple software upgrades; others may require complete redesign of the entire software system configuration. The latter example is typical in an outdated software operating system with too many add-ons over a lengthy duration, which makes the system operation totally inefficient and causes frequent application breakdowns or malfunctions. The following presents the methodology and industry best practices for software dependability improvement.

5.5.2 Software Complexity Simplification

Software dependability can be improved by means of design simplification.

- *Structural complexity* describes the logic paths for connection of software units in the software design. Each software unit can be programmed or coded to provide an executable unit of software function in the software structure. Structural complexity is related to the testability of program codes that affect fault detection, hence influencing the reliability and maintainability of the software architecture. The more complex the structure, the more difficult it becomes to test the software. The software design rules should establish a level of complexity to facilitate design for dependability.
- *Functional complexity* describes the required functions that the software unit or segment of code in the unit must perform. In an ideal design concept, one software unit should be designed to perform one function to achieve simplicity, with one set of cohesive inputs and outputs to facilitate software fault isolation and removal. In practice, both structural complexity and functional complexity should be considered for software design evaluation. Software design strategy for complexity is directly linked to the number of test cases needed for complete software verification.

5.5.3 Software Fault Tolerance

Software fault tolerance design is the prevention of software faults from causing system failure during system operation. Fault tolerance design permits the software to continue functioning and preserve the integrity of data with certain faults present in the software. Software designed with fault tolerance capability exhibits graceful system degradation and continues to operate properly for a period of time in the event of system failures. Fault tolerance is of particular importance for safety-critical systems that depend on high-availability system performance in the presence of faults or operating under adverse conditions. Software fault tolerance is constructed to have a low probability of exhibiting common-mode failure from a number of diverse system designs, including the following recommended practices:

- *Fault confinement*: software is written to restrict a fault to a local domain such that it cannot contaminate other software domains.
- *Fault detection*: software is written to respond to test when a fault arises.
- *Fault recovery*: software is written so that, after detecting a fault, it takes sufficient steps to allow the software to continue to function successfully.
- *Design diversity*: software and its data are created so that fallback versions are available.

Redundancy is a common approach to improve system reliability and availability. Multi-version programming is used for fault tolerance design of safety critical systems. The method engages multiple functionally equivalent programs that are independently generated from the same initial software specifications. The independence of separate programming efforts will greatly reduce the probability of identical software faults occurring in two or more versions of the program. Implementation of these programs utilizes a different algorithm and programming language. Special mechanisms are built into the software to allow these separate programs to be controlled by a voting scheme in the decision algorithm for program execution in application. The concept is based on the assumption from a redundancy viewpoint that output from multiple independent versions is more likely to be correct than the output from a single version. In practice, the improvement benefits of the multi-programming effort will require justification of the additional time and resource requirements to warrant cost-effective implementation.

5.5.4 Software Interoperability

Software interoperability is the ability of diverse software systems to work together to exchange information and to use the information that has been exchanged. In an open system such as an Internet protocol network, it is important to achieve interoperability of diverse software systems to establish communication links. Failure of the communication link will affect dependability and disrupt services in network performance operation. One practical approach recommended to enhance interoperability in communication networks is to incorporate a specific feature in the software system design to monitor the status of the established connection. A signal processing and synchronization scheme is incorporated in the design to send signals back and forth between the communication nodes to monitor that the communication link is established. If the link is broken or interrupted due to changes in the operating environment, the software system will automatically attempt to re-establish the link to maintain continuity in network communications such that dependability performance is not degraded. Software interoperability relies on standardizing a common set of exchange formats to read and write the same file formats and to use the same protocols. Standardization ensures compatibility of software interfaces to achieve interoperability, hence enhancing system dependability performance.

5.5.5 Software Reuse

Software reuse is the use of existing software to build new software. Software can be reused at different levels, from simple software functions to complete software application systems. The benefits of reusing software include:

- increased dependability with proven software exercised in operating systems;
- reduced process risk with less uncertainty in the costs of reusing existing software than new software development;
- effective use of software resources by reusing existing software while allowing the software designers to focus on developing reusable software that would retain and capture their special knowledge;
- compliance with standards already embedded in the reusable software;
- expedited software development.

A software design pattern is a general reusable solution to a commonly occurring problem in software design. The design pattern serves as a template on how to solve a problem that can be applied in different situations. Design patterns reside in the domain of software modules and interconnections. The application of design patterns can speed up the software development process by providing proven tried and tested solutions. Reusing design patterns can help the software designer improve code readability and prevent problematic issues that may not become visible until later in software implementation.

Reusing existing application software is common in engineering/procurement/construction contracts in which available COTS products or systems are selected and used for system design and implementation. The reuse of the entire COTS application system is achieved by configuring a system for a specific project application environment or by integrating several COTS products and systems to create a new application for the project.

Reusable software is a reusable asset. The most well known reusable asset is code. Programming code written at one time can be used in another program written at a later time. The reuse of programming code is a common technique that is used in an attempt to save time and effort by reducing the amount of repeated work. Software reusability is the degree to which a software asset can be used in different software systems or in building other assets.

From a dependability perspective, the application of software reuse in projects and its reusability characteristics should be controlled to achieve dependability improvement. Reusability is directly dependent on the software structure and modular design. For a software module or unit to be reusable, it should be confined to performing only one function in its entirety. This restriction is essential because if the intended reuse of the software unit is performing less than one function, or it is able to perform more than one function, it will be difficult to implement or to maintain for its intended reuse purpose. Deviation from this restriction will decrease the usefulness of the reusable software. Reusing software that does not perform exactly one function could have an adverse effect on dependability due to the possibility of errors introduced into the software during implementation or maintenance.

The reuse of software should be implemented only if the functional requirements of the new software unit are in line with those of the reusable software for very similar application and operational environment. Otherwise, it will diminish the cost-effectiveness of the software reuse objective and possibly decrease reliability when implemented. Reusable software should be well documented for traceability to facilitate configuration management of software assets.

5.5.6 Software Reliability Growth

Software reliability growth is the condition characterized by a progressive improvement of a reliability performance measure of the software system with time. Software reliability improvement is achieved by design, and progressive reliability enhancement is verified by means of reliability growth testing. A software program needs to be executed to discover bugs and expose software failures. The software failures uncover faults, and the removal of these faults results in reliability improvement. Software reliability growth trends are based on the fault removal rates with respect to the cumulative software execution time. For scheduling purposes, execution time can be converted to calendar time to establish the software failure rates for reliability estimation. A reliability growth program [19] can be established for the combined hardware/software system. Reliability growth models and estimation methods for assessments, based on failure data captured in the reliability growth program, are described in the statistical methods for reliability growth [20].

Reliability growth for software is the positive improvement of software reliability over time, accomplished through systematic removal of software bugs. The rate of reliability growth depends on how fast the bugs can be uncovered and removed. A software reliability model applicable to growth conditions allows project management to track the software reliability progress through statistical inference to establish trends and forecast future reliability targets. Appropriate management actions can also be taken if the trend indicates a negative reliability pattern.

Measuring and projecting software reliability growth requires the use of an appropriate software reliability model that describes the variation of software reliability with time [21]. The reliability model inputs are derived from prediction-based experience data or from an estimation of test data collected during system testing. The selection and use of any software reliability models for growth estimation should be validated. The estimation process is based on the times at which the failures occur with a sufficient data sample for significant execution time accumulation. This is to establish a reasonable degree of statistical confidence to validate the reliability growth trends. The objective is to forecast software maturity and release targets.

It must be recognized that reliability growth will be different during testing and operational phases. Thus a software reliability growth model (SRGM) developed for the testing phase may not be applicable to the operational time period. A generalized SRGM has been proposed to account for the differences between these two environments with respect to the efficiency of fault removal efficiency [22].

5.5.7 Software Maintenance and Enhancement

Software maintenance is the modification of a software product after delivery to correct faults, to improve performance or other software performance characteristics, or to adapt the product to a modified environment. There are four main categories of software maintenance:

- *Corrective maintenance*: reactive modification of a software product performed after delivery to correct discovered problems;
- *Adaptive maintenance*: modification of a software product performed after delivery to keep a software product usable in a changed or changing environment;
- *Perfective maintenance*: modification of a software product after delivery to improve performance or maintainability;
- *Preventive maintenance*: modification of a software product after delivery to detect and correct bugs in the software product before allowing further propagation into real failure occurrences.

The management issues associated with software maintenance include alignment with customer priorities, maintenance resource planning and allocation, skill training of maintenance personnel, contract maintenance work and customer satisfaction survey feedback. The technical issues in software maintenance include incident reporting, technical problem resolution, impact analysis, standardization of application procedures and testing practices, software maintainability assessment and test efficiency measurements.

Software enhancement is a part of the software evolution process. A software system in field operation is noted for its increasing complexity with time of operation due to modifications and enhancements implemented to meet customer needs. This is unavoidable due to continuous changes in maintenance support strategies in response to competitive service offerings and the need to develop the skills and techniques to accommodate the changing business environments. The ever-increasing complexity in a software system will eventually affect the interoperability in service operation, resulting in ineffective software maintenance support. The extent and achievement of software maintenance and enhancement effort should be regularly monitored, verified, validated and documented. The cost of software maintenance should form part of the dependability assurance strategy.

Software dependability data collection activity is essential for field tracking to assess the dependability of software performance operation in customer premises. This is to ensure and confirm that the accepted level of dependability performance in operation is sustained for the software deployment. The in-service field dependability information is collected together with relevant customer feedback information. The information is used to justify changes for new software requirements and initiate development of new software releases.

Due to the dynamics of application environments and technology evolution, the decisions on software new releases are often influenced by market competition and driven by business strategies.

5.5.8 Technical Support and User Training

Technical support provides a range of services to assist users with software operation or software products in use. The objective is to help users solve specific problems with product operation or application. Technical support takes on various forms, including telephone query, online service, email, remote access repair and on-site visit for problem solving. There are increasing growth and use of outsourced call centers by technology product development organizations for business, economic and geographical reasons to facilitate real-time response to technical support service needs. These call centers serve as centralized technical support for a broad range of technology products, such as computer systems and software requiring technical assistance around the clock with worldwide, toll-free user access. Technical support services form part of the maintenance support to sustain product operability and reliability performance contributing to dependability improvement.

Software user training is an important aspect of software dependability improvement. The objective is to enhance or familiarize the skill level and understanding of the software product applications from the users' perspective. Software user training exists in various forms, including online access of the product supplier's tutorial database, call center assistance or dedicated technical expert service to address unusual problems encountered.

There are many training organizations and academic institutions equipped with training specialists and laboratory demonstration facilities to train application users, operators and maintainers of specialized systems and products, including software systems. Users often engage in hands-on experience and learn directly from the expert trainers or educators in a range of training programs from system fundamentals to specific techniques and skills. There are also technical support and technology forums open for user participations. The continuing learning process promotes knowledge enhancement and greatly influences dependability improvement.

REFERENCES

[1] Balamuragan, S., Xavier, R.J. and Jeyakumar, A.E., *Control of Heavy-duty Gas Turbine Plants for Parallel Operation Using Soft Computing Techniques*, Electric Power Components and Systems, 37:1275–1287, 2009.

[2] IEC 62508, *Guidance on human aspects of dependability*

[3] ISO/IEC 12207, *Systems and software engineering – Software life cycle processes*

[4] ISO/IEC 42010, *Systems and software engineering – Recommended practice for architectural description of software-intensive systems*

[5] IEC 60300-3-15, *Dependability management – Part 3-15: Guidance to engineering of system dependability*
[6] IEC 62628, *Guidance on software aspects of dependability*
[7] Yamada, S. and Kawahara, A., *Statistical Analysis of Process Monitoring Data for Software Process Improvement,* International Journal of Reliability, Quality and Safety Engineering Vol. 16, No. 5 (2009) 435-451.
[8] *CMMI-DEV (Version 1.3, November 2010)*, Carnegie Mellon University Software Engineering Institute. 2010
[9] Kuhn, D.L., *Selecting and effectively using a computer aided software engineering tool*, Annual Westinghouse computer symposium; 6–7 Nov 1989; Pittsburgh, PA (U.S.); DOE Project
[10] IEC 60300-3-3, *Dependability management – Part 3-3: Application guide – Life cycle costing*
[11] IEC 62347, *Guidance on system dependability specifications*
[12] Ram Chillarege, *Orthogonal Defect Classification – A concept for in process Measurements*, IEEE Transactions on Software Engineering, 1992
[13] Lyu, M. R. (Ed.): *The Handbook of Software Reliability Engineering*, IEEE Computer Society Press and McGraw-Hill Book Company, 1996
[14] IEEE-1633: *Recommended Practice on Software Reliability*, 2009
[15] ISO/IEC 20926, *Information technology – Unadjusted functional size measurement method*
[16] IEC 62198, *Project risk management – Application guidelines*
[17] Schneidewind, N.P.F, *Risk-Driven Software Testing and Reliability*, International Journal of Reliability, Quality and Safety Engineering Vol. 14, No. 2 (2007) 99-132.
[18] IEC 62429, *Reliability growth – Stress testing for early failures in unique complex systems*
[19] IEC 61014, *Programmes for reliability growth*
[20] IEC 61164, *Reliability growth – Statistical test and estimation methods*
[21] Kapur, P.K., Aggarwal, A.G, Shatnawi, O. and Kumar, R., *On the Development of Unified Scheme for Discrete Software Reliability Growth Modeling,* International Journal of Reliability, Quality and Safety Engineering Vol. 17, No. 3 (2010) 245–260.
[22] Kapur, P.K., Gupta A. and Jha, P.C., *Reliability Analysis of Project and Product Type Software in Operational Phase Incorporating the Effect of Fault Removal Efficiency*, International Journal of Reliability, Quality and Safety Engineering Vol. 14, No. 3 (2007) 219-240.

Chapter **6**

DEPENDABILITY INFORMATION MANAGEMENT

6.1 UNDERSTANDING DEPENDABILITY INFORMATION

Information is intelligence that pertains to satisfying specific needs for learning and enlightenment. The specific dependability information needed depends on the relevancy of the information and the circumstances of use. Timing and availability of information are key factors of its usefulness, and sometimes they may impact decision outcomes due to inadequate information input or resulting from too much information, which can lead to unnecessary confusion. Dependability information is derived from data collected from various dependability-related sources, especially on items used in the product or the system of interest. The classification of data plays an important role in managing a credible dependability database. Sound decision-making helps attain organizational goals and achieve dependability value.

In dependability application, data are typically the results of measurements. Raw data are unprocessed numbers or noted observations of an item's condition. The item can be a component, a product, a system or an event situation. Information is the result obtained from processed data that may contain the specific intelligence sought. The extent of useful information that can be extracted from the set of raw data depends on the approach of information processing relevant to the objectives in data analysis.

Knowledge is an understanding of the facts and figures presented by the information available at hand. The ability to interpret and utilize the intelligence in the information is based on specific know-how and relevant expertise. Knowledge is acquired through study and learning, and it is gained from practical experience. The differentiation of data, information and knowledge with respect to its intelligence contents can be interpreted by the level of detail: data presents the least value and knowledge provides the most, while information offers something in between. For example, an error discovered in a computer program is presented as data; running the computer program for an account billing application is information; and designing a fault-tolerant computer program is providing knowledge. Similarly, having a thousand customers is data; serving a thousand satisfied customers is information; and retaining a thousand valued customers year after year is knowledge.

Information management is user-oriented. The information system delivering the processed data outputs to meet varied user needs is an enabling mechanism or system. The concept of the enabling mechanism is described in Chapter 3, Figure 3.3. The enabling system complements the system of interest during its life cycle stages but does not necessarily contribute directly to its function during operation. The enabling system objective and operation structure for the collection and processing of dependability data, and the dissemination of relevant information and knowledge capture, establishes the framework for dependability information management.

6.2 FRAMEWORK FOR DEPENDABILITY INFORMATION MANAGEMENT

Dependability information management provides an organized approach to management control and data processing for acquisition, analysis, retention, retrieval and distribution of dependability information. The framework for dependability information management is focused on establishing a credible system with a relevant database, capable of handling the organizational and project-specific needs for capturing and processing dependability data and disseminating relevant information. The system that facilitates dependability data processing is an information processing system. It should not be confused with management information systems, which have a broader scope and mandate for applications involving technology, information and human resources to provide strategic and operational information needed for effective management of the organization's business.

The information system established for dependability information management should be compatible with the existing information management system structure and processes implemented and used within the organization. This will simplify the information and data transfer process and facilitate the sharing of common nomenclature, cross-referencing identification, asset traceability and record management. Timely response to management query, access and retrieval to support documents and reporting structure hierarchy should conform to established quality assurance procedures. The scope of the information system database should be populated with sufficient current dependability data and historical performance records to support in-house projects and assist product deployment in field service operation.

Management in all businesses and organizational activities is the act of using available resources to accomplish the desired goals and achieve business objectives. In the context of dependability management, the goal is value creation harnessed by effective use and results-oriented applications of the information management process. The technology and tools utilized to support the dependability information system have become more complex and sophisticated, capable of mimicking humans in making decisions. Examples include various forms of expert systems and artificial intelligence that can provide convincing ready-made solutions when presented with a difficult and complex problem or question. It should be cautioned that there are no engineering cookbook answers coming out of any machines created by humans. Prudent judgment and human intuition is often necessary to override conflict resolutions based on real-life practical experience.

Figure 6.1 illustrates the process relationships of the dependability information management framework.

The process relationships show a scenario where there is full integration of process information flows from initial data source captured through information extraction to the development and utilization of dependability knowledge base. This is an ideal situation for an integrated information system. Difference organizations have different information management processes. Different businesses have different industry-specific technology information needs. For practical application, relevant parts of the dependability information process may be selected and adapted for specific use. They should be tailored to suit the specific information requirements of the organization and business needs.

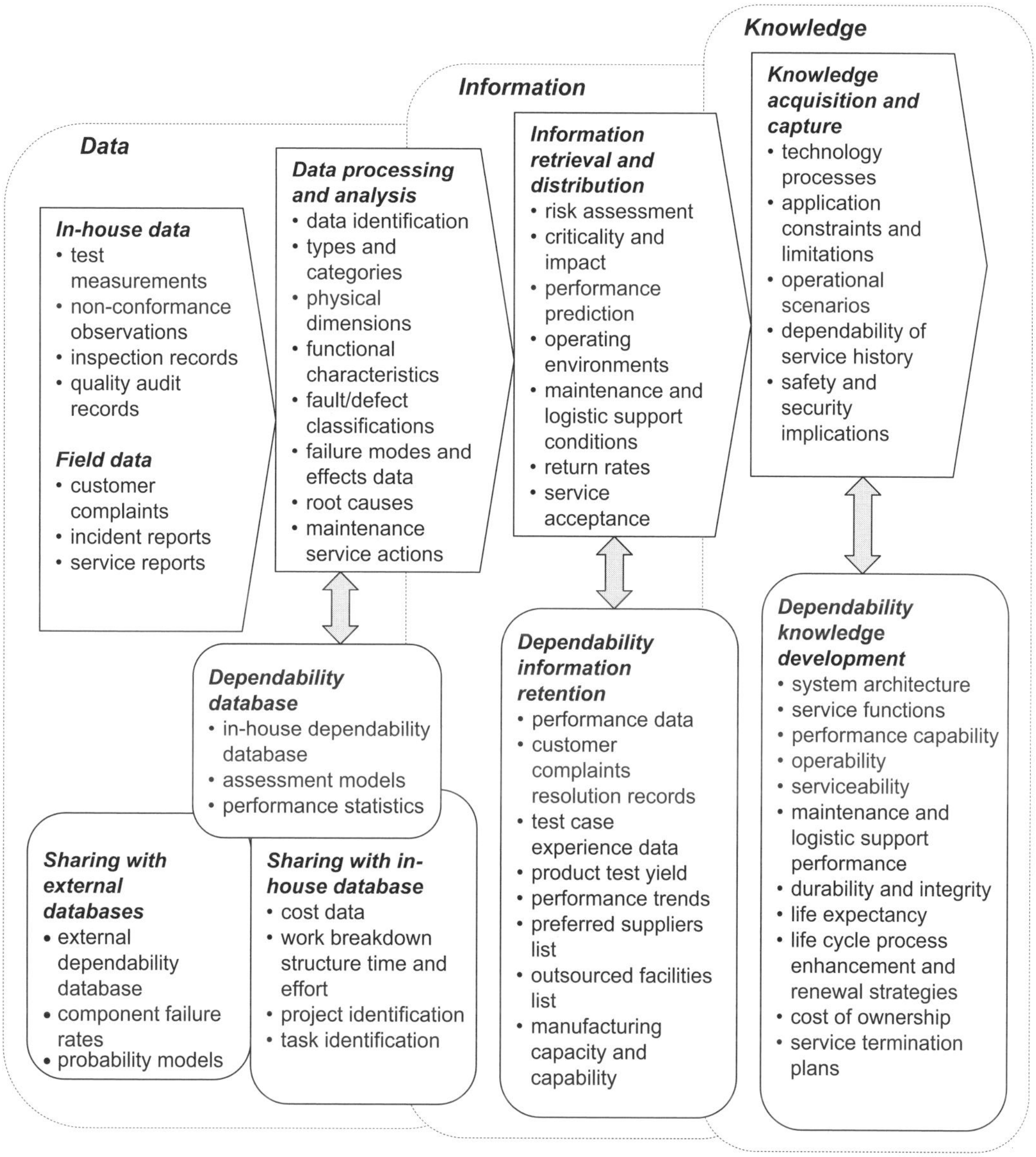

Figure 6.1 Process relationships of the dependability information management framework

6.3 ESTABLISHING A DEPENDABILITY INFORMATION SYSTEM

6.3.1 Dependability Information System Requirements

For technological systems, the requirements for a dependability information system should include:

(a) a failure reporting, analysis and corrective action system (FRACAS) to record failures, usually hardware, and establish the organization's procedures and database essential for project dependability applications;

(b) a maintenance and logistics support system to sustain system service operation;
(c) a fault management system to capture errors and faults, typically network- and software-related, to facilitate implementation of fault prevention and mitigation schemes, and development of alternate designs for fault tolerance;
(d) dependability assessment methods and techniques to capture design characteristics for performance improvement in system dependability prediction, analysis and evaluation;
(e) information retention to facilitate retrieval and distribution of relevant dependability records and data;
(f) knowledge acquisition and capture to encourage dependability knowledge development.

The basis for a comprehensive dependability information system is developed over time with the experience and knowledge gained. The development process takes into account business changes, technology advancements and continuous data capture as the organization's products are being developed for service implementation. A centralized database has become an essential information depository and knowledge base support requirement. For dependability application, most data collected come from product testing and field service records. These data provide the needed inputs to initiate population of the dependability information system database. Processed information derived from data analysis and evaluation, dependability assessment on various design functions and system configurations, and test case verification experiments will provide helpful insights for decision-making.

Establishment of the dependability information system advocates for the collaboration of information and data derived from three separate systems: FRACAS, the fault management system and the maintenance and logistic support system. The convergence of information and data from these system outputs greatly supports the implementation of a cost-effective centralized database. With the advent of communications technology, access, retrieval and distribution of electronic information and data have become real-time transactions with significant improvement on delivering data accuracy, simplified data collection procedures and searching of dependability knowledge for acquisition from all available sources. Separate paths for information flow and data flow can be identified with their convergence towards a centralized database to facilitate progressive dependability knowledge development. Figure 6.2 presents an overview of the information and activity flow of the dependability information system.

6.3.2 FRACAS

The starting point for the dependability information system is the development and implementation of a FRACAS system. FRACAS was developed for use in the United States defense industry as enunciated in the military standard MIL-STD-2155 [1]. The FRACAS framework has been adapted by industry for development of reliability database systems [2]. For expediency, some organizations have opted to acquire a similar system from outsourced vendors that provide ready-made information systems under licensing agreements. This approach is quite common in start-up business when dealing with products that are primarily electronic components and systems. Specific in-house data can be added to populate the database as they become available. This approach saves time and avoids initial development effort, but it lacks the specific procedures and practices in collecting organizational data of its own products to provide the relevant experience information most essential to the business.

Table 6.1 presents typical examples of FRACAS data collection.

FRACAS is intended to provide management visibility and control for reliability and maintainability improvement of hardware and associated software by timely and disciplined

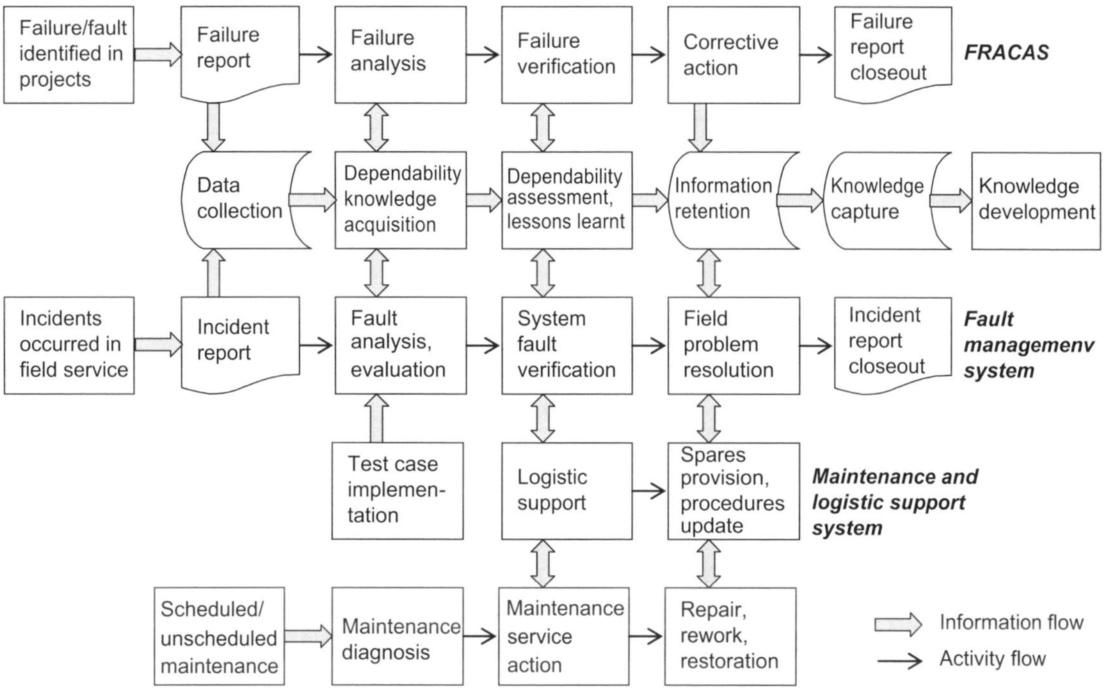

Figure 6.2 Information and activity flow of the dependability information system

utilization of failure and maintenance data. The purpose is to generate and implement effective corrective actions to prevent failure recurrence and to simplify or reduce the maintenance tasks. FRACAS is designed to be a closed-loop system where the reported failures are analyzed to determine the failure causes. The analysis process identifies and verifies the failure problem and recommends corrective action for implementation to prevent further recurrence of the failure. In practice, the failures are first diagnosed in terms of failure

TABLE 6.1 Examples of FRACAS data collection

Problem reporting data	• Failed item identification • Date and time failure detected • Failure discovered during design, production, operation or maintenance • Detected failure description • Failure detected in specific location • Person who detected the failure • Failure symptom and status • Severity and priority	Data collected should be used for reporting the problem on failure identification and occurrence.
Corrective action data	• Date failure corrected • Person who corrected the failure • Maintenance action taken • Description of modification • Identification of modules modified • Version control information • Time required to correct failure • Date verified as failure corrected • Person who verified the correction	Data collected on corrective actions taken and verified as the failure is corrected and should be used for reporting problem resolution.

patterns to determine common causes. Selective failures in accordance with the severity of the failure problems are followed up with root cause analysis to determine a corrective action for initiation. Not every failure will require an immediate correction. Some corrective actions can wait until the next design modification for practical reasons.

Not every failed item can be fixed. A typical example is the analysis of no-fault-found (NFF) on field-returned items, where the failure reports indicate that the items were returned as failed items, but upon testing and analysis, NFF situations are discovered. The percentage of NFF on field-returned items over time might provide some indication of the probable problems: a low (<10%) NFF will indicate normal system service operation; a high (>50%) NFF would signify a serious design problem. A 100% NFF would likely be a return policy problem during the warranty period, where the customers have decided to return all replaceable items for new ones. The treatment of NFF is an important dependability issue often occurring in the introduction of new products. NFF may sometimes be viewed as the cost of doing business. The NFF information and causal diagnosis should be retained in the database as lessons learned. History often repeats itself. The wisdom is to learn from past mistakes. It is prudent to take appropriate action before it becomes a real problem.

6.3.3 Maintenance and Logistic Support System

The maintenance and logistic support system sustains system operation and serves as the front end for customer service in conjunction with the FRACAS application. Maintenance diagnosis and service actions are essential information to accomplish scheduled or unscheduled maintenance efforts. Maintainability and testability designs of the system often expedite the time required for completion of maintenance service tasks. The levels of maintenance provide guidance for maintainability designs and development of test/diagnostic facilities such as the built-in test equipment. In general, there are three levels of maintenance for technological systems application.

(a) Organizational maintenance – This is also known as first-line maintenance where maintenance is performed at the operation site. The diagnostic tasks are carried out by using the organization's own test equipment. For example, an airplane has landed safely with an on-board instrument malfunction, such as an altimeter, without affecting the inflight operation. The malfunction is detected inflight by the aircraft built-in central computer and found by the pilot on visual inspection. A backup altimeter is used for landing. After landing, the altimeter is replaced as a first-line maintenance action on the tarmac to get the airplane flight-ready for its next flight. The information captured indicates the maintenance action and the diagnostic condition of the failed altimeter being replaced.

(b) Intermediate maintenance – This is also known as second-line maintenance where maintenance is performed by a nearby facility, such as an aircraft hangar workshop closed to the operation site. The failed altimeter is tested to determine if it can be fixed. In one case, the failed altimeter can be repaired and tested and calibrated for reuse. In another case, the altimeter cannot be fixed and is returned to the factory or junked. The information captured indicates the maintenance actions taken and the disposition of the failed altimeter with model identification and serial number for the record.

(c) Depot maintenance – The depot maintenance is usually the manufacturer's facility where the field-returned failed item is serviced to determine the cause of failure. Depending on its failed condition and the criteria established for restoration of high-cost items, the decision could be to repair or to throw away. The information captured indicates the disposition of the failed item and detailed repair and rework activities to restore the item for reuse.

Not every system requires all three levels of maintenance. Some maintenance systems work well with only two levels; organizational maintenance and depot maintenance. The maintenance criteria depend on the system of interest requiring service. For example, many software-intensive systems consist of self-diagnostic algorithms and health check routines programmed into the system for execution on a regular basis. Software updates can be done automatically on-line without interruption of regular business services. Service calls are only made if major changes require special on-site installation. The maintenance service information in such situations can be automatically captured and transmitted to the data center for information retention.

The logistic support system plays a significant role to support the required system maintenance service. Logistic support engages primarily the supply and demand process for replacement of items used in system operation requiring maintenance service. Logistic support criteria depend on the carrying cost of replaceable items for on-site spares inventory and the efficiency of the transportation and distribution network for spares replenishments. Information captured includes tracking the movements of inventory assets, the transportation and distribution of replacement items to various depots and system operation sites. The turn-around time for spares replenishment is an important factor influencing the process of inventory control and optimization of the entire network configuration for the maintenance and logistic support system.

6.3.4 Fault Management System

For technological systems application, the fault management system complements FRACAS. The fault management system is focused on software-intensive system operation and network performance issues, while FRACAS mainly addresses failure-related activities of hardware products. The major difference is the time for restoration, which is driven by the expectation of technology expediency and the limited patience of customer social tolerance. While normal corrective action for hardware repair or rework such as overhauling an automobile engine would likely take days or even weeks waiting for replacement parts, the recovery of a network communications system from malfunction is entrenched in the customers' perception of dependability to be almost within seconds for instantaneous gratification. For example, according to e.Week.com, Mobile News 2011-10-16: "RIM's (communication network) infrastructure went haywire Oct. 10 in its Slough, U.K. data center, only to have it spread across Europe, the Middle East, Africa, and Asia, and now includes Brazil, Chile, Argentina, Canada, and the U.S. Service was fully restored across all countries by Oct. 14. The core switch (of the infrastructure) failure left its (70) millions of BlackBerry users with slow or inaccessible e-mail for up to 3 days." It does not take much for an uninitiated customer to speculate about the dependability of the service provision. For BlackBerry addicts, RIM is "chipping away our faith," according to CNN Tech on Oct. 14, 2011.

Faults are inherent weaknesses of a technological system. Faults can lead to complete system breakdown or cause network performance degradation. Fault management is essential for system dependability improvement. In some critical situations, test cases are developed to verify methods for effective fault isolation and removal. For example, the fault insertion test is an application method to verify the effect of system failures by deliberate insertion of faults to test the consequences of the results. The purpose is to verify the effectiveness of redundancy designs, protection mechanisms and the overall fault management capability of the system. The fault management system uses the design analysis and review process to determine the means for achievement of fault prevention, fault mitigation or development of alternate designs for fault tolerance.

Classification of faults, their occurrence situations and the appropriate corrective actions provide the database and framework for fault management. The fault data collected is processed along with other system performance information collected from other sources, such as the specific on-site operation service information provided by the customers, design change notices of system functions or the new installation procedures of replacement items provided by the suppliers or vendors. Table 6.2 presents typical examples of fault information for retention.

Fault management technology directly provides the means to attain the system or network dependability characteristics. They are grouped into the following categories [3]:

(a) *fault detection*: checking the operational status of nodes and links, setting various performance thresholds, and monitoring the changes of performance to timely detect faults;
(b) *fault localization and diagnosis*: locating the fault and identifying its cause according to the information collected. For a fault that cannot be located directly, analysis can be made according to the relationship of the fault information to determine the specific location and cause of the fault;
(c) *fault isolation*: isolating the failed unit and preventing it from interfering with other working units;
(d) *fault recovery*: recovering service functions by resetting the unit or restarting part of the unit in case of a fault. For a redundancy design unit, using protection switching or resource reallocation to avoid service failure;
(e) *fault removal*: replacing the failed unit or correcting software fault by using the maintenance support system, including further in-depth analysis and follow-up recommendation for improvement actions and solution;
(f) *fault alarm*: setting an alarm to alert the system control or network management system in the event of a fault for immediate corrective action. Different alarms can be set to indicate different fault levels for priority actions. The management system collects the related fault alarm information for fault localization and preparing for further analysis;
(g) *fault prognosis*: analyzing and evaluating fault probabilities and likely consequences of faults by using methods such as trend analysis;
(h) *fault monitoring*: observing and checking the fault conditions to alert maintenance actions.

TABLE 6.2 Examples of Processed Fault Data for Information Retention

Accumulated faults detected	Accumulated faults detected data should be used to determine fault rate and reliability trend over a period of time.
Accumulated faults corrected	Accumulated faults corrected data should be used to determine known faults that require corrective actions and track the effectiveness of maintenance actions.
Fault detection rate	Faults detection rate is used to indicate trends to facilitate planning of maintenance strategy and resource management.
Fault correction rate	Faults correction rate is used to indicate trends to facilitate planning of maintenance strategy and resource management. Priority setting for maintenance action is based on the severity of the fault.
Faults per location	Fault tracking according to system functions to identify specific area of problem.
Criticality of faults	Classifying the degree of impact of faults to set priority for maintenance actions.
Number and percentage of severe faults	Indications for planning maintenance strategy.

6.3.5 Dependability Assessment Information

Information associated with dependability assessment is focused on capturing and retaining the results derived from application of dependability methods to analyze and evaluate specific dependability issues encountered during the system life cycle. Assessment is an appraisal process that provides an informed status of the specific dependability situation. Table 6.3 presents typical methods commonly used for dependability assessment.

Information retention of the dependability assessment results is needed due to practical reasons for project control and assurance purposes. The assessment process and application methods are often iterative because of updating design changes and incorporation of reused standard system design functions to avoid redesign effort. The assessment results and subsequent updates provide the sequence of assessment records for verification of design progress and system performance improvement. Dependability assessment is based on a specific system design configuration, which is subject to configuration management requiring the maintenance of design change records.

6.3.6 Information Retention, Retrieval and Dissemination

The dependability information system is established to facilitate retention, retrieval and dissemination of relevant dependability records and data. An organized database structure is essential to realize effective service application. Information processing is crucial for identification of service requirements, control of authorized/unauthorized access, and retrieval and disclosure of proprietary information. Information and data security protection, as well as the physical protection of the information-related assets, have become increasingly critical to doing business in today's competitive environments, where everything can

TABLE 6.3 Methods commonly used for dependability assessment

Method	Description of Analysis Techniques	Inputs	Outputs	When Used
(a) Failure rate prediction [4]	Estimate failure rate contributions from system elements using established prediction models with given application conditions	System element failure rates from database or experience data of similar elements used	System reliability estimates to provide MTBF or total system failure rate	System definition and design
(b) Reliability block diagram (RBD) [5] analysis	Determine probability of success or system availability or reliability based on logical block representation of system configuration	Failure rate of each block assigned	System availability or reliability	System definition and design
(c) Fault tree analysis (FTA) [6]	Analyze system failure conditions due to undesirable faults identified based on construction of a fault tree starting at the system level with a top-down analysis approach	Failure rates assigned to each undesirable fault identified based on the FTA diagram	System outage or downtime estimates with probable causes identified	System definition and design
(d) Failure modes, effects and criticality analysis (FMECA) [7]	Analyze system failure modes and effects on the next higher level of the system element in the system configuration with a bottom-up analysis approach	Failure modes and possible failure causes identified for each building block system element	Criticality of system outage or downtime due to failure modes identified	System definition and design
(e) Software reliability models [8]	Analyze software reliability by estimating the number of residual faults in the software that are acceptable for release based on initial failure rate estimation when software is put in operation	Number of detected faults at some point in time in software operation	Number of residual faults in the software	Software version release and acceptance

be exposed to risks and is vulnerable to hackers and cyber intrusions. The physical assets for information processing include building facilities, control rooms, computer systems, servers, data storage, desktop and laptop computers, cell phones and other personal communication devices, all carrying important information and data that might be misdirected or misused. Data integrity is paramount in terms of confidentiality and security to ensure that data transmission and reception are not interrupted, corrupted, intercepted, tampered with or altered in the information delivery process.

Data management is essential when the dependability information system is employed to provide distributed network services to clients. Network failure or outage breakdown can disrupt information traffic flow or cause data corruption. Loss of key data may result in the inability to restore network service functions or affect economic losses to the network carriers. For example, a JPMorgan Chase news release on Oct 14, 2011, reported that the "BlackBerry outage may cost RIM over $100 million," which reflects the magnitude of losses that may be possible.

The following methods are commonly used to manage data in order to improve network dependability:

(a) Storing duplicate data in different locations to protect the media of data storage from faults or damage.
(b) Periodic backup to protect key information and data. Important data should be protected by multiple backups and remote backup. Proper use of the backup and recovery mechanism can reduce downtime and improve performance.
(c) An encryption, acknowledgment and re-transmission mechanism, and a redundancy transmission mechanism that can be used for the transmission of important service data via the network. For example, protection and assurance of financial data transmission for banking services to avoid the loss of data packets.
(d) Use of different media for storage of program data and information data, such as "cloud" computing, could be an economical means of information storage and efficient backup.
(e) Redundancy of important computer program data can be used to ensure normal running of the system even when the running software program is incorrectly modified.
(f) Synchronization of information between the active and standby units of the redundancy system can be used to avoid data loss or abnormal behavior during protection switching.
(g) Protection, detection and verification of program data during loading or transmission to ensure incorrect data are not recorded.

Table 6.4 presents typical examples of fault information captured that could be utilized for possible knowledge development.

6.3.7 Knowledge Development

Knowledge is intellectual property. It can include information, facts, descriptions and skills acquired through experience or education. Knowledge can be theoretical or a practical understanding of a subject. Knowledge development focuses on processes such as acquiring, creating and sharing knowledge and researching the concepts and principles that support the technical practices and cultural foundation. Knowledge management is a multi-disciplined approach to achieving organizational objectives by making the best use of knowledge. The management of knowledge involves utilizing available resources that enable knowledge development for advancement of technology and application. Successful

TABLE 6.4 Examples of fault information for possible knowledge development

(a) Faults introduced by life cycle stage	Indication of when and at what stage the faults were introduced and to take appropriate actions.
(b) Faults detected by life cycle stage	Indication of when and at what stage the faults were detected and justification for delaying corrective actions for fault removal.
(c) Total time spent in analysis	Indication of the time spent on analysis for problem identification and isolation for corrective action, and the associated resources required.
(d) Total time spent in system testing	Indication of the time spent on system testing and the associated resources required.
(e) Total maintenance time	Indication of the time spent on maintenance activities and the associated resources required.
(f) Average maintenance administration time	Indication of the time spent on maintenance administration and the associated resources required. Maintenance administrative duties include before and after the fault is corrected, such as time spent in assigning maintenance personnel, system procedure updates and release of correction in a new version.
(g) Average corrective action time	Indication of the time spent on corrective actions and the associated resources required. This reflects the cost-effectiveness in the maintenance activities.
(h) Reason for corrective action	This is used to determine the source of faults. Typical reasons include: • previous maintenance action • new requirement • requirement change • misinterpreted requirement • missing requirement • ambiguous requirement • change in software environment • change in hardware environment • code/logic error • performance error
(i) Cost of corrective action	Indication of the total cost of corrective action, including fault isolation, problem resolution and administration for effective maintenance action.
(j) Percentage of functions tested and verified	Indication of test coverage, test efficiency and completeness.
(k) Classification of system failures	Indication of the different types of failures and their respective sources causing failure to plan and prioritize appropriate corrective actions.
(l) Frequency of system failure occurrence	Indication of total number of incidents per year or per month where the failure occurrences cause system outage or degradation in performance.
(m) System restoration time	Indication of the total elapsed time from the first detected incident occurrence to the complete restoration of user services when the system is restored back to normal operation.
(n) Total number of users affected per incident	Indication of the magnitude of an incident affecting the number of users in provision of system services.
(o) Historical data	Provision of data history on problem areas related to design, process and product issues.

knowledge development is highly influenced by encouraging personal aspiration, technology leadership, long-term strategy and business vision on the timing for change, and forecasting the viable adaptation of the organization's new product rollouts to societal wants and customer needs. A creative working environment with supporting facilities is essential to nourish people resources and nurture their creative minds. In such cases, technology has become a crucial enabler to harness knowledge rather than a means to provide the solution. The business processes and organizational structures should facilitate sharing of knowledge, innovations, discoveries and ideas and enlightenments on value creation. Knowledge management can be viewed from various aspects contributing to the knowledge development process. Examples include:

(a) people – job motivation, career development and leadership challenge to encourage creativity and innovation;

(b) processes – cross-functional development, team work experience sharing and a mentoring program to enhance internal and external expertise participation and contribution;
(c) technology – joint venture research, design patents and awards, conference participation and networking to develop technology platforms for next generation products;
(d) culture – change management, setting common goals and objectives, adaptive and progressive means to cultivate a knowledge-driven organization and influence the societal quality of life.

Design incorporation of dependability characteristics in technological systems will instill customer confidence and user trust and sustain system performance value over time. Organizations that embrace the dependability principles and practices are in the value creation business. Successful knowledge development is often accomplished by joint collaborative work of industries, governments and academic institutions. Numerous collaborative works have been accomplished and recorded from time to time as the "book of knowledge" in specific disciplines. These cooperative endeavors represent the legacy of human insight and wisdom gained with the passage of time that is intended for the enlightenment of future generations of mankind.

6.4 LESSONS LEARNED

6.4.1 Learning from Experience

Lessons are learned from experience. The most valuable lessons are those that are experienced in practice. The telecommunications industry has had a long history of technological innovations [9] since telephones were first connected for commercial service applications. Progress in telecommunications over the past century has presented a rich legacy of technology advancements that shape the lives of many who are dependent on various forms of communications services. The social adaptation to technological changes and the resulting influences have propagated the fast-moving events in commercial ventures and altered the divisive political and cultural landscape. The following information summarizes the selective highlights that may provide background experience to help pursue dependability values associated with technological systems.

6.4.2 Case Study on Network Dependability

Since 1992, telephone companies in the United States that provide customer services through the distributed Public Switched Telephone Network (PSTN) have been required to notify the U.S. Federal Communications Commission (FCC) of service outages affecting more than 30,000 customers. The PSTN consists of thousands of switches operating in different regions and serving households and businesses across the country. The switches were designed with the latest (1990s) technologies, incorporating redundant hardware and extensive self-checking and recovery software. For several decades, AT&T (American Telephone and Telegraph) had expected its switches to experience no more than two hours of failures in 40 years.

This case study was conducted by the U.S. National Institute of Standards and Technology (NIST) to study the sources of failure in PSTN [10]. PSTN outage data were collected from April 1992 to March 1994, beginning with the earliest FCC reports. PSTN performance statistics indicated a total of 303 network failures causing outages during this period of network service operation. The outages affected over one million customers, with a total outage duration of 3,196.5 minutes for the two-year study period. The service outages were determined by the sum of telephone service outages. They were grouped by various failure categories. The service impact was determined by the total downtime measured in customer-minutes by category.

Customer-minutes = *Number of customers affected* by failure category x *Outage duration* of the category in minutes.

Table 6.5 summarizes the PSTN failure sources and categories, the number and duration of outages, the numbers of customers affected, and the service impact during the two-year study period based on data collected from the FCC reports.

Figure 6.3 presents the PSTN outage contributions and the resultant service impact by failure categories.

The data showing the number of outages related to the magnitude of service impact differs significantly for most failure categories. For example, although network overloads caused only 6% of the total outages, they accounted for 44% of the total customer-minutes. Human error caused 49% of the outages, but only contributed 28% of the downtime. The complexity of the PSTN relies heavily on its hardware and software elements. An unexpected finding was that software errors only contributed to 2% of the system downtime and hardware failures attributed to 7%. The PSTN dependability performance provided an average availability of better than 99.999% during the time period studied.

6.4.3 Looking Beyond the Numbers

One of the main tasks in dependability assessment is reliability prediction, where the hardware and software elements are judiciously analyzed and evaluated for proper design applications. This used to be the major task assignment for reliability engineers, when they devoted much of their time in system development projects. In the PSTN case study, the combined hardware and software failures indicated a mere 9% of the total system downtime. It is prudent to address the remaining 91% of the downtime problems by looking beyond the numbers.

The PSTN is a distributed network. It performs and operates as a complex, integrated system of systems. The dependability of a network is influenced by many factors. The case study noted that other aspects of the network properties were exploited from a technology design perspective to compensate for the PSTN design complexities introduced by the high dependability requirements. The approach to PSTN software design was focused on error

TABLE 6.5 Summary of PSTN failure data [10]

PTSN Failure Source	Failure Category	Number of Outages	Outage Duration (minutes)	Number of Customers Affected	Service Impact in Customers-Minutes ($\times 10^6$)
Facility-related	Hardware failure	56	159.8	95,690	1,210.8
	Software failure	44	119.3	118,200	355.5
Traffic-related	Overload	18	1,123.7	276,760	7,527.2
Disaster-related	Acts of nature	32	828.2	159,000	3,124.0
Security-related	Vandalism	3	456.0	85,930	110.5
Schedule-related	Maintenance	0	0	0	0
Human-factor-related	Human error – procedure	77	149.4	182,060	2,349.3
	Human error – accident	73	360.1	83,936	2,415.8
Total		303	3,196.5	1,001,576	17,093.1

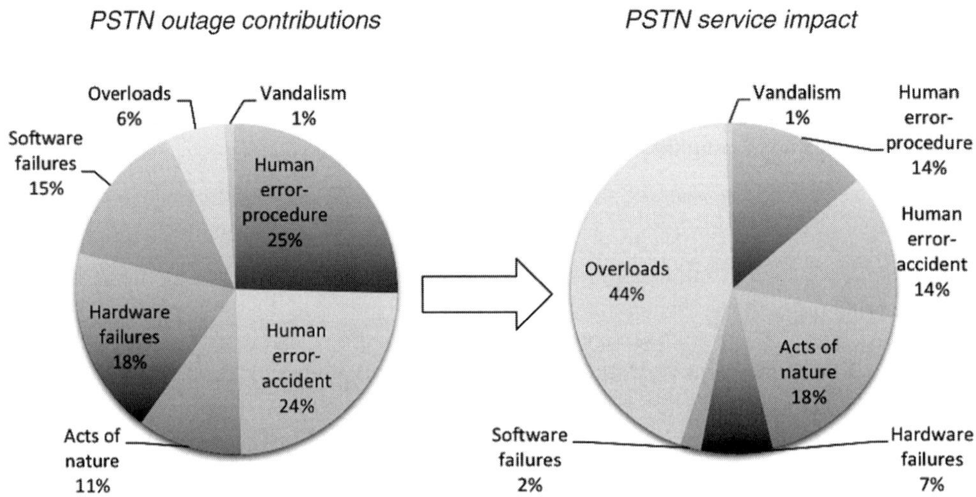

Figure 6.3 PSTN outage contributions and service impact [10]

detection and correction. Some dependability issues related to network technology designs were addressed in this case study.

(a) Dynamic rerouting
The network is highly distributed where localized failures are more likely to exist. The network switches can reroute traffic dynamically to circumvent a failed network node. Intermittent failures are usually not catastrophic. A brief failure in one network component has relatively little impact on the availability performance of the entire network. However, for the PSTN to reroute calls, it must retain a good deal of information globally. Maintaining consistent distributed databases can require complex interactions among system components. This is a backup approach to leverage risk avoidance for continuity in network operation to sustain dependability performance.

(b) Loose coupling
The PSTN is a loosely coupled network because it can dynamically reroute calls along many paths. The loose coupling is achieved at the cost of some complex interactions between system components. The network designs include the need for end-to-end acknowledgments, interactions among many systems, and the maintenance of some globally consistent databases. Major switching centers store information on alternative paths and exchange data on traffic patterns and switch status for data access throughout the day. The loosely coupled PSTN allows for flexibility in recovering from failures and enhancing dependability performance. The recovery includes automated and manual mechanisms to exploit the PSTN's loose coupling advantage to achieve high-integrity system.

(c) Human intervention
The PSTN is designed with built-in self-test and recovery mechanisms and is continuously monitored for service operation. Traffic routing benefits from automated and human operations. Information on switch status and traffic patterns

permits the software within a switch to automatically select an alternative route if the preferred route becomes overloaded or unavailable. If the switch exhausts all alternative routes, human intervention can reconfigure the network, sometimes solving the problem in a few minutes. Status data exchanged regularly between switches makes automated and human operations to reconfigure routing possible. This improves the serviceability of dependability performance.

The lessons learned from this classic case study conclude that software is not the weak link in the PSTN dependability. Extensive use of built-in self-test and recovery mechanisms in major system components (switches) contribute to software dependability. They are the significant design features in the PSTN. The network's high dependability indicates that the trade-off between dependability gains and complexity introduced by built-in self-test and recovery mechanisms can be positive. Likewise, the tradeoff between complex interactions and loose coupling of system components has been positive, permitting quick human intervention in most system failures and resulting in an extremely dependable system.

Technology has evolved since the PSTN case study in 1997. It should be noted that additional dependability approaches for engineering of communication networks [3], such as network routing optimization, traffic congestion control, and fault tolerance design have been introduced and successfully deployed for dependability performance enhancements.

6.4.4 Adapting to Change and Changing Environments

Fast forward to the present, the ICT (Information and Communication Technologies) industry has been advancing in great strides over the past two decades. The technology evolution is driven by human needs and changing environments worldwide. The lines between wireline, wireless, cable and satellite services have become increasingly blurred as communications services continue to converge. The International Telecommunications Union (ITU) estimated in 2009 [11] that there were more than four billion mobile cellular subscriptions worldwide. In comparison, there were only 1.3 billion fixed telephone lines worldwide. Almost a quarter of the world's population of 6.7 billion people (2009 statistics) was using the Internet. In the United States, more than 276 million people subscribed to mobile phone services, with over 20% of U.S. households subscribing only to a wireless service plan. The U.S. wireless customers also consumed more than two trillion air-time minutes per year and sent more than 1.3 trillion text messages annually [12]. The dynamic growth of the ICT industry is fueled by increased and diverse communications needs in the information society that is spreading over all countries and regions around the world.

The Network Reliability Steering Committee (NRSC) [13] was formed in 1993 under the auspices of the Alliance for Telecommunications Industry Solutions (ATIS) for the purpose of monitoring network reliability in the United States on an ongoing basis [14]. This is achieved with collaboration among U.S. telecom companies, service providers and the FCC by reporting telecom network outages across the country. The coordinated effort in identifying and understanding the underlying causes of reported outages serves as basis to establish network outage trends. Industry-government cooperation is recognized as an important part of learning from past experiences and preparing for future challenges as networks evolve. A framework for communications infrastructure [15] has been established for systematically reviewing the network elements by identifying possible influences on the national network outage trends. These influencing factors reflect the energy efficiencies of network performance experiences associated with the relevant network elements in terms of impacts of technology evolution and application changes. The basic communications infrastructure is used by the NRSC as a standard methodology to identify network elements essential for the operation of communication networks. Each network element includes intrinsic vulnerabilities that must be proactively prepared for and addressed.

Table 6.6 presents a systematic review of network reliability influencing factors. Examples of these influencing factors for each network element are summarized [16].

6.4.5 Exploiting Green Technology

One of the most aggressive worldwide deployments of the ICT system is the wireless network. The following presents process steps of the wireless network design approach [17] by exploiting green technology. The approach engages the network design process for environmental sustainability [18] to reduce energy consumption, heat generation and equipment footprint. The objective is to increase network performance efficiencies and reduce operational cost to achieve dependability value.

(a) Network design impact on energy efficiency
An efficient wireless network design accommodates a configuration that offers optimized performance for different network services and minimizes the energy consumption of the network operation. The relationship between network performance

TABLE 6.6 Systematic review of network reliability influencing factors

Network Elements	Possible Influencing Factors on Energy Efficiencies and Impacts to the Network Elements
Power	• increasing dependence on power capabilities for distributed remotes • increased reliance on alternating current electricity, which has more components • decreasing number of subject matter experts • increased back-up power need for cooling during commercial power failures
Environment	• increased concentration of hardware packaging increases cooling challenges • increased physical security affects access • increased distributed mesh network topology potentially reduces significance of any single site
Hardware	• increasing use of common hardware across equipment suppliers • increased outsourcing by equipment suppliers • increased capacity of single elements • increased rate of technology turnover
Software	• increased outsourcing by equipment suppliers and network operators • increased use of artificial intelligence • increased deployment of service-oriented architectures • increased presence of worms and viruses
Networks	• decreasing dependence on silicon chips for control (shift to software) • decreasing prevalence of deterministic availability and path control • increased complexity of interconnections with other entities • increasing exposure to wireless interfaces
Payload	• increasing diversity of services running on networks (video, gaming ,etc.) • increasing variation in traffic levels due to service types • decreased segregation of traffic with control messages • increased use of "always on" sessions
Policy	• increased number of connected network entities and elements • increased number of relevant standards • increased global divergence on the expected role of regulation • decreasing preparation for turn up of new capabilities
Human	• decreased time allotted for learning curve advances for new technologies • increasingly competitive environment increases overall workloads • increasing electronic authentication dependence to support virtual worksites • decreased social cohesion with proliferation virtual work teams

and energy efficiency should be fully understood in the network design process and the wireless network operation.

(b) Network traffic modeling
The network traffic modeling includes analysis of the voice, data and video services, user behavior in the network and user mobility. This is to provide a means to estimate various communication traffic patterns and application parameters. The wireless network energy efficiency can be expressed as a function of all the traffic model parameters. The energy consumption and performance of a network can be optimized with implementation of specific network resource management schemes.

(c) Network topology
The network topology design has an impact on energy efficiency in network operation. Network topology design identifies the network elements architecture, the relationships between the network elements and the underlying assumptions of end-to-end connectivity solution. The network topology design also includes the performance, distribution, scalability and reliability of the network, which may have significant impact on energy consumption and performance efficiency.

(d) Access network technology
The access network technology has an effect on energy efficiency. Examples of wireless access technologies include GSM (Global System for Mobile communications), W-CDMA (Wideband Code Division Multiple Access), and LTE (Long Term Evolution). They are access standards for wireless communications. The latest generation of wireless access technology, such as LTE, provides high capacity and optimized configuration. It has the advantage of lower energy consumption per system configuration and provides rich service offerings.

(e) System parameter optimization
Parameter optimization is a method for optimizing energy efficiency of a network. The optimization process includes on-site field parameter adjustments to antenna height, tilt, orientation, and transmit power. The objective is to achieve the maximum coverage with the minimum amount of interference. During parameter optimization, several aspects of wireless characteristics, such as radio propagation, frequency and channel allocation techniques, coding schemes and priority schemes should be considered.

(f) Service distribution architecture
Energy efficiency requires constructive service distribution and management framework. The management framework defines the distribution of application servers to centralize user services such as voice, web and streaming services. The accommodation of user access closer to the edge of the access network would minimize traffic and reduce the overall network energy consumption. The framework may require network and technology upgrade to optimize the energy efficiency and handling of multiple advanced services and content delivery.

(g) Redundancy implementation
Implementation of redundancy mechanisms in the service provider's network has major impact on the power consumption of a network. Redundancy

is needed to meet network reliability requirements, achieve end-user quality of service (QoS), and comply with regulatory requirements. Service providers may use different redundancy techniques to achieve high availability in their network to fulfill regulatory requirements. The redundancy techniques could be hot standby, warm standby, or cold standby. The effectiveness of redundancy is gauged by how quickly a standby element can be brought into service to enhance energy efficiency. The energy consumption associated with the different standby models may vary with system implementation and application software. The effects are dependent on the characteristics of the access, core and backbone network being used.

(h) Dimensioning and optimization
Dimensioning and optimization consolidate all of the previous process steps of the network design to optimize an energy-efficient network. Design trade-offs can be iterative and are carried out to reduce the number of needed network resources and optimize the associated energy consumption. During dimensioning and optimization, it is important to consider uplink and downlink traffic load separately as the traffic load in uplink and downlink directions are not identical. For example, streaming services for the uplink direction is used mostly for signaling to communicate with streaming servers in the network, but in the downlink direction the streaming payload requires higher bandwidth for transmission. The downlink direction always requires more transmit power of a radio base station for most telecommunication services.

This wireless network example shows the lessons learned from exploiting green technology to achieve energy efficiency as a way to capture technical know-how and practical experience. These valuable lessons learned should be candidates for information retention to facilitate further knowledge development of industry best practices. It should be noted that the application of ICT plays a significant role in various environmental sustainability projects, such as the application of monitoring and control systems used in Smart Grid for efficient delivery of reliable, economical and sustainable electricity services. Dependability advocates the ability to perform. In this respect, any enhancements in technological system performance imply the achievement of dependability value.

REFERENCES

[1] MIL-STD-2155, "Failure reporting, analysis and corrective action system," 24 July 1985.
[2] Smith, R. and Keeter, B., 2010. "FRACAS; Failure Reporting, Analysis, Corrective Action System," Reliabilityweb.com Press; 1st edition (December 14, 2010), ISBN-13: 978-0982051764.
[3] IEC 61907, Communication network dependability engineering
[4] Mil-HDBK-217F, "Military Handbook, Reliability prediction of electronic equipment," 2 Dec 1991.
[5] IEC 61078, "Analysis techniques for dependability – Reliability block diagram and Boolean methods"
[6] IEC 61025, "Fault tree analysis"
[7] IEC 60812, Analysis techniques for system reliability – Procedure for failure mode and effects analysis
[8] Lyu, M. R. (Ed.), 1996. "The Handbook of Software Reliability Engineering," IEEE Computer Society Press and McGraw-Hill Book Company, 1996
[9] Coe, Lewis, 1995. "The Telephone and Its Several Inventors: A History," Jefferson, North Carolina, McFarland & Co., Inc. ISBN 0-7864-0138-9
[10] Kuhn D.R., 1997. "Sources of Failure in the Public Switched Telephone Network," IEEE Computer, Vol.30, No.4 (April 1997), National Institute of Standards and Technology, Gaithersburg, Maryland 20899 USA

[11] International Telecommunication Union (ITU), Measuring the Information Society: The ICT Development Index 2009
[12] CTIA – The Wireless Association, Wireless Quick Facts – Mid-year Figures, Web 5 December 2009
[13] Network Reliability Council, Network Reliability: A Report to the Nation, June 1993
[14] ATIS-0100029, NRSC 2008-2009 Biennial Report
[15] Rauscher, Karl. F., Protecting Communications Infrastructure, Bell Labs Technical Journal Homeland Security Special Issue, Volume 9, Number 2, 2004
[16] NSTAC NGN Task Force Report, Systematic Assessment of NGN Vulnerabilities, Appendix G, March 2006
[17] ATIS Report on Wireless Network Energy Efficiency, *www.atis.org/Green/index.asp*, January 2010
[18] ATIS Report on Environmental Sustainability, *www.atis.org/Green/index.asp*, March 2009

Chapter 7

SUSTAINING DEPENDABILITY PERFORMANCE DURING OPERATION

7.1 INTRODUCTION

The operation and maintenance stage of the life cycle process represents the longest duration in the system life for most organizations engaged in providing a service. Service providers maintain a viable business operation primarily to deliver customer service for revenue generation. To sustain long-term operation and durability of systems and equipment, they must be properly maintained. This is to provide an adequate level of service acceptable to the customers. To provide dependability performance during operation, it is essential to properly manage system operation and maintenance support functions. This will result in achieving customer satisfaction and enhancing system performance and efficiency by controlling system outages and reducing service downtime.

To achieve dependability performance during operation, system availability and equipment reliability must be ascertained. This is accomplished by means of harnessing the effort of maintainability in the system design, implementing strategic system supportability plan and providing timely and effective maintenance support actions to prevent unnecessary incidents from happening that might otherwise create operational problems.

From a technological system operation perspective, the essential measures should address their impact on dependability performance. This chapter presents principles and practices for dependable system operation. Also included are some examples of dependability measures deployed and utilized in industrial systems to sustain dependability performance during system operation.

7.2 OPERATIONAL CONSIDERATIONS

It is worthwhile to first define the fundamental expectations of system operation from a dependability perspective. All systems have an inherent level of reliability. This is only achieved if proper operation and maintenance practices are followed. They consist of:

- ensuring operation within design limitations as they relate to functional, process or equipment operating conditions, such as pressure, temperature, fluid properties, contaminants and electrical aspects (e.g., current and voltage);
- staying within limits for environmental conditions such as temperature, humidity, airborne contaminants and precipitation;
- following proper startup and shutdown procedures;
- carrying out operational tasks such as routine cleaning and lubrication;

- updating system application and operating software as and when required;
- avoiding excessive stress and loading.

Reliability cannot be achieved or sustained if these factors are ignored or inadequately applied in system operation. However, it is quite possible that actual operation does not match the intended design requirements. In such cases, the options are a) to redesign or modify the system by replacing degraded or faulty components with more suitable ones, or b) to change or restrict operating practices to accept the reduced reliability limit.

Availability performance is dependent on reliability and is also affected by the level of maintenance support and the logistics provided. System availability performance is generally under the direct control of the user or operator. The system performance outcome can be adjusted based on the allowable downtime limits in order to optimize system usage and the cost of operation. Systems designed with redundancy improve reliability at the expense of added cost of duplicated equipment or system components. From a system operation perspective, redundancy provides greater flexibility with planned maintenance. With the advent of software technology applications in the system architecture and the decreasing trends of mature component costs, there are ample opportunities in optimizing a suitable configuration to further reduce the time and effort associated with maintenance support costs in the long run.

Understanding the operational constraints and appreciating the impact of infrequently stretching the existing operating limits are important. Data collected during system operation reflects the causal reactions of reported or observed incidents. This will enlighten the knowledge gained of the system operational profile and provide an indispensible information source for interpretation of availability and reliability data for use in system improvement.

7.3 OPERATION FRAMEWORK

7.3.1 Operational Objectives for Dependability

The measurement of availability and reliability is tied closely to performance requirements as well as organizational objectives and customer expectations. System operational objectives for dependability performance include:

- ensuring service levels are being achieved, such as with commercial transportation;
- satisfying customer expectations for dependability of products;
- monitoring safety performance to make sure the safety of the system meets expectations;
- measuring reliability growth of products in order to reach reliability targets;
- gathering information for operational cost analysis and dependability value assessment;
- minimizing downtime to facilitate production availability and capacity for manufacturing and industrial process enhancement;
- monitoring environmental performance to satisfy regulatory, societal and other application specific requirements;
- Identifying opportunities for operational service improvements.

7.3.2 Overview of System Operation Process

To achieve these system operational objectives, it is essential to have an infrastructure with well-established organizational processes to support successful dependability performance in system operation. Operational process activities vary widely depending on the type of industry, as summarized in Chapter 1, section 1.1.3. The operation process involves both technical and administrative support from a management perspective. This is to facilitate infrastructure development and continuous improvement to sustain business viability in system operation. Figure 7.1 shows an overview of the system operation process. It presents the relationships of process activities and support functions to enable attainment and sustainment of system dependability performance.

The infrastructure is associated with the organization that operates the system based on sound business principles and practices. Mature business operations are generally customer-focused with strong human resources equipped with technical skills and business knowledge as inputs to system operation. The operation process adheres to the planned activities and performs tasks according to the target commitments supported by management reviews and decisive actions. The resultant system outputs deliver the needed customer expectations on product quality and dependability of service. The system operation is highly dependent on its supporting systems, which act as enabling agents that are shared across the organization. It is crucial to note that efficient operation relies heavily on the availability of the supporting systems and the effectiveness in timely task implementation. A common problem often experienced is the lack of spares for replacement of a failed item during system operation, which causes delays. A more serious problem may be a seemingly simple problem of an improperly calibrated measuring instrument used in a manufacturing process, which will register incorrect indications and cause all previous measurements to be repeated for assurance purposes. In the worst-case situation, the entire lot of production items may be quarantined or junked, resulting in the total loss of the production lot.

To sustain dependability performance during a production-oriented operation, it is imperative that the information support function provides timely feedback to advise the maintenance support function on immediate needs for process adjustments to assure and

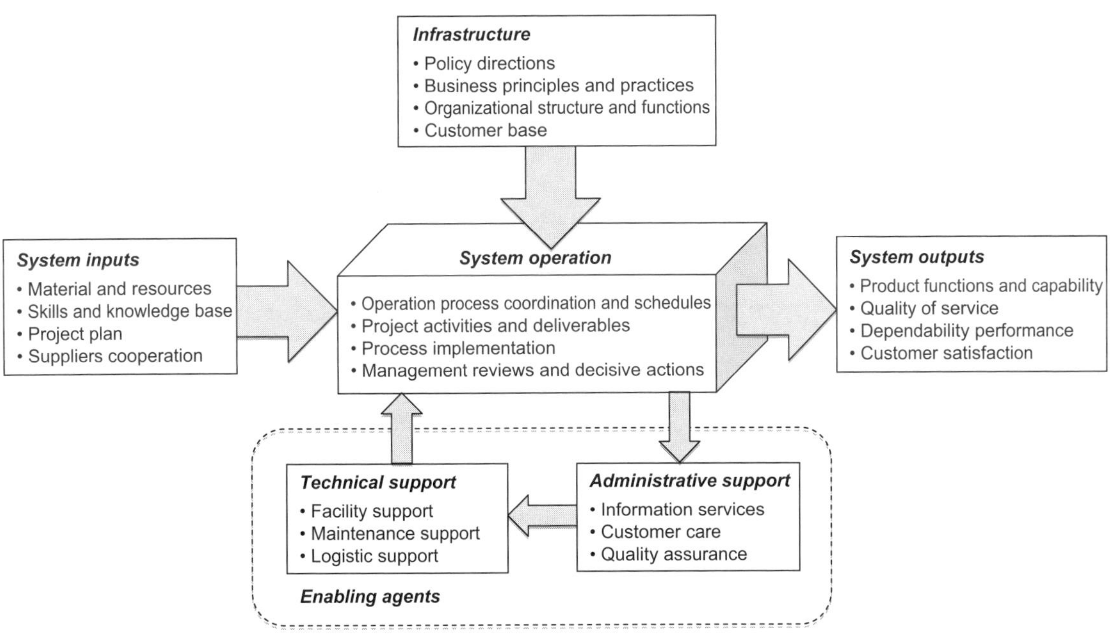

Figure 7.1 Overview of system operation process

regain full control of system operation. In this respect, it becomes critical to invest in an appropriate supervisory control system to monitor the status of all key-measuring points during the entire operation process. The supervisory control system should meet the essential needs for practical information processing. Dependability measures on system faults and failures are important for effective management of system operation. The investment will pay off in the long run in terms of time and cost savings and the reduction of maintenance support time that may otherwise result in system outage or degradation.

7.3.3 Operation Process Implementation

The duration of the operational stage of a system depends on the nature of the operation process. For many products, the operation process covers the time frame from the introduction of the new product, through market growth to maturity and saturation, followed by a gradual decline from a business life cycle viewpoint. For the energy industry, it commences with the production of the energy source and ends when that energy source is no longer available or economical to provide.

The performance of systems and equipment employed for normal routine operation degrades with time. Hence the system supporting functions kick in to mitigate the situation, attempting to bring it back to normal operating conditions. This situation also presents an opportunity for resource realignment. The supporting functions are the crucial enabling agents designed to provide maintenance support and service restoration in times of need. Technical support for the operation involves facility usage, maintenance and logistic support functions, and administrative support in managing productivity throughput in system operation.

Business operations suffer from occasional set backs due to major facility malfunctions or sudden market downturns. Organizations may sometimes face unexpected large purchase orders that they are unable to fulfill with existing capacity limitations or may be incapable to take on profitable new venture projects due to resource constraints. Business risks may result in either negative outcomes or positive benefits subject to the business management's ability to arrive at prudent, informed decisions. Administrative support for the operation deploys information services to link market inputs, customer feedback data and quality assurance process information. Dependability performance data captured during operation and subsequent information analysis provides valuable insight for process improvement and presents viable approaches for operating budget control and support cost reduction. Dependability information management plays a pivotal role in guiding effective system operation and provides critical data to support informed management decisions to address operational issues, including dependability performance impact.

7.3.4 Process Approach to Sustain Dependability Performance

Figure 7.1 presents the key drivers of the system operation process to enable effective operation and sustain system dependability performance. The application of the key drivers concept represents a process approach for assessment of system operation performance. These key drivers are the essential process functions, established in conjunction with cohesive objectives that foster an integrated system approach to facilitate performance evaluation. There are six main key drivers that can be used to describe system operation:

- infrastructure to develop and support the framework for system operation;
- system inputs to initiate the operation process;
- system operation to deliver planned throughput;
- system outputs derived from the operation;

- administrative support for the operation;
- technical support for the operation.

Each key driver represents a specific group of collaborating process activities with a unique purpose and application capability to facilitate process implementation. The applicable process activities are designed to facilitate the implementation conditions on selective process deployments. The process approach provides a credible method to determine the effects of processes on system operation and assess the resultant impact on dependability performance.

Sustaining dependability performance during operation is one of the main process objectives. The key drivers of the set of process activities that influence the outcome of dependability performance during operation are identified for the purpose of implementing their specific roles in the system operation process. The achievement of desirable performance outcomes hinges on the effective implementation of the relevant process activities of the key drivers.

(a) Infrastructure to develop a framework for system operation
Key driver: Infrastructure
Purpose and description: The infrastructure establishes a framework for system operation equipped with appropriate assets and capabilities for the operation of required functions. It facilitates process implementation, activities planning, project coordination and resource allocation to reflect the organizational hierarchy and reporting structure for management control of the system operation.
Implementation conditions: The infrastructure lays the foundation for the system operation with policy directions, business principles and practices, customer service base, supplier liaison, administrative guidance and technical procedures for running the day-to-day operation.
Effects on operation: Progressive development of business maturity in system operation, provision for learning and knowledge acquisition, and value creation to ascertain operational effectiveness and efficiency to meet market demands and satisfy customer requirements.
Impact on dependability performance: Enhanced brand value and improved reputation for quality in delivery of dependable products and services.

(b) System inputs to initiate the operation process
Key driver: System inputs
Purpose and description: A system requires resources as supplied by materials, parts and energy to activate the designed process functions. Appropriate skills and knowledge of operators and maintainers are necessary to perform the tasks for system operation.
Implementation conditions: The system inputs rely on well-established plans for initiation of the process or service. In manufacturing, the material and resource inputs are transformed into a work process for the planned system tasks. In the energy industry, the inputs consist of sources such as oil and gas. In some situations, the cooperation of the suppliers with specific knowledge of the supplied equipment and parts is needed to support timely completion of the system tasks.
Effects on operation: The availability of the needed materials and the readiness of resources are critical to sustain continuation of system task performance. The skills and knowledge of the operators and maintainers will improve the efficiency in system operation.

Impact on dependability performance: A systematic approach to managing system inputs will facilitate continuous system operation and successful end results.

(c) System operation to deliver planned throughput
Key driver: System operation
Purpose and description: System operation engages a sequence of relevant process activities to transform input material and resources into desirable output of the planned throughput. The relevant process activities are coordinated and supported by the organizational infrastructure to achieve efficiency in process task performance.
Implementation conditions: The process activities implement the appropriate application procedures with available resources, facility utilization and technical support to achieve the desirable outcomes.
Effects on operation: Regular management reviews on the status of the operation and target commitments for deliverables will provide insight into efficient and productive process outcomes for continuous improvement.
Impact on dependability performance: Process efficiency and confidence in the operation process will influence the productivity of throughput and enable successful dependability performance during operation.

(d) System outputs derived from the operation
Key driver: System outputs
Purpose and description: The delivery of the required product functions and capability, provision of quality of service and sustainment of dependability performance in system operation will achieve customer satisfaction.
Implementation conditions: Customer liaison on the quality of service being provided will facilitate continuous service improvement to meet customer needs.
Effects on operation: Successful system outputs are highly dependent on well-coordinated efforts to optimize the operation process.
Impact on dependability performance: The quality of products and services resulting from the operation process is assured by dependability performance.

(e) Administrative support for the operation
Key driver: Administrative support
Purpose and description: Provision of administrative support to enable achievement of system task performance and contribution towards successful operation process outcomes.
Implementation conditions: Coordination of information services, customer care and quality assurance functions in provision of the needed administrative support.
Effects on operation: Efficiency and effectiveness in supporting the administrative needs by management of the operation process.
Impact on dependability performance: Gained employee confidence and service recognition through timely information feedback of collaborative achievements of organizational value.

(f) Technical support for the operation
Key driver: Technical support
Purpose and description: Provision of technical support to enable achievement of system task performance and contribution towards successful operation process outcomes.

Implementation conditions: Coordination of facility support, maintenance and logistic support to provide the needed technical support.
Effects on operation: Efficiency and effectiveness in technical support needs by management for maintenance and restoration of operation process tasks.
Impact on dependability performance: Reduction in system operation downtime and decreased frequency of unplanned service interruptions.

The key drivers concept and an integrated system approach can be adapted for use in performance effectiveness assessment of system operation processes used in many industry system applications.

7.4 DEPENDABILITY MEASURES DURING OPERATION

7.4.1 Overview of Dependability Measures

In system operation, dependability measures are addressed in terms of downtimes or outages caused by system failures. A failure occurs when a function is lost or degraded. The loss of a function is not necessarily observable to the user or operator of the system. Hidden or latent faults are generally undetectable, as is often experienced with electronic systems. A fault is a state describing the manner of an item's inability to perform its designed function. A failure is an event that is observable representing the loss of a performance function. The failure identification process may require special testing to trigger the latent fault for its precipitation into an observable failure. These uncertainties directly impact how failures are measured, recorded and analyzed. This has led some industries to the development and application of highly structured and formal systems to assure data quality and analysis effectiveness associated with the dependability measures.

Once good data has been gathered, the challenge is to analyze it properly and reach useful conclusions that can be used for effective improvement. There is a range of possible analytical approaches, some of which are fairly straightforward, such as MTBF calculations, while others such as Markov analysis are complicated and require specialized expertise to rationalize the analysis results. The temptation is to use the simplest possible method, but it is important to realize that dependability analysis is fundamentally of a statistical nature and simple averages can easily hide what is really happening.

It is useful to segregate dependability analysis into two categories:

- at the top level relating to organizational or end-user objectives such as production, service levels and safety;
- at the supporting level applicable to dependability performance of assets.

The top level is specific to the dependability application and can broadly be divided into four categories of production/manufacturing, service, product and safety performance. Dependability-related aspects of these categories can be summarized as follows:

- product performance: measures are product-specific and vary because a product can be at different levels of the asset hierarchy, ranging from a simple component to a complex system. The focus is usually on monitoring failure rates in a manner appropriate to the product;
- service-level performance: measures are situation-specific and relate to failure to perform the required service in a timely manner;

- production/manufacturing performance: availability and reliability of production are the main measures, and production may vary from 0 to 100% of maximum capacity or production;
- safety performance: this is associated with a variety of negative outcomes for people and assets and is tied to incidents or accidents.

Figure 7.2 illustrates the hierarchy of dependability analyses generally used for these two levels. Details are described in the following sections.

7.4.2 Organizational or End-user Measures

End-users naturally are concerned with how dependability performance affects and impacts the system operation directly. The management of the organization has to consider the experience and perception of customers in addition to measures related to the organization.

For a situation involving a service, an end-user will be concerned with dependability performance based on general measures such as:

- availability of that service when required;
- reliability of that service once it has been acquired;
- ability to complete the service on time as promised;
- ability to complete the service without a safety-related accident or incident.

For example, with a rail service, availability means that the train arrives on time, within a specific time period of the posted schedule, a certain number of times as compared to the total number of arrivals. If the train arrives full and it is not possible to board the train, the passenger or service user may consider this incident to be a failure of dependability performance, although the actual reason is one of insufficient capacity. Reliability can be considered as a function of the number of interruptions to the train journey due to equipment failures, although interruptions due to other causes may also be perceived as

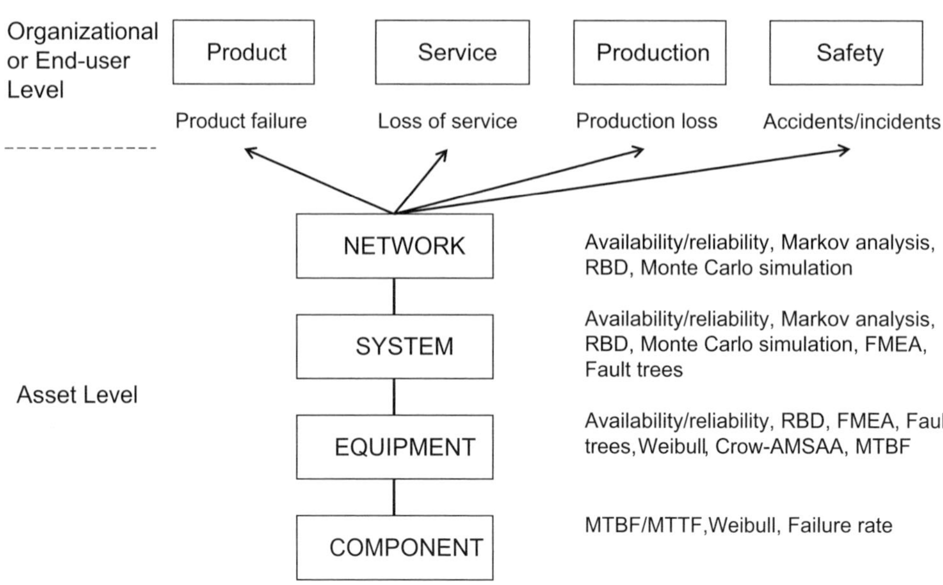

Figure 7.2 Overview of applicable dependability analyses

applicable to the train service. The third measure applies to delays caused by degradation in the performance of the train that would cause a delay in the final arrival time.

Closely related to these dependability measures are concerns related to the aspect of safety. Even though safety is often thought of as a functional concern and not a direct dependability requirement, the perception by many people is that a safety-related incident or accident is more important than the loss of dependability performance. The two are indeed closely tied together since equipment failures can cause an accident or incident affecting safety.

For products, expectations of dependability by the customer include the following:

- ability of the product to be available when needed;
- reliability of the product;
- ability to perform the functions provided by the product;
- usage of the product without a safety-related accident or incident;
- ability to operate without undue environmental impact.

As an example of what is expected of a product, a family vehicle is expected to start and be able to drive on demand, complete the journey without interruption, and do so safely. This example can be extended to all types of industrial equipment and systems.

Some parts of our modern society can be thought of as associated with either a product or a service. Examples are energy sources we use such as electricity and natural gas, and telecommunication such as telephone and the Internet. End-user measures are very similar to those mentioned above.

From the point of view of an organization delivering services or products, additional measures from a management perspective include the following:

- ability to produce enough of the product or deliver adequate service to satisfy demand;
- maintain a high level of reliability to enhance its reputation;
- minimize returns as an indicator of quality problems;
- minimize manufacturing costs to ensure profitability.

Both the end-user measures and those important to the organization are ultimately dependent on the infrastructure and assets employed to provide the service or manufacture the product. Asset-related measures are the focus of the remaining part of this section.

7.4.3 Average Failure Rate Measures

The simplest way to characterize failures is to use failure rate (the number of failures over a period of time) or its inverse, the ubiquitous MTBF (mean time between failures) or MTTF (mean time to failure). There is much confusion about these measures, but MTTF is generally considered to apply to non-repairable components and MTBF to repairable equipment or systems. For non-repairable components, the data can be assumed to be independent and identically distributed, whereas for repairable items this assumption is often not true because a repair does not always return the system to an as-new condition. [1]

There is the further complication in the assumption that the underlying distribution should be exponential with a constant failure rate. If data covers an initial period of equipment operation where infant mortality failures had occurred or includes the later wear-out period where the failure rate is increasing, the value of MTBF may be quite misleading. If the equipment MTBF quoted by a manufacturer is based on a constant failure rate without consideration of a wear-out phase, the value for MTBF may be a very long time period, perhaps in the hundreds of years, which may not be practical for measurement. Many electronic components that are designed and built into equipment are unlikely to reach the component wear-out phase because their useful life expectancy far exceeds the actual service life in normal equipment applications. Actual in-service measurements of equipment MTBF are more likely to include components that have failed at random for the period after infant mortality and prior to wear-out. Some vendors are now beginning to talk about "service life," which might be a more useful measure for data comparison [2].

In addition, if regular replacements are occurring, the resultant failure rate will depend on the replacement interval. To properly understand the behavior of failure, it is necessary to fully understand the situation and move to a statistical analysis as is further described in the following section.

Where a large number of equipment exists at a facility (e.g., pumps at a refinery), it is very tempting to divide the number of failures for pumps in a year by the number of pumps and then by inverting to deduce the average MTBF for pumps. This ignores the fact that not all pumps will be operating and pumps may be of different types, sizes, manufacturers, type of operation (base load or standby) and other factors. If this method is used, the pumps should at least be categorized and actual operating hours used. Even then, the true picture is likely to be obscured [3].

In spite of these issues with the use of an average failure rate measure, it is often recorded in reliability databases and subsequently used in analyses.

Another example of the difficulty of using averages is a metric such as System Average Interruption Frequency Index (SAIFI), which is utilized in the electrical industry to monitor electrical outages by customers [4]. The underlying distribution is bounded by zero at the low end and is usually highly skewed to the high end by a small number of large values. For example, there may be a range of zero to twenty outages in a year but a system average of only two. That may represent only the lowest third to quarter of a utility population. In addition, there may be pockets of frequently interrupted customers that are masked by the overall average. The message is that averages need to be used very carefully.

7.4.4 Statistical Failure Rate Measures

Statistical measures or metrics for failure take into account the distribution of the data and therefore provide a much more accurate representation of the nature of the failure data. The most flexible statistical measure of failures is the Weibull distribution [5]. Its major advantage is that it can characterize a range of distributions that cover failure rates from the initial infant mortality period to end-of-life wear-out. An example of its application to wind turbines [6] found that a three-parameter Weibull model was needed to account for the fact that only incomplete data was available for many wind turbines. Another application where Weibull analysis has been applied is with production data to make it possible to distinguish between processing and reliability issues [7].

Another application of statistical techniques is the prediction of failures with scarce data, as is the case with water distribution systems, where pipe breaks are infrequent occurrences but predicting future failures is highly desirable [8]. The paper compares three models of future trends: time-linear, time exponential and generalized linear (GLM) models such as a Poisson distribution. A logistic GLM was found to provide the best fit against actual data but was actually only marginally better than a model based only on pipe length.

Reliability performance and growth can be estimated by utilizing the Crow-AMSAA (U.S. Army Materiel Systems Analysis Activity) method [9]. It is similar to a Weibull analysis in that the slope of the line indicates whether reliability is increasing, decreasing or remaining constant [10]. Its advantage is that it is easier to plot and can be used for multiple failure modes, in contrast with Weibull analysis, which requires data to be divided into individual failure modes.

7.4.5 Availability and Reliability

The use of average availability and reliability metrics is extensive because it can be easily related to downtime of production or manufacturing capacity. A number of industries have developed common methods of measuring, although there is not always consistency between them.

A generalized view recognizes that assets can have four basic states of operation and maintenance; operating, no demand, planned maintenance and unplanned maintenance (also commonly referred to as a forced outage). For any time period, these states can be converted to availability in % as shown in Figure 7.3.

The advantage of this method is that the time basis is always consistent (although the number of hours in the month does change). The disadvantage is that if operating time is low, then comparisons with high-time equipment are misleading.

It is also possible to ignore the no-demand period in the total time. The disadvantage of this method is that the time basis is inconsistent if it is being reported on a calendar basis, e.g., every month. The advantage is that the comparisons between high- and low-time equipment are more valid. In the end, if comparing only high-usage equipment, both methods are very close to each other.

Steady-state availability, mean rate of failure, mean duration of downtime and lower bound reliability of a general system can also be derived from randomly and independently failing repairable components [11].

7.4.6 Reliability Analysis Techniques

Complex systems require a more sophisticated approach to dependability analyses, and various methods are available to accomplish this. Probably the most common is the Reliability Block Diagram (RBD). A larger system or network is divided into blocks that are connected in series and/or parallel. The availability or reliability of each block is defined by average failure rate (or MTBF), or a Weibull characteristic and the resultant system availability or reliability can be calculated. It is a practical and useful technique that is widely used [12].

Another technique is FTA, which is applicable to large complex systems but is also more time-consuming [13]. Even more flexible but also more complicated is Markov

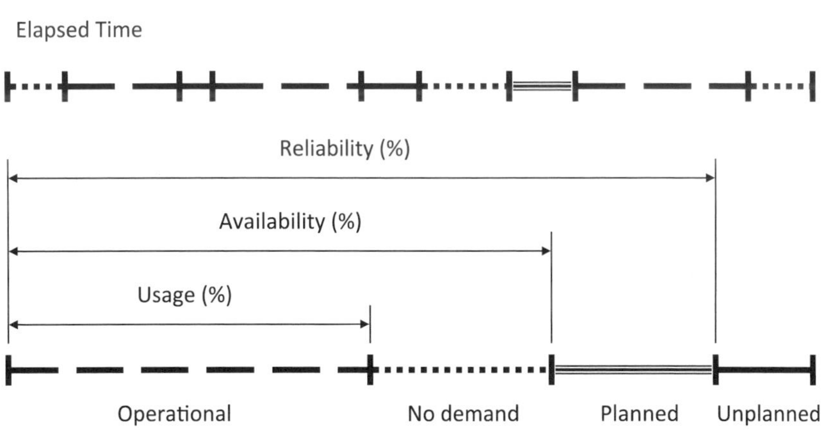

Figure 7.3 General calculation of availability and reliability based on total time period

analysis, which can handle multiple states [14]. Availability and reliability modeling is best handled by Monte Carlo simulation techniques [15, 16].

7.5 DEPENDABILITY DATA SOURCES

7.5.1 Data Acquisition

It is common with some product applications for the operators or users to take responsibility for dependability data gathering, analysis and corrective action, often without direct involvement of the original product manufacturer. Examples include operators of complex systems and users of computer networks where incident reporting and problem diagnosis are automated and reported on-line. The operators and users are in the best position to capture accurate dependability information first hand for reporting as the incidents occurred.

In some industries, independent service organizations specializing in maintenance service are established to maintain and service equipment and systems on a full-time or on-call basis. Examples include the oil and gas industry, where trained and certified maintainers are dispatched to provide the needed customer services. These independent service organizations are well equipped to enable on-site problem resolution and capturing dependability information for their own dedicated information system database for continuous client service and business improvement.

There are many efforts to gather data on availability and reliability, but no single source exists for any specific situation. There is no consistent method of data gathering, analysis and reporting across all industries, although there can be consistency within a given industry. In general, data sources can be categorized as follows:

- data gathered by individual operators or users for their own benefit;
- data collected by individual vendors or manufacturers that may be shared with users but is intended mainly for their own use;
- industry-based programs to assist members with monitoring and improving dependability performance;
- programs within an industry involving both equipment vendors and operators, generally to promote and improve the performance of that industry;
- commercial databases for general use;
- government-sponsored programs, including those run by regulatory agencies.

It will be seen that properly constructed databases incorporate a defined taxonomy or structure of the assets for which data is being gathered, procedures for collecting and recording data, and specific methods for analysis and presentation of the results. Nevertheless, there are also attempts to harvest data from sources such as the Internet [17]. Clearly for this to work, there are major challenges to ensure that the data is relevant and correct for dependability purposes.

It is not possible to document every dependability database available today. As examples, several are discussed in the following sections.

7.5.2 International Standards Information

A number of current standards address the dependability information process. IEC 60300-3-2 [18] provides a more general description of the data collection and analysis process. It describes the objectives of data collection to be:

- identification of deficiencies in product design;
- changes to maintenance and logistics support;
- determination of customer problems so they can be corrected;
- the use of root cause analysis to prevent future failure modes from occurring in the next design.

Data can be gathered at different levels of equipment complexity and at different stages of the life cycle. Data can be collected on:

- inventory or items that exist in the field or are being used;
- usage of items that have been installed, operated and removed;
- environment or the operating conditions encountered, or;
- events such as failures, faults, maintenance actions or incidents experienced by the equipment.

Approaches to data collection include:

- time-based continuous, windowed for a certain period of time, multiple windowed or a rolling window (time can be calendar, operating hours, number demands, mission-based, distance, etc.)
- complete or limited data collection of specific subsets of items;
- quantitative or non-quantitative measurements.

The quality of the data and information derived from data collection is of paramount importance since bad data leads to inadequate information and misleading analysis. Quality data and information exhibits relevance, accuracy, completeness, confidence in the source, communication to the right person, appropriate timing and sufficient detail. Quality can only be ensured by a process of data validation, in which incorrect data is identified and corrected.

A more specific standard exists for the petroleum, petrochemical and natural gas industries for the collection and exchange of reliability and maintenance data for equipment [19]. The emphasis is on standardization of data collection and analysis for the mutual benefit of plants, owners, manufacturers and contractors. The primary areas of use for this data are reliability, availability/efficiency, maintenance and safety and environment.

Data gathering is performed in a structured manner with a defined format and is generally divided into the following categories:

- a boundary description that defines what equipment is included (for example, a compressor might consist of the compressor itself, the lubrication system, the sealing system and the control and monitoring system);
- a taxonomy or systematic classification of items into generic groups from the highest level (industry) to the type of plant and down to the specific system and so on to the smallest part of interest;
- timeline issues related to the surveillance and operating periods;
- data about the equipment such as manufacturer's data and design characteristics;
- failure data concerning what has failed, when it occurred and how it happened, defined by the failure mode, cause and consequences;
- maintenance data with details on the nature of the maintenance action and resources expended.

Standard descriptions for all of these are found in the annexes of this standard for general use. In this fashion, valid comparisons can be made across this industry.

7.5.3 OREDA - Offshore REliability DAta

A prime example of an industry-based program is OREDA (Offshore REliability DAta). Focused on the offshore oil and gas sector, its objectives are to improve safety, reliability, availability and maintenance effectiveness and enhance industry reputation. It has existed since 1981, with between seven and eleven international oil and gas members from five different countries [20]. It was established as a forum for the collection and exchange of reliability data and utilization of reliability methods for the oil and gas industry. The first OREDA Reliability Data Handbook was issued in 1984, and the latest is the 5th edition [21]. As the most comprehensive reliability database for oil and gas equipment, its methods have been incorporated into ISO 14224 [19]. Detailed data is available in an electronic database, but this is available only to members.

OREDA has a detailed taxonomy or structure to standardize data collection and analysis. The four main types of equipment cover rotating machinery, mechanical equipment, control and safety and sub-sea equipment, and these categories are further broken down into systems and equipment classes.

Figure 7.4 presents the database structure divided into installation, inventory, failure and maintenance records. A large number of failure modes are covered and are divided into critical, degraded, incipient and unknown categories. The number of equipment and installations, number of demands, number of failures and both calendar and operating time incurred are noted to assist with understanding the validity of the information. Failure rate is recorded as lower, mean, upper, standard deviation and MTBF. Mean and max data is also included for active repair hours and man-hours.

Software developed by OREDA has been created to automate data collection, acquisition and analysis. It features advanced data search and selection, as well as commonly used reliability analyses. The main modules are shown in Figure 7.5.

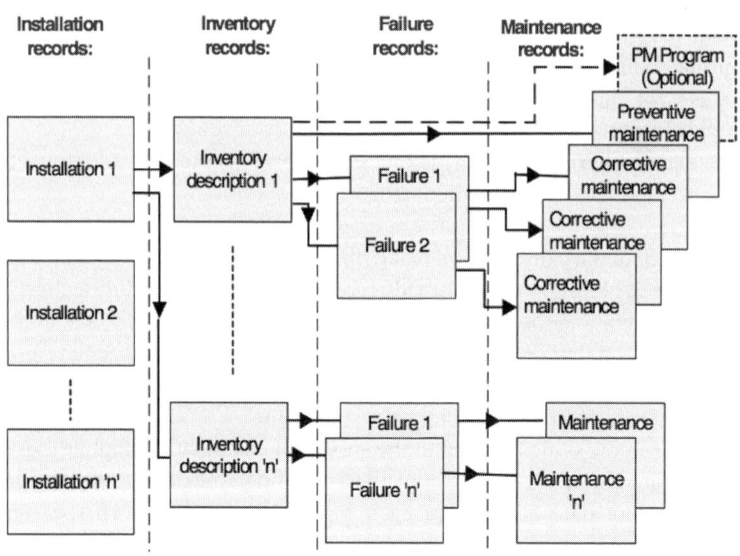

Figure 7.4 OREDA database structure [20]

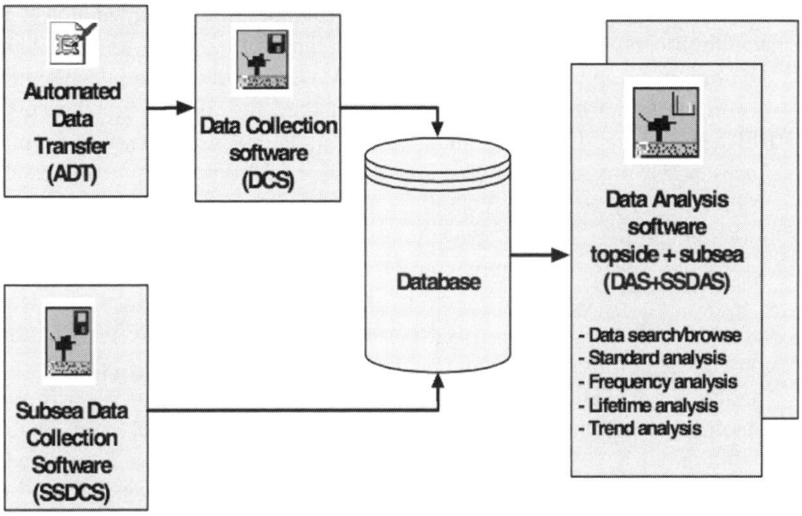

Figure 7.5 Main software modules for OREDA [20]

7.5.4 Gas Turbines

7.5.4.1 Background

Since its inception, the gas turbine market has been driven by an emphasis on improved operating efficiency and greater megawatt output, with reduced environmental impact through technology advancement [22]. The technology advances that have taken place include higher firing temperatures and pressure ratios, improved cooling schemes with advanced metallurgy and coatings for hot gas path parts, consideration for tighter clearances, low-emission combustion systems, and fuels flexibility. At the same time, there has been a commensurate focus on ensuring that the achievable levels of availability and reliability of these evolving product offerings would be optimized for various duty cycles, applications and plant arrangements. In fact, as product evolution and advancement has taken place, there has been an expectation that availability and reliability performance would be consistent with the best achievable levels experienced by existing and more mature technologies.

Since the mid 1970s, there has been a significant focus on improving the perceived and actual reliability of gas turbine power plants, both in simple and combined cycle operation. A major investment of engineering resources and efforts was undertaken either solely by various equipment manufacturers or with the collaborative support, both financial and technical, of the Electric Power Research Institute (EPRI). These efforts established the process of using field data with standard analytical methods and techniques with the intent of improving product reliability performance, a programmatic focus on reliability, availability, and maintainability (RAM). At this time, EPRI-funded programs, like the "High-Reliability Gas Turbine Combined-Cycle Development Program," placed an emphasis on improving the "perceived and actual" levels of equipment availability and reliability performance, with a specific emphasis on the gas turbine, control and accessory systems, as well as the balance of plant systems and components. The primary questions to be addressed were: what is the current actual level of equipment performance, and is it acceptable? If it is not acceptable, what must be done to improve reliability performance to meet market expectations for equipment performance? This program provided a basis for ensuring that availability and reliability characteristics and performance metrics would be considered by the original equipment manufacturer (OEM) during design and development efforts.

In the mid 1980s, these efforts culminated in a further collaborative effort between EPRI and General Electric, the "Design of High-Reliability Gas Turbine Controls and Accessories" program. The goal and objective of this program was to establish availability and reliability goals and expectations for "F" class gas turbine plants, to identify areas of improvement for reliability growth on a system and component basis (with an emphasis on control and accessory systems), and to quantify the reliability impact of identified design alternatives. The objective was to treat reliability as a design parameter to drive component and system selection. This program established a design for reliability approach that coupled reliability methods and analysis tools, with actual field data available through the Operational Reliability Analysis Program (ORAP) system, along with the OEM's design principles or practices, culminating in a design driven by a focus on reliability.

In the early 1990s, in a report from the U.S. Department of Energy to Congress, which is entitled "Comprehensive Plan for Advanced Turbine Systems (ATS)," the emphasis on maintaining the reliability, availability, maintainability and durability of the combustion turbine at a sufficiently high level and consistent with current "expectations" continued. This Department of Energy initiative continued the process of driving technology advancement by establishing aggressive operational goals: 60% combined-cycle thermal efficiency (LHV) with minimum NOx emissions and a 10% reduction in the cost of electricity, while maintaining the achievable level of availability and reliability in line with state-of-the-art expectations. As the industry has evolved from the 1970s to today through continued technology advancement and driven by market demands and challenges, as well as with the increasingly competitive market environment, the value of availability and reliability performance is abundantly clear. Profitability goals of various projects depend on pro forma levels of availability and reliability being realistic and achievable. Unless these availability and reliability performance expectations can be achieved, the financial performance of the project will be at risk.

7.5.4.2 Metrics

The key performance indicators, in priority order, include: Service Factor (%), Service Hours per Start (Ratio), Availability Factor (%), Reliability Factor (%) or Forced Outage Factor (%), and Scheduled and Unscheduled Outage Factors (%). While several other RAM metrics exist and provide valuable information, in essence these metrics are sufficient to characterize the performance of today's gas turbine fleets at the highest level. The precise definition of these metrics can be found in two industry standards: IEEE 762 [23] and ISO 3977 [24]. It should be noted that these metrics are time-based not energy-based, as typically used by the World Energy Council (WEC). It should be further noted that both the North American Electric Reliability Corporation (NERC), with their Generating Availability Data System (GADS), and SPS' ORAP system follow these industry standards when reporting performance metrics.

A brief overview of the metrics follows:

- The service factor and the service hours per start metrics are indicative of the duty cycle, or the mission profile, that the gas turbine (either in simple or combined cycle) must meet. Both metrics are based on service hours (the time that the turbine generator is synchronized to the grid at any load). These parameters reflect the economic mission that should have been established in the pro forma.

- The capacity and output factors provide an indication of the MW output contribution of the operating assets.
- The availability factor represents the percent of time that the turbine is available for service, either actually operating (i.e., service hours), or in a state of ready reserve. The complement of availability is unavailability (the percent of time the unit is out of service).
- The reliability factor is the complement of the forced outage factor. The forced outage factor represents the percent of time that the turbine plant is forced out of service. Therefore, the reliability excludes forced outages.
- The scheduled and unscheduled outage factor is the percent of time that the turbine plant is out of service for maintenance (whether scheduled in advance or not).

7.5.4.3 NERC GADS Data

To appreciate the performance levels of gas turbine plants, whether in simple or combined cycle operation, an important starting point is the NERC GADS database. The NERC GADS data is shown for five-year periods, providing both a historical and current perspective: 1984-1988, 1989-1993, 1994-1998, 1999-2003, and 2004-2008. The NERC GADS data is further segmented into three specific gas turbine classes: 50MW+, combined cycle block, and combined cycle GT units. The data are not segmented by fuel (i.e., gas or oil). The objective is to show the operational difference between simple cycle and combined cycle performance.

The data for the 50MW+ units shows that these simple cycle gas turbines operate primarily in a peaking duty cycle with a service factor between 3% and 6%. Therefore, the relatively high level of availability of 91-93% is driven by the significant amount of reserve standby time reported. It should be noted that the capacity factor is consistent with the service factor, and, when operating, MW output is between 40% and 70% of nameplate-rated capacity. As a further point of reference, during the previous ten-year period, 1975 to 1984, NERC reported availability of 82.6% with a service factor of 8.2% for gas turbines in the 50MW+ class. Consequently, reliability growth is evident in the later five-year reporting periods.

The combined cycle plants as reported by NERC are operating at substantially higher number of operating hours on an annual basis. Consequently, their availability factor is driven by the substantially higher number of operating hours per year (typically over 3,500 service hours per year). The NERC data in Table 7.1 shows substantially higher service factors for combined cycle plants. This is also supported by the significantly higher mission profile on a SH/ST (service hours per start) basis, and thus it would be expected that the availability and equivalent availability factors would be higher. Both availability and equivalent availability are approximately 90% (an unavailability of approximately 10%). A further review of Table 7.1, looking at the forced outage factor, suggests that the driver of combined cycle unavailability is both scheduled and unscheduled maintenance.

7.5.4.4 ORAP - Operational Reliability Analysis Program

ORAP is an automated system for monitoring RAM of both combustion (gas) and steam turbine driven plants, with the emphasis on the total plant, including condensate and feed water systems, power distribution, heat recovery steam generators, electrical generators, driven equipment and all mechanical and electrical balance of plant systems [25]. Standard equipment codes, developed by SPS under an EPRI contract, are the basis for the uniform

TABLE 7.1 NERC GADS data for combined cycle by MW range [22]

Category	Period	Availability Factor	Service Factor	SH/ST	Capacity Factor	Output Factor	Forced Outage Factor	Equivalent Availability Factor
Combined Cycle Block	1984-1988							
(100-199 MW)	1989-1993							
	1994-1998	92.1	62.8	60.7	37.7	59.2	0.7	87.7
	1999-2003	88.2	51.0	62.5	33.7	67.7	1.5	85.6
	2004-2008	86.9	35.4	37.2	25.1	72.0	5.4	84.5
Combined Cycle GT Units	1984-1988							
(100-199 MW)	1989-1993							
	1994-1998							
	1999-2003	91.4	51.6	58.4	41.3	80.8	2.9	90.1
	2004-2008	90.0	55.6	46.9	44.6	81.3	2.4	88.9
Combined Cycle Block	1984-1988	88.9	58.0	72.0	31.6	55.2	1.3	80.5
(200-299 MW)	1989-1993	86.3	65.3	86.3	38.8	59.4	1.1	78.1
	1994-1998	89.0	55.3	57.8	37.3	67.4	1.9	82.7
	1999-2003	88.3	54.9	69.9	39.1	72.3	3.7	84.5
	2004-2008	90.2	47.6	63.6	36.3	76.6	2.7	88.1
Combined Cycle GT Units	1984-1988							
(200-299MW)	1989-1993							
	1994-1998							
	1999-2003							
	2004-2008	87.2	62.5	62.0	51.2	81.6	5.5	85.9

reporting system across all product lines and OEMs. These codes provide reporting uniformity across all equipment types, providing a basis for combining or segmenting data at a component level across equipment manufacturers, size ranges or other valid criteria. Additionally, ORAP is capable of obtaining data using the European KKS standard.

Data from participating plants are submitted to SPS on a monthly basis, or in some cases on a "real-time" basis, for engineering review and data validation, data acceptance according to relevant industry standards and incorporation in the ORAP database. This process is illustrated in Figure 7.6.

An important aspect of the data collection process is to obtain event data (forced and scheduled outages) at the right level of detail to support an understanding of the causes of unavailability/unreliability. Standard event types, based on IEEE Standard 762 [23], were created and used by ORAP. ORAP data are collected and reported from the "bottom up." In other words, ORAP allows the engineer to see the impact of a component failure, including shared equipment, up to the system, up to the plant. This level of detail is essential for effective reliability analyses.

The ORAP information system then converts the data into RAM statistics such as, detailed system and component outage factors, failure rates, starting unreliability, service factors, time to repair and other outage factor information. Additionally, ORAP provides outage description details, outage causes, failure modes and corrective actions taken (as supplied by the plant operator). This information provides the basis for assessing plant, system and component RAM performance, as well as for developing RAM values to assess improvement. It should be noted that the data reported to the SPS ORAP system from various utility and co-generator participants are reviewed for accuracy and verification before being entered into the database. The information is not modified by SPS unless the participating customer concurs with and accepts the recommended change. SPS engineers

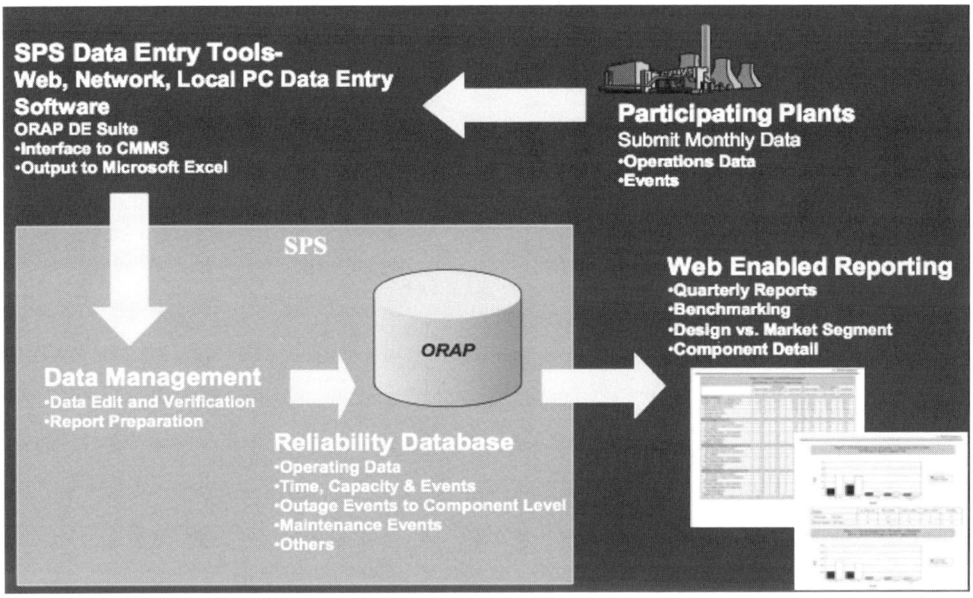

Figure 7.6 Data transformation and transfer process for ORAP [25]

work with each participant to ensure data accuracy and completeness. This ensures that SPS ORAP data reflects the specific operational, failure and maintenance history for each component in the database, and therefore the availability and reliability performance measurements are valid indicators of unit experience and capability.

A proprietary and non-disclosure relationship establishes the basic rules for how SPS will process unit- and site-specific data for each participant, as well as combining "like" data to develop fleet-level performance. Participation in ORAP provides members with quarterly reports of statistical comparisons of their equipment with units in similar applications and MW range for industry assessment, fleet benchmarking and up-to-date operational information. Moreover, the ORAP data are used to perform reliability assessments that can effectively measure and demonstrate reliability performance.

The ORAP data [22] is further stratified by technology and other important parameters; duty cycle, application and single-shaft and multi-shaft combined cycle plants. It should be noted that the trends that will be shown reflect annual unit year data. Year-over-year variation may be driven by such things as sample size (generally increasing), the affect of significant downtime contributors (in terms of downtime hours), and the changing service factor (increases/decreases in service hours on an annual basis).

Figure 7.7 provides a review of the availability and service factor metrics for both "E" and "F" class units across all OEMs. It should be noted that these classifications are based on the specific technology, not the MW range that was used to characterize the NERC GADS data shown previously. The data in Figure 7.7 shows a significant performance difference when compared to the NERC data seen in Table 7.1. The "E" and "F" class units reporting to ORAP show a substantially higher service factor as a group when compared with the NERC GADS data. This is because the ORAP participants cover all of the major duty cycles, from peaking, to cycling, to base load. The metrics suggest that, for both "E" and "F" class units, availability is in the range of 92% to 94%, and that over time the expected performance for both classes has converged. As would be expected, due to improved thermal efficiency, the service factor of the "F" class is ultimately higher than

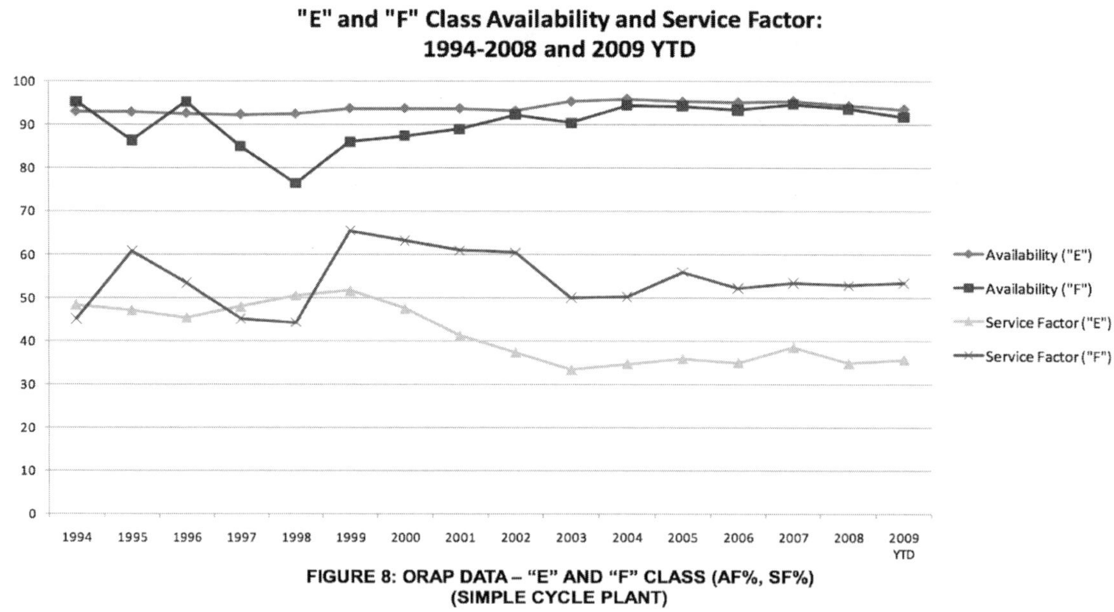

Figure 7.7 OPRAP availability data - "E" and "F" class for simple cycle plant [22]

that experienced by "E" class units by about 16% (or approximately 1,400 service hours per year).

Figure 7.8 shows the forced outage factor trend over time. Since 2004, the forced outage factor has been relatively consistent for both classes. The data suggest (similar to what was seen in the NERC GADS data) that both scheduled and unscheduled maintenance are the primary sources of equipment unavailability for these mature technologies.

Figure 7.9 provides a review of the major downtime contributors (top 10) for the "F" fleet monitored by SPS in both single-shaft and multi-shaft combined cycle plants. These charts include all gas turbine data irrespective of class, MW size, or OEM, and are presented only to demonstrate downtime contributors.

As technology advancement continues with a focus on improved output and thermal efficiency (for both simple and combined cycle), as the use of gas turbines expand with the ability to burn unconventional fuels (including clean coal through IGCC), and as new lifetime issues associated with rapid start and stop capability increase, the focus on availability and reliability performance must be maintained and demanded.

7.5.5 Infrastructure

Our society is now very dependent on infrastructure that includes a wide variety of systems and networks such as roads and bridges, water supply and distribution, pipelines, electrical generation, transmission and distribution, and communications. As this infrastructure ages, reliability becomes more critical.

As an example, water systems are receiving increasing attention as water sources are being impacted by both burgeoning populations and the negative effects of climate change in some areas of the world. Dependability for water systems is characterized not only by the reliability of the infrastructure itself, i.e., the water mains, pumps, valves, desalination plants and storage reservoirs, but also the availability of water sources, i.e., rainfall, rivers, water bodies and recycled water. Government and regulatory agencies are taking a major

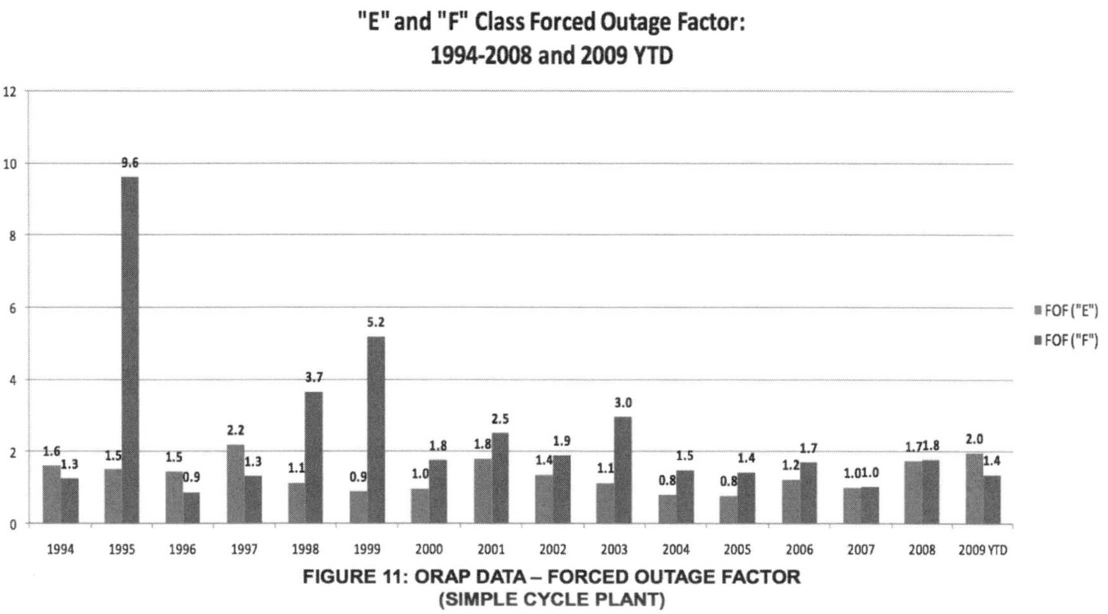

Figure 7.8 OPRAP forced outage data - "E" and "F" class for simple cycle plant [22]

role in monitoring water supply reliability as well as water quality for safety purposes. One example of government involvement is in California where water supply is a critical issue, one that is exacerbated by decreasing water supply availability. This has led to a program on water supply reliability called the CALFED Bay-Delta Program, which emerged from the water crises of the 1990s and funding crises of the early 2000s [26].

7.5.6 Telecommunications and the Internet

Telecommunications have made great strides in technological advancement in the past decade. The Internet has evolved extensively due to consumer demands. Media convergence has

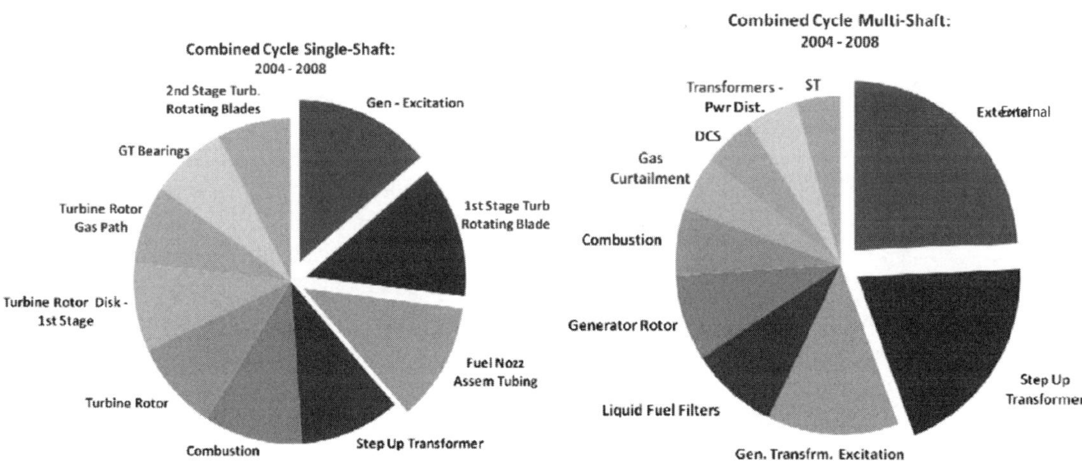

Figure 7.9 ORAP data on downtime contributors for single-shaft and combined cycle multi-shaft gas turbines [22]

resulted in multi-media merging that offers versatility and flexibility with voice, data and video communications within a single network infrastructure. The achievement of dependability in enabling the performance of telecommunications and the Internet has propagated into widespread applications for on-line e-commerce services and influenced the frenzy of social communications such as texting, Web surfing, VoIP (voice over IP), and streaming media. User demands have, to a large extent, been satisfied with high quality of service, robustness in operation at moderate cost and ease of modification and upgrading in services, although security and privacy issues are often outpaced by intentional intrusion of hacking and malware.

From a dependability vantage viewpoint, technological advancement in every stage of human existence also comes with a legacy of the good, the bad and the ugly. The good is the modernization in the quality of life in that the voices of many more people are being heard in changing the political landscape of nations; the bad is the deliberate hacking in cyberspace for fun and mischief, and the ugly is the invasion into personal privacy and financial security for despicable profit gain and unethical actions of corporate greed. This often raises doubts for those who are not totally convinced of the benefits and long-term reliability of technological advancement.

A report developed by SIFT Information Security Services on behalf of the Australian Department of Communications, Information Technology and the Arts [27] names a number of factors that influence Internet reliability, including:

- physical damage to the infrastructure directing and delivering Internet data;
- interoperability issues with the protocols used to carry Internet data;
- vulnerability of key Internet systems;
- commercial or political issues leading to service disruption;
- factors influencing consumer confidence;
- congestion through electronic bottlenecks;
- lack of quality of service guarantees.

7.6 EXAMPLES OF ANALYZING DEPENDABILITY DURING OPERATION

In the appendices, two examples are presented to illustrate how dependability is measured, analyzed and applied during operation. It is interesting to note differences and similarities and the varying involvement in these activities by customers, operators, vendors and other organizations such as government and regulatory agencies.

In Appendix E, a study was performed on a re-liquefaction system for liquefied natural gas (LNG) carriers in order to verify design improvements intended to increase reliability. RBDs were used to determine the reliability of the system and various alternatives. The result was incorporation of redundancy for the three most unreliable sub-systems.

Appendix F presents an example of a study of the availability of compressor stations to establish the need for standby units to meet contractual requirements. A Monte Carlo simulation was utilized to estimate the available capacity and test several scenarios. Based on an economic analysis, it was determined that it was cost effective to install a standby unit at each of ten compressor stations.

REFERENCES

[1] Unknown, 2008. "MTTF, MTBF, Mean Time Between Replacements and MTBF with Scheduled Replacements," Reliability HotWire, Issue 94, December 2008.

[2] Unknown, 2002. "The Bathtub Curve and Product Failure Behavior Part Two - Normal Life and Wear-Out," Reliability HotWire, Issue 22, December 2002.
[3] Sutciffe, F., 2007. "Lies, damned lies and statistics," E&P Magazine, July 2007.
[4] Kram, E.A., 2003. "Leveraging Operational Data to Improve Asset Management and Maintenance Decisions," www.bluearcenergy.com/reports, March 5,2003.
[5] IEC 61649. Weibull analysis, Edition 2.0, 2008.08.
[6] Guo, H., Watson, S., Tavner, P. and Xiang, J., 2009. "Reliability analysis for wind turbines with incomplete failure data collected from after the date of initial installation," Reliability Engineering and System Safety 94 (2009), pp. 1057-1063.
[7] Roberts, Jr., W.T. and Barringer, H.P., 2001. "Consider using a new reliability tool: Weibull analysis for production data," Hydrocarbon Processing, October 2001, pp. 73-82.
[8] Yamijala, S., Guikema, S.D. and Brumbelow, K., 2009. "Statistical models for the analysis of water distribution system pipe break data," Reliability Engineering and System Safety 94 (2009), pp. 282-293.
[9] Sun, A., Kee, E., Popova, et al, 2005. "Application of Crow-AMSAA Analysis to Nuclear Power Plant Equipment Performance," 13th International Conference on Nuclear Engineering Beijing, China, May 16-20, 2005, pp. 1-6.
[10] Comerford, N., 2005. "Crow/AMSAA Reliability Growth Plots," Vibration Association of New Zealand, 16th Annual Conference, Rotorua. Pp. 1-22.
[11] Kiureghian, A.D., Ditlesen, O.D. And Song, J., 2007. "Availability, reliability and downtime of systems with repairable components," Reliability Engineering and System Safety 92 (2007), pp. 231-242.
[12] Wang, W. Et al, 2004. "Reliability Block Diagram Simulation Techniques Applied to the IEEE Std. 493 Standard Network," IEE Transactions on Industry Applications. Vol. 40, No. 3, May/June 2004, pp. 887-895.
[13] Volkanovski, A., Cepin, M. and Mavko, B., 2009. "Application of the fault tree analysis for assessment of power system reliability," Reliability Engineering and System Safety 94 (2009), pp. 1116-1127.
[14] Unknown, 2003. "The Applicability of Markov Analysis Methods to Reliability, Maintainability and Safety," START Selected Topcs of Assurance Related Technologies, Volume 10, Number 2, START 2003-2, pp. 1-8.
[15] Ge, H. and Asgarpoor, 2011. "Parallel Monte Carlo simulation for reliability and cost evaluation of equipment and systems," Electric Power Systems Research 81 (2011) pp. 347-356.
[16] Borgonovo, E. Marseguerra, M. and Zio, E., 2000. "A Monte Carlo methodological approach to plant availability modeling with maintenance, aging and obsolescence," Reliability Engineering and System Safety 67 (2000), pp. 61-73.
[17] Dussault, H., Zarubin, P.S., Morris, S. and Nicholls, D., 2008. "Harvesting reliability data from the internet," Proceedings of the 2008 Annual Reliability and Maintainability Symposium, Las Vegas, Jan 28-31, 2008, pp. 322-327.
[18] IEC 60300-3-2 Edition 2.0, "Dependability management - Part 3-2: Application guide - Collection of dependability data from the field," 2004-11-10.
[19] ISO 14224:2006, "Petroleum, petrochemical and natural gas industries -- Collection and exchange of reliability and maintenance data for equipment".
[20] OREDA Brochure, www.oreda.com/, accessed May 7, 2011.
[21] OREDA Participants, Offshore Reliability Handbook, 5th Ed., 2009.
[22] Della Villa, S.A. And Koeneke, C., 2010. "A Historical and Current Perspective of the Availability and Reliability Performance of Heavy Duty Gas Turbines: Benchmarks and Expectations", Proceedings of ASME Turbo Expo 2010: Power for Land, Sea and Air, June 14-18, 2010, Glasgow, UK, GT2010-23182, pp. 1-11.
[23] ANSI/IEEE Std. 762-1987, IEEE Standard Definitions for Use in Reporting Electric Generating Unit Reliability, Availability, and Productivity.

[24] International Standard ISO 3977-9:1999, Gas Turbines – Procurement, Part 9: Reliability, Availability, Maintainability and Safety.

[25] Steele Jr., R.F., Paul, D.C. And Torgeir, R., 2007. "Expectations and Recent Experience for Gas Turbine Reliability, Availability, and Maintainability (RAM), Proceedings of GT2007 ASME Turbo Expo 2007: Power for Land, Sea and Air May 14-17, 2007, Montreal, Canada, pp. 1-8.

[26] CALFED Water Supply Reliability, calwater.ca.gov/calfed/objectives/Water_Supply_Reliability.html, accessed on May 4, 2011.

[27] SIFT Information Security Services. Future of the Internet (FOTI) Project, Reliability of the Internet, www.dbcde.gov.au/_data/assets/pdf_file/0004/75676/FOTI-Reliability-FinalReport.pdf, accessed May 4, 2011.

Chapter 8

MAINTAINABILITY, SUPPORTABILITY AND MAINTENANCE ENGINEERING

8.1 INTRODUCTION

Maintainability, supportability and maintenance support performance are dependability characteristics exhibited in system performance. Maintainability and supportability are essential measures to determine the efficiency and efficacy of maintenance, which is a process to preserve the system in good operating condition. Maintenance is part of the system support infrastructure to sustain dependability performance in system operation. This chapter addresses the maintainability and supportability measures and their relationships in maintenance engineering to reinforce achievement of successful operation in technological systems.

All systems deteriorate in use and degrade in performance with time. Machines do wear out, electronic devices do fail, and software programs for system application do require attention more than once in their lifetime. Maintenance support plays an important role in keeping the system running smoothly when engaged in most business operation.

An overview of recent trends in maintainability and maintenance support reveals substantial changes that include:

- increased use of condition-based maintenance;
- increased outsourcing of maintenance support and the use of long-term service agreements;
- widespread implementation of structured techniques for determining the optimum maintenance program, especially reliability centered maintenance (RCM);
- more sophisticated methods for maintenance optimization;
- emphasis on enhancement and life extension;
- continual cost pressures;
- spare parts agreements with vendors/suppliers and reduction of inventory by operators.

Figure 8.1 presents an overview of maintenance-related factors for maintainability and supportability as they support availability. The inherent maintainability characteristics together with supportability jointly determine the maintenance of equipment and systems. Supportability can be further divided into management aspects and required resources to address maintenance support and logistic support issues.

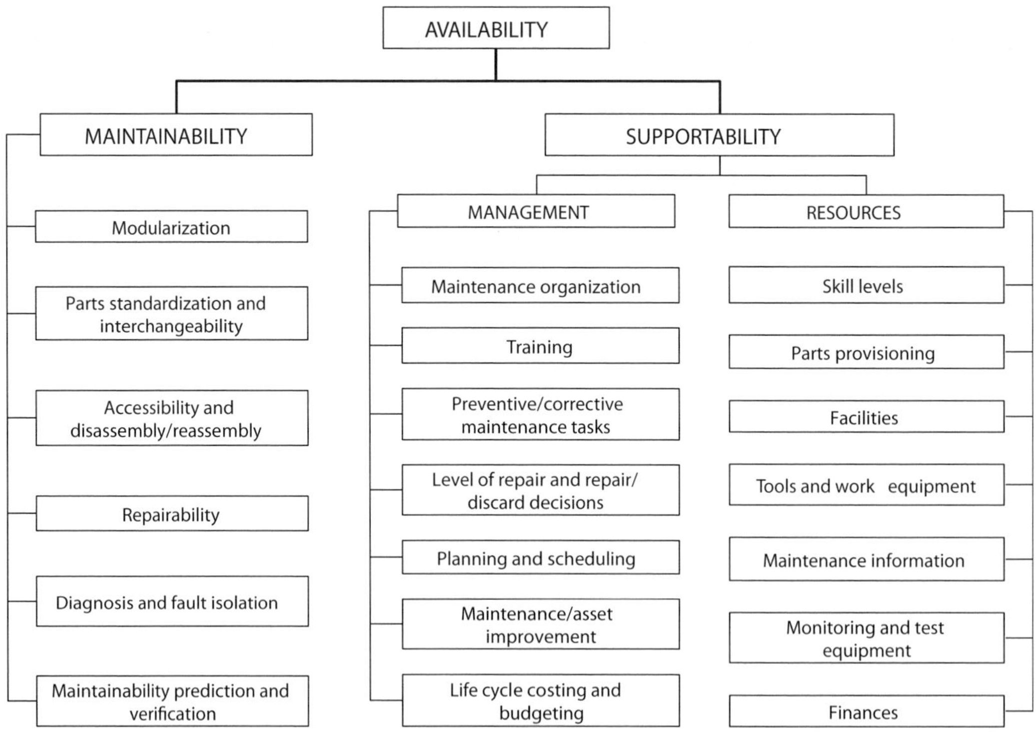

Figure 8.1 Overview of maintenance-related factors

8.2 MAINTAINABILITY

8.2.1 What Is Maintainability

Maintainability is an intrinsic characteristic of a product or system that is determined by its design. It can be defined in general terms as the ability to carry out maintenance actions intended to either retain or restore required performance under given conditions of use and maintenance. It is often described as the ease of doing maintenance. In other words, something is considered to be maintainable if:

- its components are readily accessible for removal or repair;
- the skills required to perform the work are as basic as possible given the technology of the product or system;
- diagnostics, either built-in or externally provided by specialized methods or condition monitoring, are available to isolate faults or identify failures;
- standard tools are able to be used;
- comprehensive maintenance procedures are available;
- spare parts are defined and easily acquired;
- repair time is relatively short compared to the scope of the work;
- reassembly is as foolproof as possible;
- software code is properly organized and documented.

Designing for maintainability has to compete in projects with other priorities such as cost, schedule and design resources. It is critical to projects such as the construction of nuclear power plants [1], where maintainability designs that are inadequate or neglected will result in significant long-term negative consequences with immense difficulty to rectify at later stages. More recent developments have placed much more emphasis on maintainability and the importance of considering human aspects in maintenance.

8.2.2 Maintainability during Design

The main purpose of considering maintainability during design is to optimize the time to perform maintenance, both preventive and corrective, or even eliminate the need for maintenance, thereby also reducing the cost of maintenance. There are several maintainability design objectives [2]:

- modularization;
- parts standardization and interchangeability;
- accessibility and disassembly/reassembly;
- repairability;
- diagnosis and fault isolation;
- maintainability prediction and verification.

Maintenance is greatly assisted by packaging components into functional self-contained units or modules that can be removed and replaced if failure occurs or maintenance action is required. This is now standard practice in electronics with plug-in circuit boards, which can easily be removed and replaced to facilitate testing for failure. It may be desirable to package multiple boards into a larger assembly module that can then be replaced as a whole with environmental protection. These are sometimes called line replaceable units (LRU), intended for quick replacement to maximize availability with module repair being done at a repair facility. The same philosophy is used for mechanical equipment, such as aero-derivative gas turbines and pumps. Electrical equipment interfaces such as wiring assembly connections and circuit board connectors should be handled with care since these may become the most critical failures in connectivity affecting system performance operation.

The use of standard parts with maximum interchangeability minimizes the amount of inventory required. Parts standardization increases the availability of parts for inventory and reduces the cost of inventory holdings. It also reduces the amount of training time for the skills required to perform maintenance and simplifies the maintenance test methods, test equipment and technical manuals. Functional interchangeability implies identical connectors and interfaces to simplify design. The corollary is that standard parts that are not functionally equivalent should not be physically interchangeable. This will avoid human error.

Accessibility is concerned with how components, parts or modules are configured so that more frequent maintenance or failure items can easily be removed and replaced. It takes into account the ability of the maintenance personnel to access the workspace and perform the required maintenance tasks. In some situations, it may be necessary to disassemble other components in order to reach the failed part. Other associated maintenance issues should be considered such as human limitations, accessibility of tools, adjustments, alignment and servicing and the use of lifting equipment. Care must be taken to ensure accessibility in the final installation to ensure that safety of maintenance personnel is not compromised by poor accessibility. For reassembly, it is important to ensure that accidental damage does not occur and that the reassemble methods and practices are foolproof.

Each part or component can be repairable, partially repairable or non-repairable by design. Repairability is enabled by the previous design activities of modularization, parts standardization and interchangeability, and accessibility. The decision to repair or discard a failed item is an economic issue related to maintenance support and must be made by the operating organization.

The most difficult and time-consuming step is often the one to uncover the reason for the failure and isolate its cause, a process usually referred to as troubleshooting or diagnostics. Built-in test equipment (BITE) can effectively facilitate the detection and diagnosis of the fault and identify the failed item, module or component to be changed. Where multiple redundant modules are installed, it can also initiate an automatic switchover with the ability for removal of the faulty module or component during operation so that the system availability performance is not affected. The reliability of the BITE must be very high so that false reports are not made or unnecessary time lost spent on diagnosing and repairing the BITE.

Another solution is to use specialized test equipment that can be used to diagnose a problem or possibly to read a cause code stored in the module at time of failure. This automatic test device may be utilized at the location of the equipment or the module may be replaced and transported to a repair facility for diagnosis. The other option is to install condition monitoring equipment, such as vibration detectors, monitors, and pressure and temperature sensors, which can provide inputs into a condition monitoring system.

It is worth emphasizing that reliability and maintainability are complementary requirements that must be considered together. Increases in reliability may drastically reduce the need for maintainability and the resultant need for maintenance. Conversely, lower than desired levels of reliability place a greater emphasis on maximizing maintainability. Trade-off analyses may need to be done to select the best solution.

8.2.3 How to Measure Maintainability

The basic measurement of maintainability is the mean time to repair (MTTR), also known as mean time to restoration. Since it is easy to confuse between repair and restoration, it is worthwhile to clarify the meaning of these terms. As can be seen from Figure 8.2, although the term "repair" is very commonly used, restoration is actually more appropriate since repair in its strict sense is more restrictive than restoration. It is important to recognize that MTTR covers both preventive maintenance (maintenance actions taken prior to failure) and corrective maintenance (maintenance carried out after a failure has occurred). Some of the maintenance activities are similar, but the basic processes of preventive maintenance and corrective maintenance are different.

The active maintenance time starts with technical delay for both preventive and corrective maintenance. This consists of time for the equipment to be rendered safe for maintenance, including safety checkouts, lockout and isolation, equipment cool down, electrical grounding and setup of test equipment and tools. For preventive maintenance, the main task is then to perform the prescribed tasks of gaining access to the equipment/component, disassembly, component replacement or repair, calibration, testing, balancing or other relevant maintenance activities followed by proper reassembly. For corrective maintenance, the main task is to first diagnose the fault and then to carry out tasks similar to those for preventive maintenance. The final step for both preventive and corrective maintenance is to check out the equipment in order to verify its successful completion. This may require

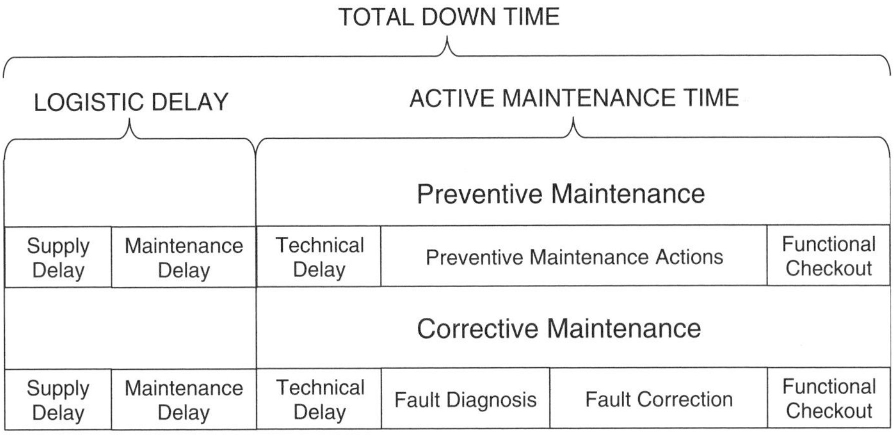

Figure 8.2 Breakdown of maintenance times

checks prior to resumption of operation, such as re-alignment or functional verification after startup of adequate pressure or temperature, or condition-related parameters such as reduced vibration.

In addition to active maintenance time, there are possible logistic delays, especially for corrective maintenance. These logistic delays can be broken down into supply and maintenance delays. Supply delay entails provisioning of spare parts consisting of administrative lead times, procurement lead times, repair of components and transportation times, as well as special tools and work equipment such as lifting equipment. Maintenance delay refers to notification time, waiting for personnel with the necessary skill sets and travel time to the site. In theory, there should not be any logistic delays for preventive maintenance, but in practice they may still occur, for example, if an unexpected problem is found during maintenance. The logistic delay is shown as the first activity, but in fact it may occur at any time in the maintenance process.

It can be seen that the logistic delay is not a direct consequence of the maintenance activity but the result of the available resources. Therefore, it is not considered to be part of the repair time, although it is only the combination of the two that determines the actual availability and downtime of the equipment. The repair or restoration time used for specifying maintainability in terms of MTTR during design is thus the inherent repair time, which excludes all logistic delays. The total time to recover from a failure is also called the time to recovery.

From a mathematical point of view, the simplest measurement is the mean time as it relates to downtime, restoration or recovery times. It can be calculated by dividing the time (calendar or running hours or some other measure such as number of cycles) by the number of maintenance actions. This is however, merely an average, while the actual times will follow a statistical distribution that the mean only serves to hide from view. For situations where the maintenance tasks apply to a simple system and are similar in length and quite repeatable, a normal distribution will be an adequate representation. Where it is applied to a more complex system with a larger variance in the types of tasks, the most common and applicable statistical distribution for maintenance mean times is log normal, which can represent a skewed distribution. Therefore, if the MTTR is specified or determined for a simple system with a limited number of tasks, the variance in the maintenance times will be relatively small. For larger systems with maintenance tasks ranging from routine tasks to major overhauls, the distribution will be skewed accordingly, and it can be concluded that the value of MTTR will be less meaningful. Details on the statistical calculations associated with maintainability have been described by Ebeling [2] and can be found in other textbooks.

An example of how maintainability principles can be applied is provided in Appendix G.

8.3 SUPPORTABILITY

8.3.1 What Is Supportability

Supportability is the ability to provide support to sustain the required system availability with a defined operational profile and specified logistic and maintenance resources. When a system is designed for supportability, this facilitates the enhancement and extension of its service life. It is dependent on the maintainability of the system, combined with factors external to the system that affect the provision of maintenance and logistic support. It can be viewed as the time to provide and use the technical data, skill sets, tools and spare parts so that maintenance can be effected and may be quantified using appropriate performance measures.

Supportability factors are generally determined by the operating organization and often change over the life of the equipment. The exception is the process of integrated logistic support, discussed in the next section where the entire maintenance support concept is set up prior to operation.

Referring to Figure 8.1, the most important concern with the required resources for supportability is the skill sets of the maintenance staff. Basic skills are needed for general maintenance activities. Some advanced maintenance tasks may require training and certification. The provisioning of spare parts and consumables is also important, without which most maintenance activities are severely constrained. Facilities refer to maintenance-related buildings and yards for maintenance and office staff, repair shops, inventory warehousing and storage of tools and work equipment. Tools and work equipment are another important maintenance resource along with specialized monitoring and test equipment such as portable vibration monitors. Maintenance information may not be an obvious resource but is critical for work order management, planning, scheduling and evaluating maintenance history. Money has to be allocated for all of the above resources and constraints in providing finances to support maintenance.

The management of maintenance is equally important and dependent on resources. Unless the number of staff involved in maintenance is very small, it will be organized into different departments or sometimes even added to another department such as operations. These organizational units may be geographically split although modern communications technology has done much to counteract physical distances. There is no ideal maintenance organization, and the critical management issue is to promote cooperation and communication. The maintenance organization and its supervision and management have the responsibility for ensuring adequate training of staff, planning and scheduling of maintenance activities, the determination of a proper and updated maintenance program of preventive and corrective tasks, policy on the level of repair and repair/discard decisions, improvement and life cycle costing and budgeting.

8.3.2 Supportability During Design

In certain industries it is common for maintainability and supportability to be established prior to the system operation stage, a process called Integrated Logistics Support (ILS). This approach is standard for a lot of military equipment and also public transportation. In many ways, ILS is the ideal way to ensure that maintainability and maintenance support are considered early enough to be able to influence the design. In addition, it is possible to understand the availability that is achievable and match that to the level of service expected. Its ultimate goal is to minimize life cycle cost. ILS works well where the equipment is the major component of a system, such as a train or a military aircraft, but it becomes less feasible where there is a facility such as a gas plant that consists of many different equipment and its design is realized by an EPC (engineering/procurement/construction) company whose major role is to integrate existing equipment into a specific production system.

The defense establishment is the primary user of the ILS approach, with two major documents available, one by the U.S. Army [3] and the other as a U.K. Defence Standard [4]. There is an international standard by the IEC as well [5]. The major elements of ILS according to IEC 60300-3-12 are:

- maintenance planning;
- spares/materials;
- support equipment (including tools and test equipment);
- technical documentation;
- manpower and personnel;
- training;
- packaging, handling, storage and transportation;
- facilities;
- software support.

The initial basis for ILS is guided by customer requirements and a general logistic support strategy that lays out both constraints on what will be allowed and the expected framework for satisfying logistics. Reliability and maintainability analyses supported by techniques such as FMEA/FMECA are used to establish the maintenance and support resources required. If these are deemed to be excessive and negatively impact performance, availability and cost, changes to the design can be considered and tradeoffs evaluated. Maintenance activities and the scheduled preventive maintenance tasks are normally analyzed and identified using RCM. A logistic support analysis evaluates different options for meeting logistic support requirements. This includes a level of repair analysis (LORA), which considers where line replacements should take place and provides decisions on whether replacement or repair is more economical. Detailed plans are made for satisfying the various resources of maintenance skills, spare parts, facilities, tools and work equipment and maintenance documentation and information.

8.3.3 Maintenance Support During Operation

Maintenance support activities during operation are very similar to those above for ILS, except that only some are specified during the design phase and the remaining are decided during operation with further changes occurring as the equipment ages or the operation requirements change.

Taking a petrochemical facility as an example, it can be seen that this approach makes sense and is even the most appropriate one. Smaller equipment such as pumps, instrumentation and valves will have a wide range of application and criticality, and the manufacturer is only able to indicate guidance on their operational environment and maintenance program. They will provide spare parts and various degrees of support, but outside of the warranty period they will provide no further guarantees. Many users will be able to and indeed should do their own maintenance analysis to establish the most relevant tasks for that equipment in their physical environment and how maintenance support should be carried out to meet their objectives. For larger and more complex equipment and systems such as industrial gas turbines, the organization may engage the manufacturer or a third party for support such as major overhauls or may have a long-term service agreement that essentially outsources all of the maintenance effort. The various elements of maintenance analysis and implementation that need to be considered are further elaborated in section 8.4.

8.4 MAINTENANCE ENGINEERING

8.4.1 Reliability Centered Maintenance

The foundation for every maintenance activity is an effective set of preventive maintenance tasks that fit the operating context of the equipment. The accepted method for establishing these tasks is reliability centered maintenance (RCM), which was originally developed for the Boeing 747 as MSG-1 (Maintenance Steering Group) and then expanded into other aircraft projects for the military. A seminal report by Nowlan and Heap [6] described the first use of the term "Reliability Centered Maintenance." Its applicability to other types of equipment was quickly recognized by the power generation industry and tested through several pilot projects on nuclear and fossil fuel power plants. From there it has spread to many other industries, with the most influential proponents being John Moubray [7] and Anthony (Mac) Smith [8]. There is an International Standard on RCM by the IEC [9] that provides guidance on the generic processes of RCM.

Since there are different definitions on preventive maintenance, the one used here is shown in Figure 8.3 and includes all maintenance actions carried out prior to failure, whereas corrective maintenance are those performed after a failure. Note that the terms often used in RCM analysis are shown in parentheses.

The RCM process encompasses both equipment and structures, but its application to structures and static equipment such as piping, pressure vessels and infrastructure are commonly known as risk-based inspections (RBI). It is a risk-based approach to prioritizing and planning inspection activities, particularly in the oil and gas industry, usually by non-destructive testing techniques. Guidance for developing, implementing and maintaining a risk-based inspection program for pressure-containing equipment is provided by API Recommended Practice 580 [10].

The use of the term "reliability" in RCM is actually somewhat misleading, and it can be argued that "risk" would be a better substitute [11]. By understanding the likelihood of failure and its potential consequences for failure modes of equipment, RCM determines its criticality and finally what is actually worth doing from a maintenance point of view. If no maintenance task can be found that is effective in reducing the failure rate to an acceptable level, and it is worth doing from an economic point of view, the conclusion is run-to-failure or no preventive maintenance task. If these are still not acceptable, other options such as redesign/modification or different operating procedures are considered. The core analysis process in RCM is a modified FMECA followed by a task selection logic that prefers condition monitoring. As with its use as a design tool, there are different versions and implementations of FMECA within RCM, and therefore there is no standard RCM process used by all practitioners. The basic processes are the same however, and consist of the following basic steps.

- Initiation and planning: determining the objectives and boundaries of the analysis, identifying and assigning resources for the analysis and developing the operating context.

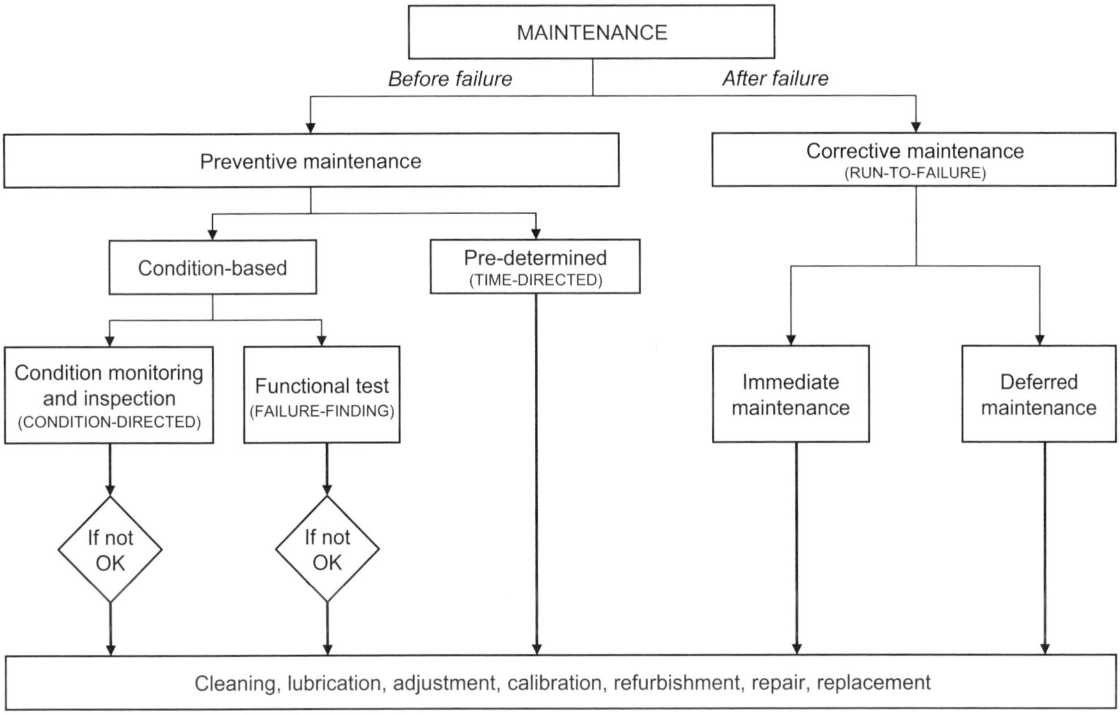

Figure 8.3 Preventive and corrective maintenance tasks

- Functional failure analysis: collecting available failure and maintenance data, partitioning the system functionally and identifying functions, functional failures, failure modes and their effects, consequences and criticality.
- Task selection: selecting the most applicable and cost-effective tasks and their interval.
- Implementation: identifying task details with respect to procedures and resources, rationalizing maintenance intervals and doing initial age exploration where needed.
- Continuous improvement: regular review of maintenance effectiveness against safety, operational and economic targets and improving the maintenance program.

In the ideal situation, RCM is first performed during the design phase where feedback on reliability and maintainability can easily be accommodated. In practice, for many industries, this rarely happens, primarily because it takes considerable effort and the expertise is often not available (e.g., in the oil and gas industry with the EPC companies, whose primary objective is to build the facilities). The RCM effort is only initiated once the warranty period is over, or even later when the organization realizes that its maintenance program is poorly suited to its operation and reliability is suffering.

Even though the basic framework for RCM is consistent with most implementations of RCM, there is a wide range of methods that are used, some being very time-consuming with detailed quantitative techniques such as Weibull analysis, and others using qualitative or semi-quantitative techniques to speed up the process and still achieve results without much available data. Although purely qualitative analysis may gain some benefits, most situations require at least a semi-quantitative approach for both the likelihood of failure and useful measures of consequences. Since acceptance and support by management in addition to buy-in by maintenance staff is crucial to the success of RCM, presenting RCM in the context of risk management is very important [12].

8.4.2 Maintenance Optimization

In addition to RCM, there are significant recent developments in what can generally be called maintenance optimization. It has long been recognized that there is a tradeoff between failures and preventive maintenance actions. In other words, not only is it not feasible to eliminate absolutely all failures (although it may be highly desirable from a safety point of view), it is also not economical. The optimum point for both can be illustrated by the graphs as shown in Figure 8.4. Two versions of the graphs are shown for optimization.

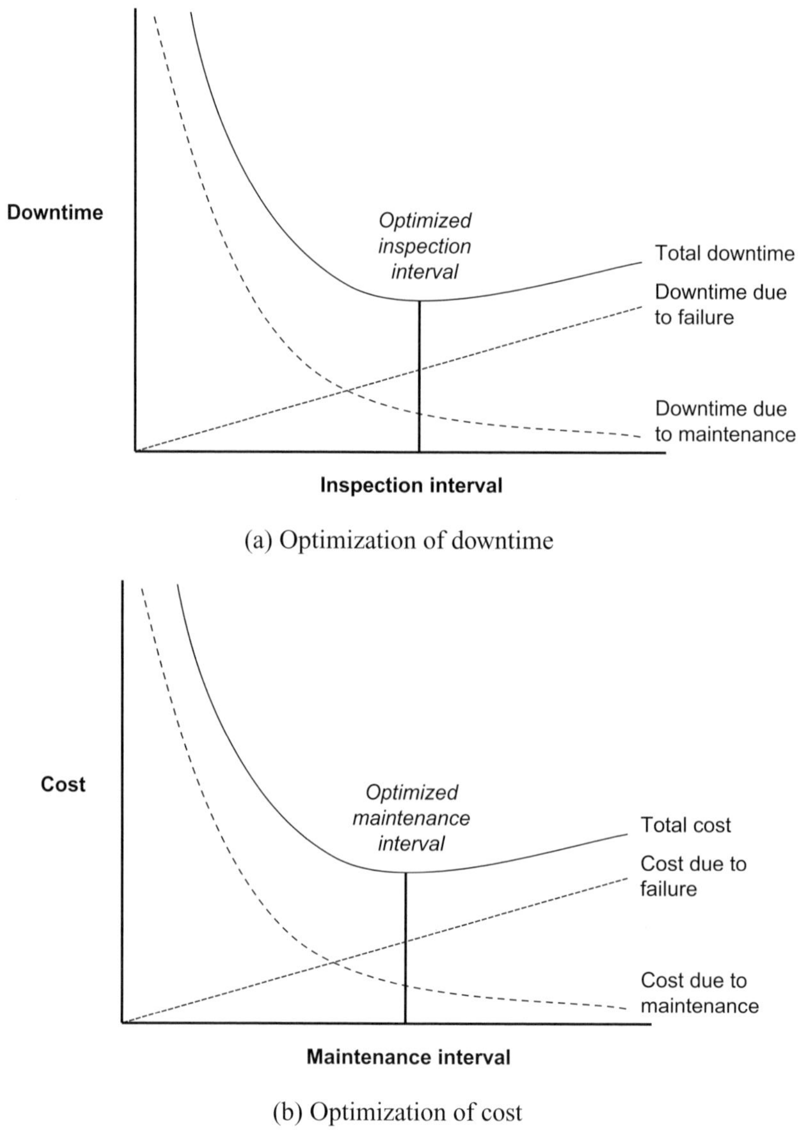

Figure 8.4 Conceptual view of maintenance optimization

In the first case, the objective is to minimize downtime as a function of inspection interval, and in the second it is cost.

For those inclined to dig into the mathematical models, a good reference is by Jardine and Tsang [13]. There are many papers in the literature on this topic, but most are too theoretical to be of practical use. For practical application, software programs are now commercially available to enable easy calculation of the optimum point of inspection or replacement, with options for handling both repairable and non-repairable components. From the failure data provided, a Weibull distribution is calculated and used to model failure rates.

8.4.3 Facility and Equipment Enhancement and Renewal

The process of maintenance optimization can be extended to calculating the optimum age for capital equipment replacement. This requires a life cycle point of view that takes into consideration the tradeoff between acquisition cost and operations and maintenance costs. This is illustrated in Figure 8.5, showing similar graphs to those in the previous section on maintenance optimization. The average annual acquisition cost decreases with time the longer the equipment is in use. The operation and maintenance costs tend to increase with time as the equipment ages. This implies an increasing failure rate, and, if the failure rate is constant, replacement is not economically effective and is actually a waste of money. Fixed costs such as routine maintenance and operations are ones that change little with time and actually have no effect on the optimum replacement age. All costs are referred back to a consistent year to eliminate the effect of discounting or the changing value of money over time (net present value).

There are different scenarios for replacement, including:

- determining the economic life for equipment that is utilized consistently each year;
- establishing economic life for equipment with varying utilization such as base load versus peak demand;
- replacing equipment with more efficient, reliable or functionally superior equipment

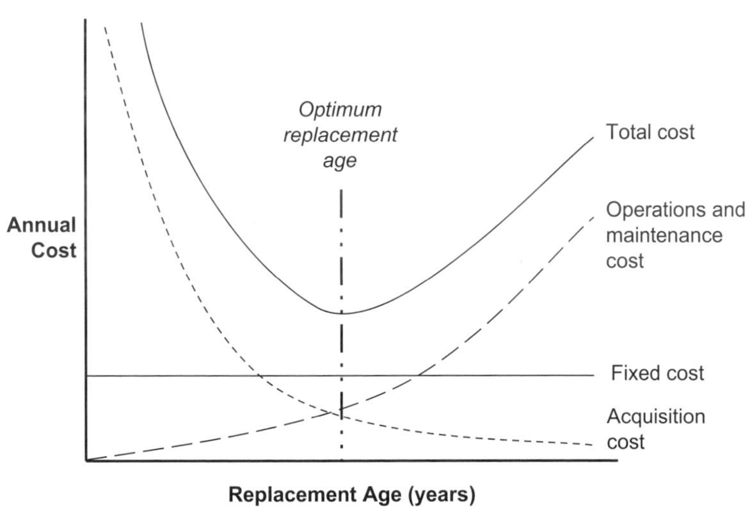

Figure 8.5 Conceptual view of optimum time for replacement

The basic mathematical model for replacement with the objective of minimizing the total discounted cost is

$$C(n) = \frac{C_1(n)}{1-r^n} = \frac{\sum_{i=1}^{n} C_i r^i + r^i(A - S_n)}{1-r^n} \qquad \text{Equation 8.1}$$

where

C(n) is the total discounted cost with replacements of n periods;

n is the age in periods, usually years;

r is the discount factor;

C_i is the operations and maintenance cost for the *ith* period;

A is the capital acquisition cost;

S_n is the resale value of the equipment at the *nth* period.

Detailed descriptions of how this equation can be applied to different scenarios can be found in Jardine and Tsang [13]. A number of software programs are available to facilitate the use of these equations for various scenarios.

Although appearing to be a simple equation, the difficulty is in the details. First, a breakdown of acquisition and ongoing costs must be made. A general cost breakdown structure is shown in Figure 8.6 as an example, but there is no cost breakdown structure that fits all applications. Each one has to be developed to fit a specific situation. Note that the acquisition cost is broken down to include all development costs even though the purchaser of equipment will only see an aggregated price which includes all sunk costs and profits by the vendor.

In addition to knowing the replacement cost, estimates must be made for future maintenance costs, both preventive and corrective. Modeling of failure rate can be done using

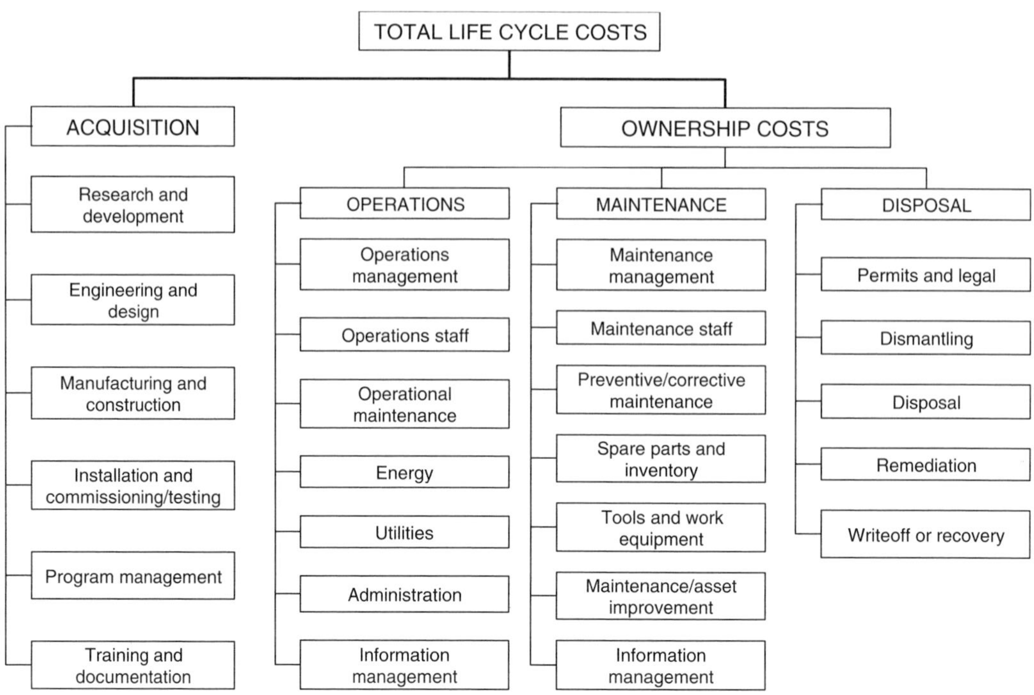

Figure 8.6 Cost breakdown structure for life cycle cost analysis

Weibull data, and the variation in repair times can be estimated by using a log normal distribution. Predicted costs can then be statistically determined by using Monte Carlo simulation method. Some useful examples for pump replacement were analyzed by Barringer and Monroe [14] using a spreadsheet, and later a similar analysis was done by using a freely available software program called RAPTOR [15].

Replacement analysis is particularly useful for managing fleets of equipment such as buses, as discussed in Jardine [16]. Sometimes energy cost is the largest component of life cycle costs, and a technological upgrade will be beneficial. For other cases, downtime is the driving factor, and the cost of lost production greatly overshadows both acquisition and operating and maintenance costs.

8.4.4 Spare Parts Provisioning

The practical and logical consequence of maintenance optimization is spare parts optimization, which has the opposing objectives of ensuring that spare parts are always available when needed against the obvious costs involved in maintaining inventory. If there was only a need for regular replacements, this problem would be more straightforward, but it is complicated by failures and replacements indicated by condition monitoring. The analysis becomes more complex when repairable parts are considered with a further lag time for repair.

Inventory levels are normally established with automatic reorder points and reorder quantities, which may be influenced by economic order quantities for purchasing efficiency. Many users rely on manufacturers or suppliers to eliminate costly inventory through special supply agreements. One key to successful spare parts policies is to properly forecast demand by modeling projected usage based on a combination of failure probability, scheduled maintenance and other possible requirements.

Some models invoke risk-based analysis to assist with spare parts optimization. Presented in Bharadwaj et al. [17] is a risk-based approach to spares inventory management aimed at establishing an optimum level of spares such that financial benefit is maximized within accepted levels of risk and the remaining (residual) risks are clearly identified. In this context, risk is the combination of the probability of a stock-out event and its consequence, where a stock-out is an event when a spare is not available on demand. In the method described here, a risk profile of the spares is obtained by considering the likelihood of the inability to meet the demand for a spare in conjunction with the consequences to meet that demand. This risk profile is then used to find the optimal level of inventory such that financial benefit is maximized given an identified acceptable risk level.

Another approach has been developed [18] in the context of long-term service agreements, where a minimum level of overall plant availability is usually to be guaranteed by the maintenance services supplier. Both parties agree upon this availability threshold, and bonus/penalty clauses are based on this value. The problem is to define both the set and levels of spare parts in a warehouse for one or more installations (pooling) and the location of the warehouse itself. The solution for this problem is based on RBD, Monte Carlo simulation techniques and an associated what-if analysis. One of the deliverables of this optimization process is the ranking of all the system components in terms of their influence on the availability of the whole system.

The different phases of this process are:

- P&ID's and specification analysis;
- gas turbine operating condition analysis;
- reliability data collection and analysis;
- maintainability program definition;

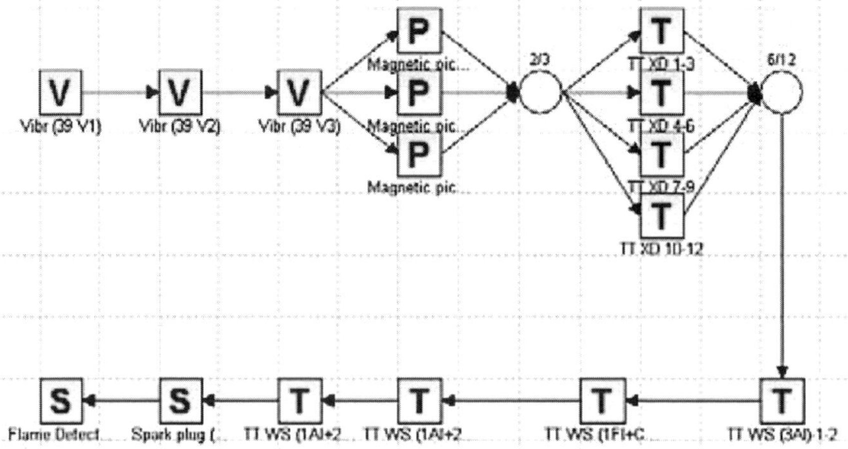

Figure 8.7 Example of a RBD [18]

- RBD of the flange-flange and auxiliary systems;
- Monte Carlo simulation and Availability Importance ranking;
- definition of optimum spare parts list according to the required level of gas turbine availability.

A typical RBD is shown in Figure 8.7, with some spare parts in series and a few in parallel or redundant.

For each block, the failure distribution, repair time and logistic time (waiting hours for a component that has to be replaced but is not in the warehouse) are derived from different sources, including actual field data analyzed through a Weibull analysis. Examples of a failure and repair distribution are shown in Figure 8.8.

The first simulation of the system is carried out supposing an infinite number of spare parts available in the warehouse. This step has two deliverables: the first one is the maximum level of system availability and the second one is the ranking of the components according to their influence on the system availability. For this latter point, the parameter "Availability Importance (AI)" is used. This parameter was invented at Nuovo Pignone and is a formula that integrates the components of availability - MTBF, MTTR and delivery time - with reliability importance relative to the other parts. A partial list is provided in Table 8.1.

Once simulated for the situation with every component available in the warehouse, the optimum spare part list can be defined. Removing the less important components up to a level of spares from the warehouse makes it possible to reach the availability level that was required by the customer.

This is but one example of how reliability is closely tied to spare parts provisioning. The excellent book by Jardine and Tsang [13] further expands on the methods for spare parts provisioning.

8.4.5 Condition-Based Maintenance

There has been a tremendous amount of progress made in condition-monitoring techniques over the past 30 years, particularly in vibration monitoring [19], performance

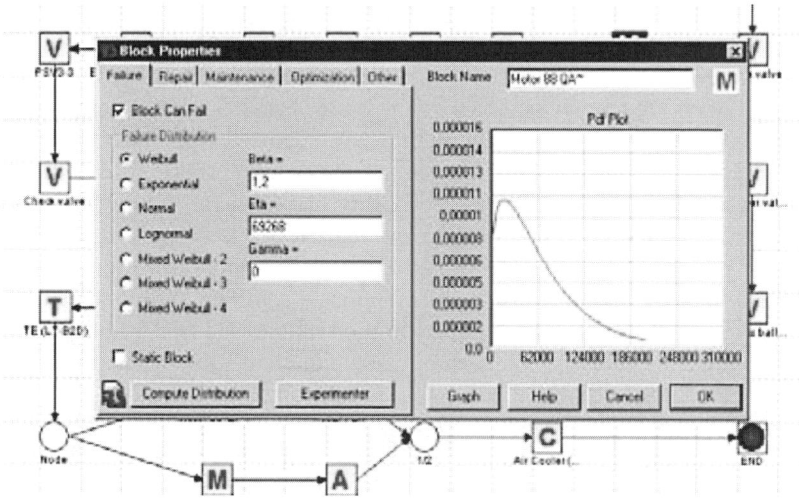

(a) Failure distribution

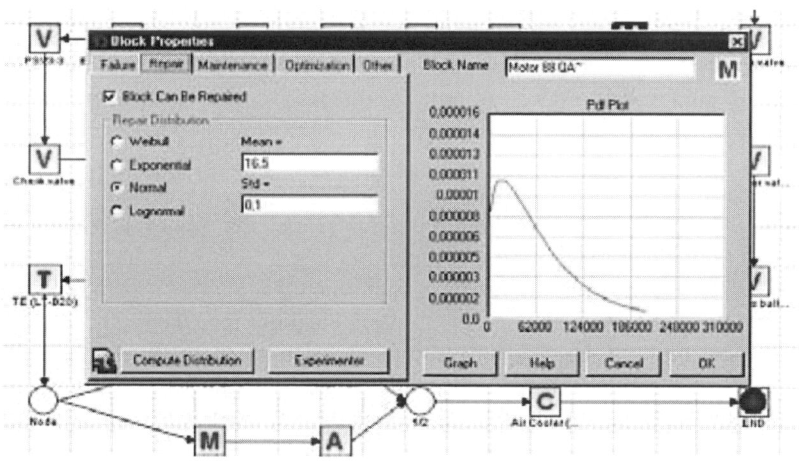

(b) Repair distribution

Figure 8.8 Examples of RBD block properties [18]

TABLE 8.1 Example of a partial list of parts with Availability Index [18]

Subsystem	Component	A.I. Norm.
Flange to Flange	Liner	1,54E-02
Flange to Flange	1st & 2nd stage shrouds +Hcomb (*)	1,13E-02
Hydraulic Oil System	Main pump gear (PH-1)	9,21E-03
Lube/Oil System	Main pump gear (PL-1)	7,12E-03
Flange to Flange	Transition Piece	6,48E-03
Flange to Flange	3rd&4th stage shrouds +Hcomb (*)	4,98E-03
Flange to Flange	#2 journal bearing (*)	4,97E-03
Flange to Flange	#3 journal bearing (*)	4,95E-03
Cooling & Sealing Air	Solenoid valve (20CB-1)	4,22E-03
Flange to Flange	2nd stage nozzle (*)	4,13E-03
Flange to Flange	2nd stage buckets (*)	4,12E-03

monitoring [20], thermography, ultrasonics and general condition monitoring. However, gathering and analyzing data is not sufficient to meet the ultimate goals of the organization such as reliable production, low life cycle cost and acceptable service levels. These objectives are quickly tied back to the basic characteristics of dependability: availability, reliability, maintainability and maintenance support. The value of condition monitoring is obtained by enabling optimization of the decision-making step. This requires analysis that combines the conditional probability of failure (or hazard rate) with the age of the equipment and condition-monitoring data. The statistical procedure that estimates the risk of failing equipment is known as the Proportional Hazard Model, developed originally by Cox [21]. The age-based component relies on Weibull data to model reliability over time. It has been applied to vibration monitoring by Jardine and Banjevic [22] and subsequently to a variety of monitoring situations implemented in a software program called EXAKT [23].

In addition to improvement in the various condition-monitoring technologies, the advent of advanced and integrated communication has enabled the development of remote health monitoring, which brings together data gathering, automated processing and analysis, and access to technical experts in a central location. Some of these remote health-monitoring systems are set up and managed by operators, while others are provided by vendors of major equipment such as gas turbines.

An example of a remote health-monitoring system is for monitoring locomotives [24]. The reliability of rolling stock equipment determines how large a fleet must be to maintain the required service level. In order to increase service without increasing capital expenditures for new rolling stock equipment, a transit agency must increase reliability of the equipment. The optimal fleet size will cost less to purchase and less to maintain than one that is overcapitalized. Locomotive reliability is critical to ensuring the most effective service for the public. Improving reliability potentially allows a transit agency to increase service without major capital expenditure.

The primary purpose of the remote monitoring system is to enable all maintenance personnel to have ready access to information describing the health and location of all electric locomotives. This requires interfacing with train subsystems, collecting maintenance event files, and analyzing the event data. The data collected is analyzed to improve system operational reliability, which will reduce the number of in-service failures of train equipment and the associated costs.

The system is called the MEAP™ system, and its architecture is shown in Figure 8.9. Onboard each locomotive or train set is a computer that connects to the built-in sensor suite and expert system. The expert system onboard the train generates faults and events based on logic rules. The faults are downloaded to the MEAP™ onboard computer where faults are evaluated against a list of select faults. If a select fault occurs, then the onboard computer immediately sends a message notification to a small group of key personnel. Normally, the system downloads are simply stored on the train and then transmitted to wayside once every hour. At the wayside server, the events and faults are stored in a relational database where further analysis and reports are generated. The wayside server allows browser-based access to the database by personnel with access privileges.

Regular reports are provided to maintenance personnel that help them perform necessary maintenance based on the actual condition of the material and systems. The cost and safety benefits of repairing items prior to failure rather than on a strictly time-based schedule are self-evident. MEAP™ enables Amtrak to fix train systems prior to failure. The system was conceptualized and developed as a solution to identify the root cause of select locomotive failures that were all related to the need for remote monitoring and predictive maintenance efforts.

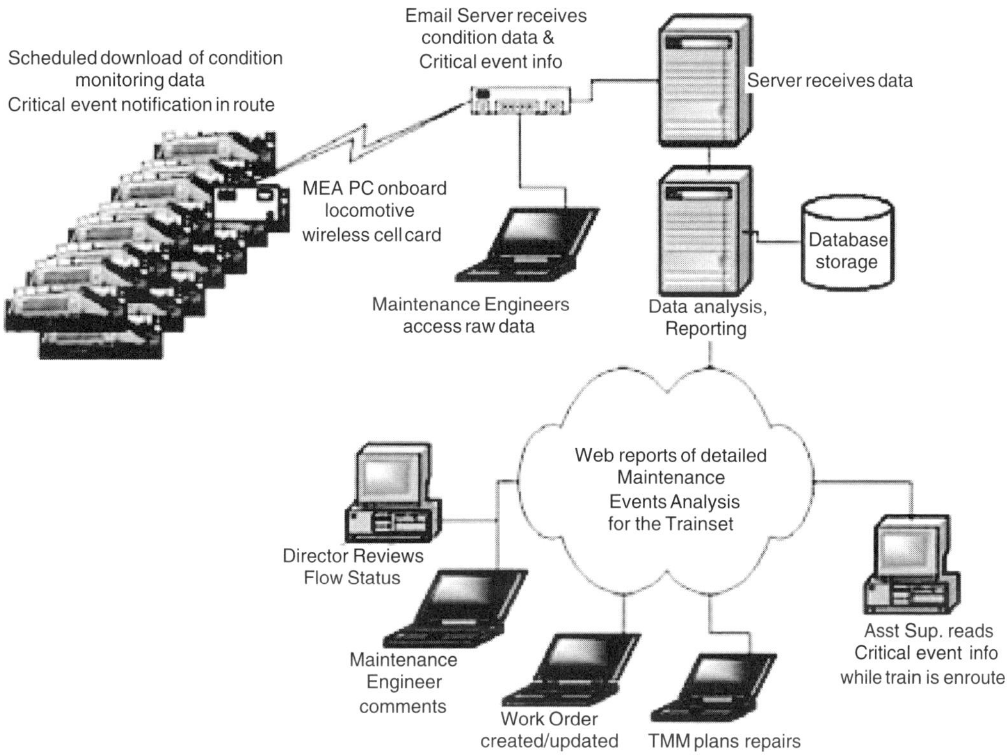

Figure 8.9 Architecture of a locomotive remote health monitoring system [24]

Condition-based maintenance (CBM) is a term for methods and methodology that, based on the actual condition and predicted future use, in theory allows maintenance to be performed at the best possible date for each component [25]. Typical applications include components that do not fail instantaneously but deteriorate in a quantifiable and, preferably, observable way over a period of time. An early failure indication enables the user to avoid the consequences of an unexpected breakdown. Early signs can be detected by the use of diagnostic equipment and/or by analytical calculations that take the actual service conditions of the equipment into account—so-called prognostics. Equipment operators increasingly use condition-based maintenance instead of, or in addition to, scheduled maintenance to reduce lifetime equipment operating costs.

However, merely having diagnostics and/or prognostics is not enough to derive all or even most benefits from CBM. In order to maximize the benefits from CBM for the enterprise, it is as important to focus on the aftermarket supply chain (i.e., the back-end of the process, including maintenance) as it is to develop better data gathering, diagnostic and prognostic techniques. Furthermore, it has been shown that optimizing the value chain results in lower costs and higher availability.

In practice, better knowledge of the actual maintenance requirements of the components of a machine should be reflected in maintenance intervals dynamically adapted to the current condition and predicted usage of the components. For gas turbines, predictions regarding future component condition and lifetime are based on factors such as load profile, quality of fuel, ambient temperature, particle levels and so on. To maximize the benefits of CBM, maintenance also needs to be re-planned whenever the current condition and future predictions, and hence the future maintenance intervals, change significantly.

With a growing emphasis on life cycle cost reduction for capital equipment such as gas turbines, equipment operators are increasingly investigating potential cost reductions. One way to minimize life cycle costs and maximize earnings is to optimize maintenance according to a customer's specific condition. Achieving an optimal or near-optimal maintenance plan, which minimizes the total cost, depends on the availability of diagnostics and prognostics, as well as on maintenance planning technologies. Successful planning also involves developing accurate and comprehensive user knowledge, in part because solutions engineered for one user can then be adapted to the specific needs of other users [26].

The maintenance process adopted in Bohlin et al. [25] combines condition information with the requirements of the operator. This is done in order to carry out maintenance as efficiently as possible, thus ensuring that potential short-term profits will be evaluated in an overriding life cycle cost perspective. To manage all relevant information, a preventive maintenance optimization tool (PM-opt) has been developed. PM-opt plans preventive maintenance for complex technical systems and maximizes earnings for a system operator. This is done through the use of an advanced prognosis process and input from an operator regarding operation profile, ambient conditions and financial data such as production value and standstill costs. This information is processed in PM-opt, generating an optimized preventive maintenance schedule adapted to an operation-unique situation, hence maximizing profit. The process is also supported by an advanced diagnostic tool to further increase reliability and availability.

The goal is to provide operating conditions that will increase availability with predictable scheduled maintenance, based on condition-monitoring assessment, with little or no downtime during deployments. Any changes in, for example, operation profile will instantly affect the preventive maintenance. Also, if an unplanned opportunity occurs, maintenance can be rescheduled if it is profitable to use this "new slot." PM-opt can deal with these situations and re-optimize maintenance if this is financially justifiable for the operator of the gas turbine.

The approach was evaluated using a real-world scenario in the oil and gas business. The gas turbine used for evaluation (a Siemens SGT-600) consists of 17 components with individual maintenance schedules. For some of the components, maintenance deadlines were determined from predictive lifetime analyses using a prognostics tool. Other components in the gas turbine were required to be maintained according to their original maintenance schedules.

The standard maintenance schedule used for comparison is based on a combined calculation of equivalent operating hours and cycles (EOH/EOC). The critical components in the gas generator stage (combustion chamber, burners, compressor turbine guide vanes and blades) were modeled and evaluated in a prognostics tool to determine suitable inspection intervals. However, at the time of writing, lifetime data was not available for the combustion chamber and burner components. Therefore, the original maintenance deadlines were used for these components. For the critical components, the relative increases obtained from using the prognostics tool, compared to the standard maintenance schedule, are shown in Table 8.2. In Table 8.2, replacements marked "n/a" were not present in the EOH/EOC schedule and are therefore not included in the prognostic schedule. Replacements marked "n/n" were not necessary in the prognostic schedule since the estimated component lifetimes were significantly higher than the standard lifetime of the turbine.

In the evaluation, four different maintenance strategies for two situations in two different scenarios were compared. The scenarios were set up to simulate maintenance planning for a new gas turbine and for a gas turbine with a non-empty maintenance history, in the cases where a seasonal stop was either absent or present. The four maintenance strategies were set up to simulate either the absence or the presence of advanced prognostics and/or maintenance optimization, respectively.

TABLE 8.2 Increases in maintenance intervals obtained from the prognostics tool [25]

	Prognostics	
Component	Inspection	Replacement
Guide Vanes Stage 1	88 %	*n/a*
Guide Vanes Stage 2	151 %	*n/n*
Combustion Chamber	0 %	0 %
Burners	0 %	0 %
Blades Stage 1	101 %	*n/n*
Blades Stage 2	41 %	*n/n*
Blades Stage 3	245 %	*n/n*
Blades Stage 4	72 %	*n/n*

Table 8.3 shows results for a simulated brand new gas turbine. Since a new gas turbine should have an empty maintenance history, all component lifetimes are set to their predicted values. Where lifetimes are obtained from the standard maintenance schedule for the gas turbine, the necessary maintenance time points are already synchronized according to the maintenance packages determined for the original maintenance schedule. This makes the planning of maintenance packages easier, especially in the beginning of the contract.

In Table 8.3, the rows "EOH" and "Progn" give the results for schedules obtained by planning maintenance activities at the last possible date, given by the maintenance intervals obtained from standard EOH calculations and the prognostics tool, respectively. This corresponds to the theoretically best possible case from a direct maintenance perspective (in other words, not considering the effect on the customer) and is obtained without using any minimization of production losses. On the other hand, the rows marked "EOH opt" and "Progn opt" provide results for schedules obtained by optimizing maintenance with regard to both maintenance and customer (loss of production) costs. Results are given for two scenarios: one where there is an already preplanned production stop of three weeks during the summer ("with seasonal stop"), and one where production is assumed to continue throughout the year, without advantageous maintenance opportunities ("without seasonal stop"). In this second case, maintenance can be freely placed. However, since maintenance stops always incur a significant cost, more focus must be placed on grouping maintenance activities together in suitable packages.

Results are reported in the form of availability ("Avail"), maintenance costs ("Maint index") and productive days spent doing maintenance ("DT days"). Availability is computed as the number of productive days when maintenance is not done (not including seasonal stops, which are assumed to be unproductive) divided by the total number of productive days for the maintenance contract. Direct maintenance costs include material and work costs. Maintenance costs are expressed using an index. In it, 100 represents the cost of

TABLE 8.3 Results of maintenance optimization for a new gas turbine [25]]

	With seasonal stop			*Without seasonal stop*		
	Avail %	Maint index	DT days	Avail %	Maint index	DT days
EOH	***97.60***	***100***	***131***	***97.60***	***100***	***131***
EOH opt	99.99	109	0.42	98.15	120	101
Progn	98.20	61	98	98.20	61	98
Progn opt	100.0	62	0	98.81	75	65

doing maintenance according to the maintenance intervals computed using the EOH/EOC calculations, the current state of practice. The highlighted line in Table 8.3 is the reference case corresponding to maintenance being done at the latest possible date.

Better lifetime estimates had a significant result on maintenance costs, availability and downtime. Adding the optimization of maintenance (with regard to both maintenance costs and production losses) yields even better results and increases direct maintenance costs slightly. This is natural, since production losses in this case are very costly and optimization is done with regard to both loss of production costs and direct maintenance costs. Table 8.2 also shows that for a schedule with no advantageous opportunities, downtime can be reduced by more than 50% using PM-opt and a prognostics tool.

The PM-opt software is currently in use for gas turbine maintenance planning at Siemens Industrial Turbomachinery AB. Calculations show a possible increase in availability of 0.5% to 1% in practice. Even when better lifetime predictions are not available and maintenance intervals are kept at the same length as before, significant reductions of preventive maintenance downtime are possible. Downtime reductions of approximately 12% due solely to the improved planning of preventive maintenance activities are expected in the general case.

8.4.6 Risk Assessment for Pipelines

Even though there are certain similarities, there are fundamental differences in maintenance approaches between machinery/equipment and static or fixed equipment. Static or fixed equipment can be infrastructure such as roads, buildings or bridges. For the oil and gas industry, the most important are pressure-containing equipment such as pressure vessels, piping and pipelines. For machinery/equipment, the primary mode of maintenance is component repair or replacement supplemented by condition monitoring while for static equipment or structures, inspection is the main maintenance activity. As well, the frequency of maintenance for static equipment is much less while the consequence can be much greater. These differences dictate that a risk assessment approach is more common for static equipment.

The use of risk assessment techniques is well developed for liquid and gas pipelines. In order to promote a positive view, this is often referred to as integrity management. The emphasis is especially on mitigating consequences related to public safety when a leak or rupture occurs. It is now commonly required for pipeline companies to prepare integrity management plans (IMP). The objective of the integrity management program, executed in conjunction with other general maintenance tasks, is to make the likelihood of a pipeline failure so remote within the life of the pipeline system that the risk of failure can be considered and controlled [27].

The principal aims of pipeline integrity management are to efficiently allocate operation and maintenance resources to assure:

- the safety of personnel (members of the public, company employees, and contractors);
- environmental protection; and
- system reliability.

If these aims are met, the operating cost and financial impact will be minimized, return on investment will be maximized, and the operating company's image will be maintained. Risk assessment in integrity management planning has the following objectives:

- prioritization of pipelines and pipeline segments for scheduling of physical integrity assessment and other risk mitigation actions;
- demonstration of risk-reduction benefits;

- determination of the most effective mitigation measures for the applicable threats;
- assessment of the integrity impact of modified integrity assessment/inspection intervals;
- assessment of the use of or need for alternative integrity assessment methodologies.

Risk assessment methods can be categorized broadly into:

- quantitative risk assessments (QRA) in which the probabilities of failure are expressed as classic probabilities (a number naturally occurring between 0 and 1.0) and the consequences are reducible to a common measure (e.g., number of casualties or a monetary value); thus, the risk value is also quantifiable (e.g., individual risk or monetary risk) and has physical meaning;
- non-quantitative risk assessment (non-QRA) methods in which likelihood (term used to distinguish the quantity from classic probability) and consequence are represented by arbitrary indices that only have meaning in relation to one another (i.e., giving a relative risk score which on its own has no physical meaning).

The QRA methods can be further sub-categorized as:

- deterministic QRA (involves the modification of baseline failure rates using deterministic modification factors);
- reliability-based QRA (standard reliability methods are used to account for uncertainty in parameters either by direct input of probability density functions or by simulation).

A quantitative pipeline risk assessment methodology is based on a set of failure probability and failure consequence models that are applicable across a wide range of pipeline network types and transported products, where failure is defined as a loss of containment event.

The failure probability and failure consequence models comprise both fully quantitative and semi-quantitative inputs. The individual threat models produce failure rates (per km-year) that are converted to annual probabilities of failure. These models incorporate historical failure data from the subject system or a peer pipeline system scaled by model-based factors and indices that are functions of the pipeline system attributes and system operating conditions.

The consequence models are a combination of quantitative models (e.g., volume release models) and correlations of various parameters with consequence severity estimates based on the industry-accepted methods in API RP 581 [28]. The models quantitatively evaluate the impact of pipeline loss-of-containment events. The resulting consequence quantities can be mapped to a common severity scale using an industry standard approach such as the Integrated Risk Assessment Process [29] to allow comparison and consolidation across different consequence types.

The resulting failure frequencies and consequences for pipeline segments, pipelines and systems of pipelines are used to evaluate risk quantities. These risk quantities can be compared against tolerable levels or shown on a risk assessment matrix (usually the preferred approach) and used to drive integrity management program actions. Due to the quantitative nature of the risk assessment models, the probabilities of failure, consequences of failure and risk of failure results produced have a physical meaning (unlike qualitative and semi-quantitative methods) and can be compared across threats, across pipelines and even across different system types. An example risk matrix is shown in Figure 8.10.

Potential Severity	Service Interruption	H&S Effect	Environmental Damage	<1E-06 Virtually impropable and unrealistic (Improbable)	1E-06 to 1E-04 Not anticipated to occur (Remote)	1E-04 to 1E-03 Occurrence considered rare (Rare)	1E-03 to 1E-01 Expected to occur at least once in 10 years (Probable)	>1E-01 Occurs several times a year (Frequent)
1 Slight	$100K to <$1M impact	Slight injury requiring first aid only	Only minor cleanup required	L	L	L	L	M
2 Minor	$1M to <$10M impact	Minor injury or restricted work only (reversible)	Clean up 100% achievable, no latent effect to environment	L	L	M	M	M
3 Moderate	$10M to <$100M impact	Major injury i.e., single LTI with permanent disabilities	Localised environmental damage	L	M	M	S	S
4 Major	$100M to <$1B impact	Single fatality or multiple LTI with permanent disabilities	Major environmental damage	L	M	S	S	H
5 Severe	>=$1B impact	Multiple fatalities	Massive environmental damage	M	S	S	H	H

Figure 8.10 Example of a risk matrix for a pipeline [27]

An example of the development of a pipeline integrity program can be seen for a 14,000-km gas transmission system [30]. Prior methods for determining maintenance expenditures and their relative priority were based on ground and aerial surveys as well as class location surveys. Worst-case scenarios were often selected for determination of the consequences. Probability of failure was not numerically evaluated in the process. Sites were selected for direct examination based primarily on soil survey results. Segments were selected for inline inspection based primarily on pipeline age, diameter, class location and failure history.

Considering the foregoing concerns, past maintenance practices and emerging trends in the industry, the pipeline company recognized the benefits of developing a system integrity program that implemented a proactive strategy using both probability and consequence as the bases for inspection and maintenance decisions. Completion of this initiative was expected to provide a maintenance program that was entirely risk-based. The risk management strategy required that two fundamental steps be taken: 1) the completion of a qualitative pipeline risk assessment for risk-ranking purposes; and 2) the subsequent completion of quantitative assessments for the purpose of identifying actual risk levels and determining the appropriate maintenance actions.

The quantitative risk assessment tool PIRAMID™ was used to calculate the failure rate, failure consequences and risk level along each pipeline and to facilitate the maintenance optimization process. The software calculates the level of risk associated with a specific pipeline segment and quantifies the expected reductions in the risk level that would result from carrying out various possible maintenance actions. It then develops cost comparisons for candidate maintenance actions in which the total annual cost for a line segment is presented. This information can then be used as the basis for integrity maintenance decisions and to develop maintenance plans for each segment in the pipeline system. The basic premise used in developing these plans is to ensure that acceptable safety standards are met and maintained (i.e., safety risk is kept at or below tolerable levels) at the lowest possible total cost.

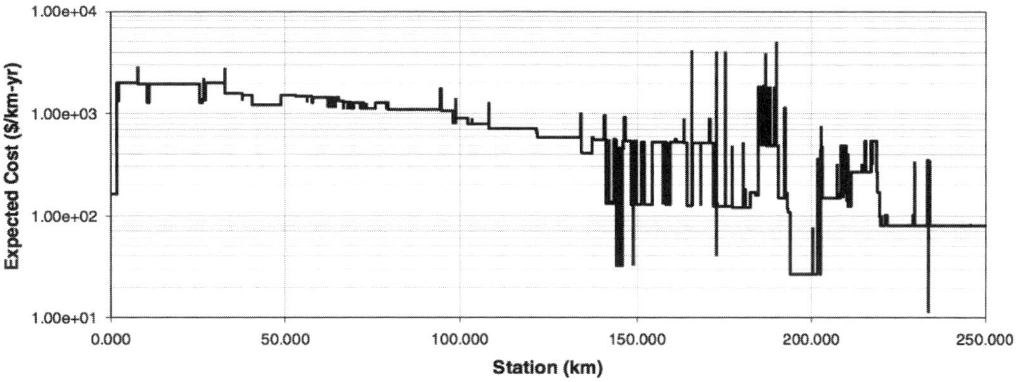

Figure 8.11 Example of an expected cost risk profile [30]

Risk profiles were used to display variation in risk along the length of a pipeline segment and identify high-risk areas. An example profile of the variation in financial risk along an example pipeline segment is shown in Figure 8.11.

The concept of an individual risk ratio (IRR), the ratio of calculated individual risk to the tolerable individual risk, was used to facilitate evaluation of life safety risk along the length of a pipe segment that includes variations in tolerable risk level. Where the IRR is greater than 1.0, the calculated individual risk exceeds the tolerable level at that location. If the IRR exceeds a value of 1.0 at any point along its length, it is said to violate its individual risk constraint.

The first step in developing the maintenance plan was to rank all segments with regard to the maximum IRR along its length. The second step was to identify segments that do not meet tolerable individual risk criteria (i.e., segments with a maximum IRR greater than 1.0). The third step was to select candidate maintenance scenarios that address the dominant failure threat (e.g., corrosion, stress corrosion cracking, equipment impact).

Selecting an optimal integrity maintenance action is a process of optimization under uncertainty. A comprehensive approach for solving this type of problem is decision theory, which provides a systematic and consistent method to evaluate alternatives and make optimal choices. One recognized decision-making approach is constrained cost optimization. This approach assumes that life safety risk, expressed in individual risk, can be treated as a constraint. The integrity maintenance option associated with the lowest expected total cost over the period of interest, without violating the life safety constraint, is considered the preferred option.

To evaluate maintenance alternatives, the total expected cost for a pipeline segment was calculated for existing conditions and for each candidate maintenance option. The total expected cost is the sum of two components, the average annual expected cost and the amortized maintenance cost. Average annual expected cost is the sum of the failure-related costs, in present day dollars, for each year during the period of interest divided by the total number of years. This provides the average annual failure related cost over the specified time period. Amortized maintenance cost is the initial capital cost of the maintenance action amortized over the period of interest. This provides the annual maintenance-related cost over the specified time period.

Calculating the total expected cost for a range of time periods (i.e., one to fifteen years) for a scenario, the cost curve will generally start at a high value, decrease with time until it reaches a minimum, and then increase again. The high initial value reflects the initial cost

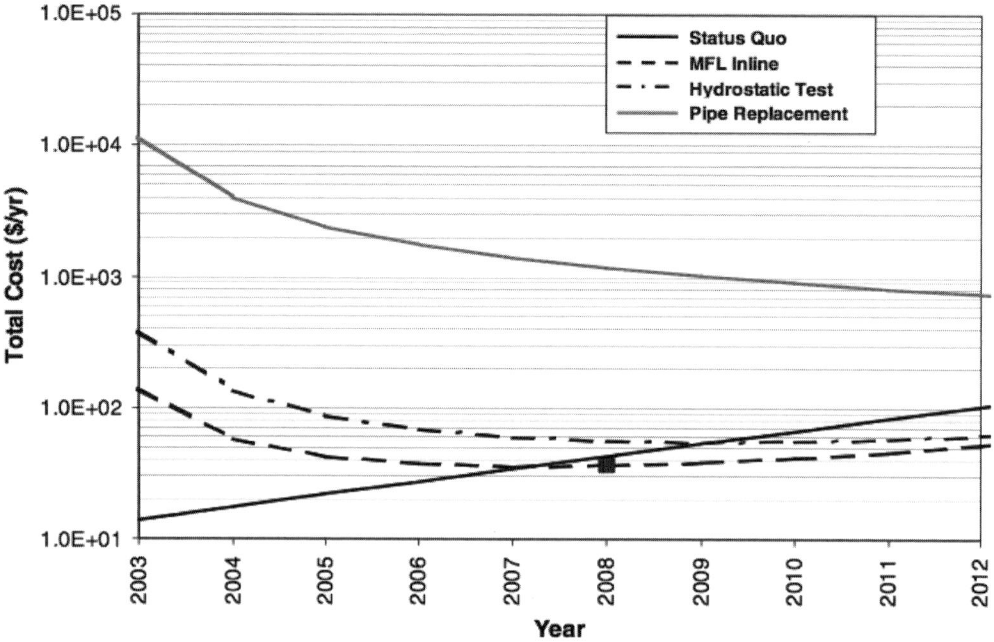

Figure 8.12 Example of cost optimization without safety constraint [30]

associated with implementing the scenario. This value decreases with longer periods of time because, as the useful life of the scenario increases, the initial cost is amortized over a longer period of time. At a certain time point, the rate of increase of the risk-related costs with time will exceed the rate of decrease of the initial cost and the total cost will start to increase again. The optimal useful life (or optimal time to next maintenance event) for a given scenario is the lowest point on the total cost plot for that scenario. For the example shown in Figure 8.12, the optimal useful life is 6 years (2008) for the MFL inline inspection option and 7 years (2009) for the hydrostatic test scenario. Since the status quo does not involve any initial expenditure for maintenance activities, the corresponding cost curve does not have the decreasing portion.

Cost optimization analyses were carried out for each line segment found to have a maximum IRR greater than 1.0. The recommended maintenance plan for each segment was the minimum cost option that at the very least meets the criteria for tolerable individual risk level. The cost optimization analysis resulted in recommended maintenance actions that were implemented through a multi-year program, since coordination of the high number of pipeline outages was seen as possible only with several windows of opportunity not normally available in a single annual cycle.

8.5 MAINTAINABILITY INTEGRATED WITH RELIABILITY AND AVAILABILITY

It must now be apparent that there is a close integration between maintainability and maintenance support, and availability and reliability. Figure 8.13 illustrates this integration.

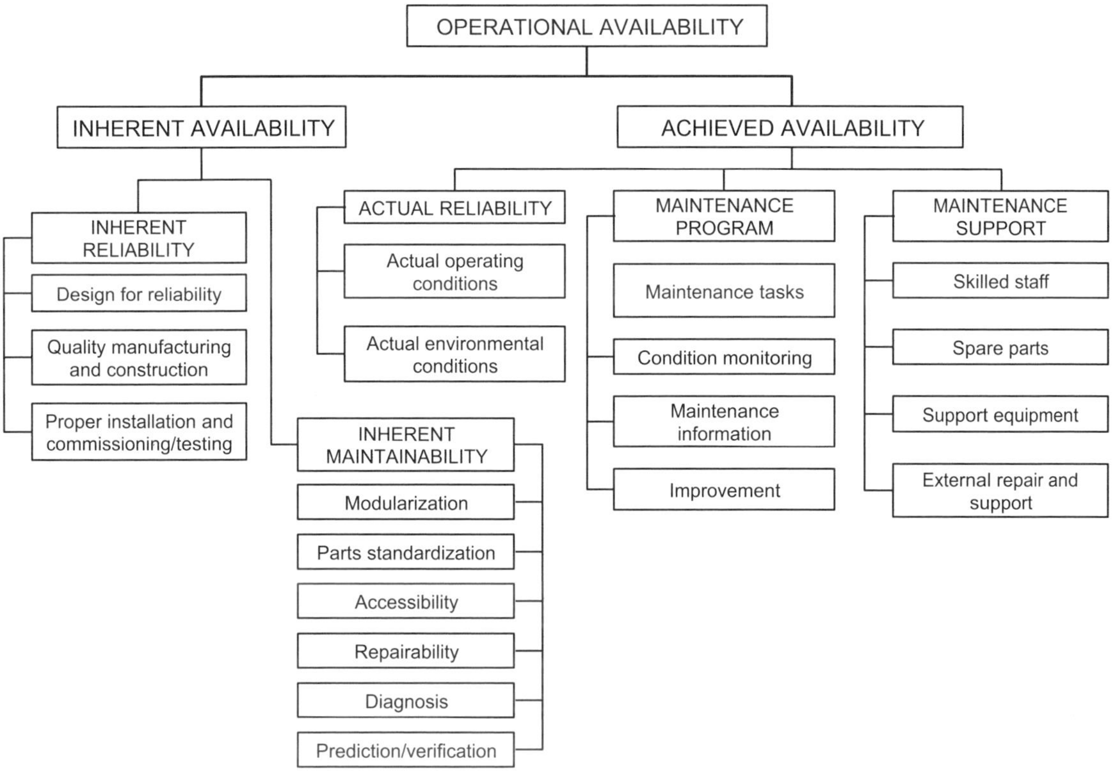

Figure 8.13 Integration between dependability characteristics

It is useful to recognize again that equipment will have inherent reliability and maintainability that results in inherent availability. The inherent reliability is dependent on design efforts to build reliability into the design by methods such as redundancy, use of highly reliable materials and components and setting environmental and operational constraints that enable long equipment life. Inherent design reliability has to be followed by quality manufacturing and proper installation. As discussed in section 8.2.2, inherent maintainability is the product of modularization, parts standardization and interchangeability, accessibility and disassembly/reassembly, repairability, diagnosis and fault isolation, and maintainability prediction and verification.

Under the heading of achieved availability are some factors that come into play once equipment becomes operational. The first item relates to reliability since there are factors that may decrease the inherent reliability, in which case actual achieved reliability will be less. A major influencing factor that will reduce reliability is operating the equipment outside of design specification, such as running the equipment at higher than design load or with a large number of starts and stops. The other significant factor is utilizing the equipment under harsh environmental conditions, such as high or low temperatures, dusty or polluted conditions, or any conditions outside of those specified. Experience has shown that underutilizing equipment is also detrimental to reliability.

The second important influencing factor for actual availability is the maintenance carried out on the equipment. This can be broken down into the type and extent of the maintenance program. This consists of the preventive maintenance tasks performed, the amount of condition monitoring being applied, the quality and amount of maintenance and asset performance information collected and analyzed and efforts made for improvement. The other aspect of maintenance is the maintenance support provided with the major elements being skill maintenance staff, available spare parts, support equipment and use of external maintenance and repair. It should be mentioned that maintainability is basically inherent in design and is difficult to change once operations are started, so it does not therefore appear as a factor under actual availability.

Another view of the relationship between maintainability and reliability can be made by correlating MTBF with MTTR. It should be noted that this could be interpreted in terms of outage as well as maintenance time. Figure 8.14 shows this relationship in the form of four quadrants. Reliability is a function of MTBF, and availability is the combination of MTBF and MTTR.

Prior to establishing this relationship, some judgment will have to be made as to what values are appropriate for the equipment being compared. The best situation is where the MTBF is high and the MTTR is low; in other words, when there are few failures and they are repaired quickly. For the quadrant where MTBF is high and MTTR is also high, reliability itself is not likely to be an issue but emphasis should be placed on reducing the repair times. Where both MTBF and MTTR are low, improving reliability should be the focus. The worst case is low MTBF and high MTTR, where both reliability and repair times have to be addressed.

A word of caution is necessary. Since these are average values, they hide the underlying distributions. If used for components or small equipment, the variance in repair times should be small, whereas for large equipment or systems there may be a few very long repair times that will skew the average repair time. For MTBF, there may be cases where there are very short times to failure that may be the result of poor maintenance quality. If these are instances that are not expected to be repeated, it may be premature to draw conclusions and it may be worthwhile to remove these data points.

In Appendix H, an example of a group of steam turbines in an oil refinery is used to illustrate an analysis where reliability, availability and maintainability each played a part in determining the effectiveness of this equipment and identifying the pieces that needed improvement.

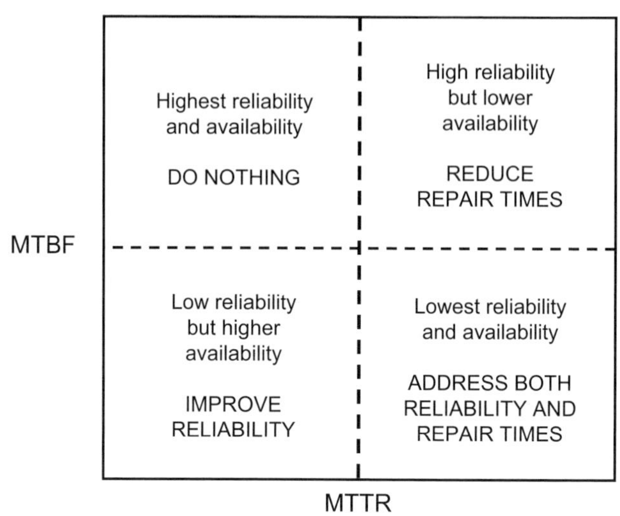

Figure 8.14 Relationship between MTBF and MTTR

REFERENCES

[1] Seminaraa, J.L. and Parsons, S.O., 1982. "Nuclear power plant maintainability," Applied Ergonomics, Volume 13, Issue 3, September 1982, Pages 177-189.
[2] Ebeling, C.E., 2005. "An Introduction to Reliability and Maintainability Engineering," Waveland Press, ISBN 1-57766-386-1
[3] Unknown, 2009. "Integrated Logistics Support," Army Regulation 700-127, April 29, 2009.
[4] Unknown, 2010. "Integrated Logistics Support. Requirements for MOD Projects," Ministry of Defence Defence Standard 00-600, Issue 1 Publication Date 23 April 2010.
[5] Unknown, 2011. "Dependability management - Part 3-12: Application guide - Integrated logistic support," IEC 60300-3-12 Ed 2.0, 2011-02-17.
[6] Nowlan, F.S. and Heap, H.F., 1978. "Reliability-Centered Maintenance," Report AD-A066-579, December 29, 1978.
[7] Moubray, J., 2004. "Reliability Centered Maintenance," Industrial Press, 2 Edition, ISBN 0831131462, April 8, 2004.
[8] Smith A.M., 1993. "Reliability-Centered Maintenance," McGraw-Hill, ISBN 0-07-059046-X.
[9] Unknown, 2009. "Dependability management - Part 3-11: Application guide - Reliability Centred Maintenance," IEC 60300-3-11 Ed 2.0, 2009-06-17.
[10] Unknown, 2009. "API Recommended Practice 580, Risk-Based Inspection," Second Edition, November 2009.
[11] Selvik, J.T. and Aven, T., 2010. "A framework for reliability and risk centered maintenance," Reliab Eng Syst Safety (2010), doi:10.1016/j.ress.2010.08.001.
[12] Yates, L., 2007. "RCM and Risk Management," The Journal of the Reliability Analysis Center, Part 1 - Third Quarter 2007, Part 2 - Fourth Quarter 2007, Part 3 - First Quarter 2008.
[13] Jardine, A.K.S. and Tsang, A.H., 2005. "Maintenance, Replacement and Reliability: Theory and Applications," CRC Press, 2005.
[14] Barringer, P. and Monroe, T.R., 1999. "How to Justify Machinery Improvements Using Reliability Engineering Principles," 1999 Pump Symposium Sponsored by Texas A&M Turbo Lab March 1-4, 1999 George R. Brown Convention Center, Houston, Texas.
[15] Barringer, P., 2005. "How To Justify Equipment Improvements Using Life Cycle Cost and Reliability Principles," North American Association of Food Equipment Manufactures Conference 2005, Miami, Florida, January 14, 2005.
[16] Jardine, A.K.S., 1979. "Solving Industrial Replacement Problems," Proceedings, Annual Reliability and Maintenance Symposium, pp 136-142.
[17] Bharadwaj, U.R., Silberschmidt, V.V., Wintle, J.B. and Speck, J.B., 2008. "A Risk Based Methodology for Spare Parts Inventory Optimisation," Proceedings of IMECE2008 ASME International Mechanical Engineering Congress and Exposition October 31-November 6, 2008, Boston, Massachusetts, USA.
[18] Ceschini, G.F., Carlevaro, F., Racioppi, G. and Masi, A., 2003. "Turbogroup Spare Part Optimization by Availability Centered Maintenance Methodology: an Application to LTSA Contract," Proceedings of ASME Turbo Expo 2003 Power for Land, Sea, and Air June 16–19, 2003, Atlanta, Georgia, USA.
[19] Mitchell, J.S., 2007. "From Vibration Measurements to Condition Based Maintenance: Seventy Years of Continuous Progress," Sound and Vibration 41 no 1, Jan 2007, pp. 62–78
[20] De Maria, R.L. and Gresh, M.T., 2006. "The Role of Online Aerodynamic Performance Analysis," Proceedings of the Thirty-Fifth Turbomachinery Symposium, 2006, pp. 55-61.
[21] D.R. Cox, 1972. "Regression models and life tables (with discussion)," J.Roy. Stat. Soc. B, 34, 187-220.
[22] Jardine, A.K.S. Joseph, T. and Banjevic, D., 1999. "Optimizing condition-based maintenance decisions for equipment subject to vibration monitoring," Journal of Quality in Maintenance Engineering, Vol. 5 No. 3, 1999, pp. 192-202. # MCB University Press, 1355-2511.
[23] OMDEC Case Studies, www.omdec.com/solutions/case-studies/, accessed Sept 22, 2011.

[24] Green, B., Hull, G., Hurtado, J. and Harvill, M., 2008. "Remote Health Monitoring Increases Reliability and Condition-Based Maintenance," Proceedings of RTDF2008 2008 ASME Rail Transportation Division Fall Technical Conference September 24-25, 2008 Chicago, Illinois, USA.
[25] Bohlin, M., Warja, M., Holst, A., Slottner, P. and Doganay, K., 2009. "Optimization of Condition-Based Maintenance for Industrial Gas Turbines: Requirements and Results," Proceedings of ASME Turbo Expo 2009: Power for Land, Sea and Air GT2009 June 8-12, 2009, Orlando, Florida, USA.
[26] DePold, H., and Siegel, J., 2006. "Using Diagnostics and Prognostics to Minimize the Cost of Ownership of Gas Turbines". In Proc. ASME Turbo Expo 2006. Paper no. GT2006-91183.
[27] Dawson, J., Colquhoun, I. Yablonskikh, I., Wenz, R. and Nguyen, T., 2010. "Deterministic QRA Model and Implementation Experience Via an Integrity Management Software Tool," Proceedings of the 8th International Pipeline Conference IPC2010, September 27-October 1, 2010, Calgary, Alberta, Canada.
[28] API RP 581, Risk Based Inspection Technology, 2008.
[29] Morrison, L., "Integrated Risk Assessment of Several Approaches for Handling Runaway Reactions," NOVA Chemicals Inc., Moon Township, Pennsylvania (white paper).
[30] Wickenhauser, P.L. and Playdon, D.K, 2004. "Quantitative Pipeline Risk Assessment and Maintenance Optimization," Proceedings of IPC 2004 International Pipeline Conference, October 4 - 8, 2004 Calgary, Alberta, Canada.

Chapter 9

DEPENDABILITY ASSURANCE

9.1 ESTABLISHING DEPENDABILITY ASSURANCE FRAMEWORK

9.1.1 Understanding Assurance in Business and Technical Performance

The term "assurance" infers confidence. It also implies certainty, assertion and trustworthiness. The concepts and principles of assurance span multiple disciplines, application domains, system properties and diverse technologies. Assurance is a management process to ensure confidence that established requirements have been or can be fulfilled. Dependability is a system property or attribute with the ability to perform as and when required, encompassing the time-dependent quality characteristics. The dependability assurance process is guided by the management policy and technical directions of an organization in doing business. From a business perspective, system developers and system service providers should ensure that customers and users are getting value for their money. The implementation of an assurance process ensures sustainable business operations. This is often summarized as Quality of Service (QoS) [1], which is influenced by the following factors:

- system service functions to satisfy user needs;
- system performance capability to meet service demands;
- security of service;
- dependability of service.

From a technical performance perspective, dependability assurance should ensure the capability of system service performance in meeting customer expectations and achievement of QoS. Dependability assurance strategies and implementation methodology should be developed to provide effective guidance for project management in business operations. This is essential for delivery of technical service needs in a safe and secure manner as well as for gaining customer confidence and user trust. The technical focus of dependability assurance is to ensure cost-effective deployment of system service functions and timely service support performance to satisfy the expected service needs and to achieve QoS objectives.

9.1.2 Dependability Assurance Framework in System Performance

The dependability assurance framework can be described by using a multi-layer performance model as shown in Figure 9.1.

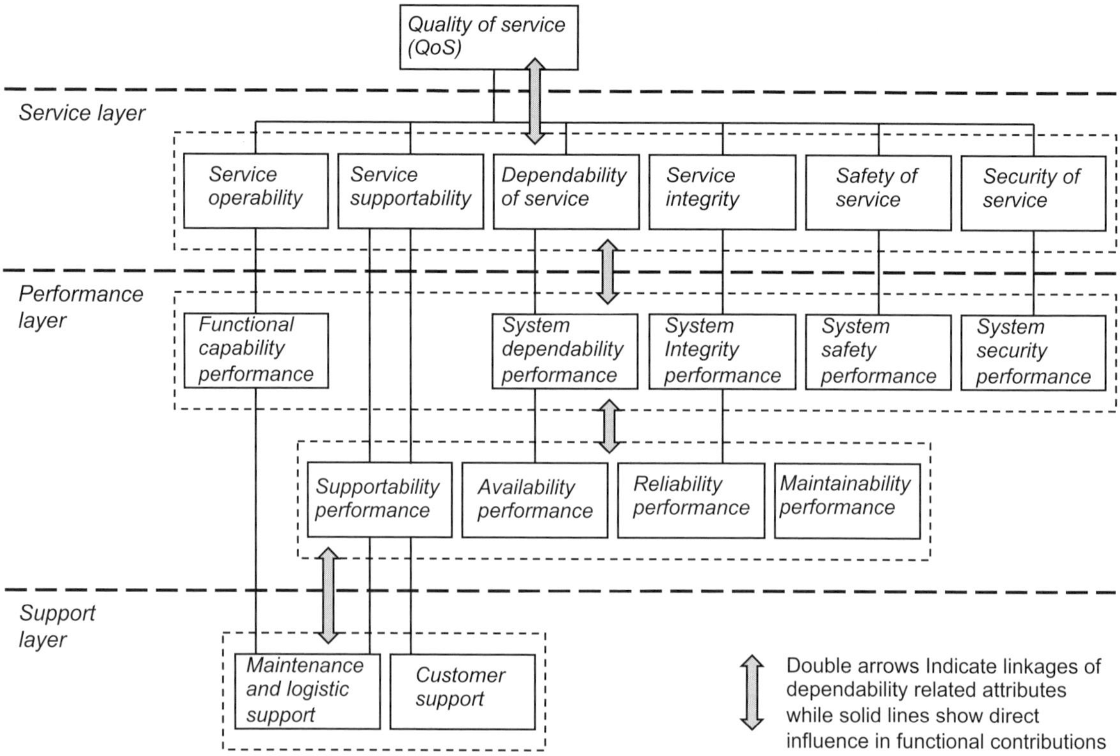

Figure 9.1 Performance model description of the dependability assurance framework

The multi-layer approach exhibits the linkages of the service-performance-support layers in a hierarchical structure. It shows the configuration and relationships of relevant system functions to achieve the dependability assurance objective. Many systems can be classified in terms of:

- a service layer to identify the service functions required to meet the QoS objectives;
- a performance layer to determine the required performance functions for system design purposes;
- a support layer to recognize the needed support functions.

A layer is a collection of similar functions that provides services to the layer above it and receives services from the layer below it. Each responsible layer provides a set of accessible system functions that can be controlled and linked to a corresponding set of system functions in the other layers. The linkages of these collaborating system functions establish their specific causal relationships. The objective of the service provided by the system is to achieve the level of QoS that the customer or user requires.

The layer approach is based on the hierarchical functional model to establish the causal relationships of system functions. The limitation of the layer approach is that it presents only a static state of configuration of the performance functions. The performance model as described in Figure 9.1 represents the system performance view in full-scale operation in delivering customer service. Additional interpretations are needed to illustrate the dynamic situations as the system evolves at different life cycle stages with different operations and changing operating scenarios. The time dependent aspects of the changing system performance situations are addressed by the system life cycle process. The layer approach can also be used to address mission objectives where the top service layer would become the mission layer with the needed functions to achieve mission success criteria.

9.1.3 Collaboration of Assurance in System Performance

The collaboration of assurance is essential to achieve cost-effective system performance. There are ample examples in industry applications such as where equipment safety is closely linked to the use of reliable components, quality material application improves product durability, safety designs increase structural integrity, and fault tolerant software enhances security and dependability of service. The collaboration of assurance attributes enables appropriate incorporation in system functions to facilitate performance applications. Table 9.1 illustrates how the collaboration of assurance attributes influences the system functional capability in system performance.

The common link between these assurance attributes and the associated system performance characteristics is the presence of risk in all their applications. Hence, systems containing such properties enabled by incorporation of assurance attributes are generally known as risk-based systems. The set of technical disciplines involved in system design and realization are known as assurance sciences.

Some or all of the assurance attributes are needed to work in cohesion with the essential service-oriented system properties. This is for delivery of service operability and supportability to achieve full-functioned capability in system operation and provision of QoS. The assurance attributes alone do not replace the functional design of the system capability, but rather act as enabling mechanisms to enhance and sustain the system performance in operation and service application.

The collaboration of assurance attributes contributing to functional capability of system performance is illustrated in the following cases.

Case (1) From a system development perspective for a system that provides a service, the assurance objective is to enable the design of system functions to achieve their specified performance requirements.

Case (2) From a system operational perspective for a system that is deployed for a service, the assurance objective is to support and enhance system performance in providing QoS to achieve customer satisfaction.

The assurance collaboration is illustrated in Case (1) for design trade-offs, and in Case (2) for joint assurance applications during operation. This can be ascertained by focusing on the extent of assurance influences in system performance for various situations. The

TABLE 9.1 Collaboration of assurance attributes influencing system performance

Collaboration of Assurance Attributes	Primary Focus of Assurance Attribute Enabling System Performance Achievement	Key Measurements of System Performance Characteristics
Dependability	Performance	Availability/reliability
Quality	Perceived value	Customer satisfaction
Integrity	Robustness	Integrity level
Security	Physical/functional protection	Security assurance level
Safety	Harm/threat avoidance	Safety integrity level
⇓		
Functional capability	System performance	QoS

extent of influences can be denoted as strong, medium or weak. Quantitative rating such as 1 to 9 can also be established to designate the priority of assurance attributes in applications and the effects on system performance characteristics. The assignment of values to reflect the extent of influences is highly subjective and may require justification in actual project application. A methodology based on quality function deployment (QFD) [2] is used to determine how assurance attributes influence system performance to achieve the respective assurance objectives. A monitoring and control system for industry application is used as an example to illustrate the methodology application in these two cases. The primary focus of this approach is to address the dependability influences in system performance. Dependability relationships with other collaborating assurance attributes are noted for their respective influences for illustration purposes.

Figure 9.2 illustrates the collaborating assurance attributes that influence the realization of system performance in Case (1).

The assurance focus on the development of a monitoring and control system emphasizes quality components for remote sensors and durable hardware assembly construction, reliable software for centralized information processing, and human interaction for remote control and monitoring of major operating functions. The collaborating assurance attributes influence the extent of design structure for realization as shown in Figure 9.2 for Case (1). Numerical values are assigned to designate the influences of dependability characteristics on the system design functions. The weighted averages of the numerical values of influences are determined as a basis for ranking the relevant factors of concerns. The

Assurance Attributes Collaboration

SYSTEM PERFORMANCE FOR REALIZATION

Major operating and physical functions

Importance of dependability characteristics for design incorporation

Assurance Attributes Collaboration	Characteristics	Hardware functions	Software functions	Human aspects	Interface connections	Structure configuration	Importance	Rank
Dependability	Availability		□				12.3%	4
	Reliability	□	○	□	○		38.3%	1
	Maintainability		○				6.9%	5
	Recoverability		○	△	○		17.8%	3
	Durability	□				□	24.7%	2
		24.7	32.9%	16.4%	13.7%	12.3%	100%	
		2	1	3	4	5		
Quality			○					
Integrity		○	○		△			
Security		○	△	△	△			
Safety		△		○		○		

Priorities of operating and physical functions for realization

Influences □ Strong = 9 ○ Medium =5 △ Weak = 3

Figure 9.2 Collaborating assurance influencing system performance realization

dependability characteristics are ranked according to their importance for design incorporation and trade-offs. The operating and physical functions are ranked according to their priorities for realization.

Figure 9.3 illustrates the collaborating assurance attributes that influence the achievement of system performance in Case (2).

The assurance focus on operation of the monitoring and control system emphasizes operability, supportability and dependability of service while sustaining the integrity, safety and security in service provision. The collaborating assurance attributes influence the extent of the achievement of system performance as shown in Figure 9.3 for Case (2). Numerical values are assigned to designate the influences of dependability characteristics on the functional capability performance objectives. The weighted averages of the numerical values of influences are determined as basis for ranking the relevant factors of concerns. The dependability characteristics are ranked according to their importance for influencing service performance. The functional capability performance services are ranked according to their priorities of performance achievement.

The QFD methodology adapted for assurance collaboration provides a viable approach for assurance project management that often encounters separate organizations or expert teams to address application-specific problems such as safety and security issues. The collaboration of assurance approach facilitates project management to identify priorities for assurance project assignments. The approach also presents a means to assess the importance of assurance characteristics relevant to the project that may share common solutions

Assurance Attributes Collaboration

ACHIEVEMENT OF SYSTEM PERFORMANCE

Functional capability performance objectives

Importance of dependability characteristics influencing service performance

	Characteristics	(QoS) Operability	Supporta-bility	Dependability of service	Service integrity	Safety of service	Security of service		
Dependability	Availability	□		□			○	21.1%	3
	Reliability	□	△	○	□	□		32.1%	1
	Maintenance		□					8.2%	5
	Restoration		○	○	○	□	□	30.3%	2
	Logistic support		□					8.3%	4
		16.5%	23.9%	17.5%	12.8%	16.5%	12.8%	100%	
		3	1	2	5	4	6		
Quality		△	△						
Integrity		○		△	□		○		
Security		○	○	○			○		
Safety		△	□			□			

Priorities in functional capability performance services

Influences □ Strong = 9 ○ Medium =5 △ Weak = 3

Figure 9.3 Collaborating assurance attributes that influence the achievement of system performance

to address the same assurance problems of safety, security and dependability concerns. The QFD methodology can work at several related hierarchical levels to determine the extent of influences of assurance efforts necessary for incorporation at the component and product levels for system development and realization. It can also apply laterally to service functions at the system level to identify criticality of service and specific support needs to enhance service operation for compliance to QoS agreements. The QFD process provides a means for knowledge base capture that would likely help to establish a common assurance procedure for ongoing and future project implementation.

9.2 EVOLUTION OF ASSURANCE STRATEGIES

9.2.1 Learning from Past Experience

Learning from experience is an age-old process for adjustment and adaptation to changes. Learning from history is an enriching experience that enlightens our understanding. The lessons learned from historical quality assurance and the quality management evolution provide valuable insights into adapting management strategies for dependability assurance to meet changing business environments and technological challenges.

The history of quality assurance and quality management spans more than six decades of industry experience. Quality management came into world-wide prominence with the publication of ISO 9000 series of standards in 1987 for global industry implementation. The ISO 9001:1987 standard [3] was developed with a set of quality requirements for certification and conformity assessment purposes. The ISO 9001 requirements focused on a quality management system, management responsibility, resource management, product realization, and measurement, analysis and improvement. The Plan-Do-Check-Act model was adopted in a later version for process improvement. The standard in practice tended to place emphasis on conformance to specific procedures rather than improving the overall processes for effective management. However, the management model presented for quality management was well suited to the manufacturing and service industry sectors, especially for product manufacturers and service providers in developing countries seeking entries for global market penetration.

The ISO 9001:1994 revision emphasized quality assurance via preventive actions and required evidence of compliance with documented procedures. This revision led organizations to compile extensive procedure manuals and documentation that often impeded process improvement. The ISO 9001:2000 revision introduced the concept of continual process improvement, involved upper management in the process and integrated quality into the business system. Process metrics were used to determine performance effectiveness and to track customer satisfaction trends. The ISO 9001:2008 revision introduced clarifications to improve consistency of quality management systems with the ISO 14001 environmental management systems [4]. There were no new initiatives in this latest revision. At the 2009 ISO/TC 176 Quality Management and Quality Assurance General Meeting in Tokyo, concerns regarding the relevancy of ISO 9001 as contributing value to business were expressed. The primary issues concerned the cost-benefit ratio to achieve and maintain the organization's certification program for compliance, the efficacy of the auditing process from a business perspective and the engagement of third-party auditors unfamiliar with the organization's business culture and technical processes to arrive at actionable recommendations for meaningful improvements. Some strategic framework changes were needed for ISO TC176 to revitalize the ownership of the ISO 9001 brand and to regain customer confidence. It appeared that industry customers were looking for quality management solutions in other management systems and not in ISO 9001. Certification to

ISO 9001 is no guarantee for producing quality products. A revival strategy is currently in the works for ISO TC176.

In the intervening years, many industry-specific quality management standards have been developed. Quality management systems and methodologies have been oriented toward specific industry needs and entrenched practices. Examples of such industry-specific standards include TL 9000 for the telecommunications industry [5], QS 9000 (now replaced by ISO TS 16949) [6] for the automotive industry and AS 9000 [7] for the aerospace industry. As new technologies have evolved, new quality-related standards have been developed, such as CCMI [8] for software capability maturity model integration and SRA [9] for software reliability assurance.

Total quality management (TQM) evolved from quality assurance. TQM is a management philosophy that seeks to integrate all organizational functions and services, such as marketing, finance, design, engineering and production, to focus on meeting customer needs and the organization's objectives. TQM empowers the total organization, from employees to top executives, with the responsibility of ensuring quality in their respective products and services and in the management of their processes through appropriate process improvement channels. TQM takes a long time, about 10 years, to administer change, and takes longer to mature the process. The decline of TQM adoption is due to a lack of top management commitment. Effective process implementation should be time-sensitive and adaptive to change.

On a positive legacy trail, the TQM strategy was well embedded in the quality process. TQM influenced the development of several industry-based business excellence models to guide business strategies and operations to align with the organization's goals and mission. Examples include the European Foundation for Quality Management Excellence Award [10], the Malcolm Baldrige National Quality Award [11] and the Deming Prize [12]. These award programs are aiming at national or international competitors to identify and recognize top-ranked quality organizations. The models used for the award examination address a broad range of quality excellence criteria including leadership, strategic planning, customer focus, measurement, analysis, knowledge management, workforce focus, process management and results. The criteria for performance excellence are used to advocate organizational self-assessment and self-improvement with respect to the organization's performance practices, capabilities and results in delivering value to stakeholders, hence strengthening the organization's competitiveness and contributing to the knowledge and continued learning within the organization. The business excellence models have been embraced by industry organizations as a proactive approach toward achievement of strategic management goals.

Six Sigma [13] is a business management strategy for process improvement via root cause identification and elimination to minimize variability in manufacturing and business processes. Established quality management methods and tools, such as control charts, cause and effect diagram, and quality function deployment (QFD), are used to facilitate the improvement process. A special group of people within an organization is designated as "Black Belt" experts to guide implementation of specific methods. Each Six Sigma project carried out within an organization follows a defined sequence of steps with quantifiable targets. These targets could be on financial budget control for cost reduction or production scheduling for product delivery commitments. Six Sigma is seen by some quality experts as a revamped version of the basic quality improvement process but with more hype and a flamboyant introduction. It has been recognized by some [14] that Six Sigma will not be successful from a customer-centric point of view unless it incorporates reliability into its processes.

A more recent development is the adoption of "lean" thinking, as evidenced by the terms such as "lean manufacturing" and "lean maintenance." Here the focus is on the systematic elimination of waste or processes that add no value. Although the basic intent is long-term improvement, it has often been seen mainly as a cost reduction exercise with a mentality of short-term results. As with Six Sigma, the key is to integrate "lean" with reliability as a primary means of achieving the objectives of lean principles [15]. An example of this is the proposed use of a failure mode and effect analysis (FMEA) approach to enhance the reliability of lean systems [16]. There are also attempts to combine Six Sigma with lean production in order to improve both the quality of products and manufacturing efficiency, presumably because no one method is complete in its effectiveness [17]. The core of both of these objectives must be the application of dependability for them to be effective.

It should be cautioned that adoption of any of these management models requires initial investment and significant resource commitment to sustain continued operation. Sometimes the cost and time required for the model's adoption might outweigh the benefits in meeting the organization's business challenges. The model adoption process might also involve the need for cultural change and staff retraining within the organization. This might affect ongoing customer services and relations. It would be prudent to call for a thorough examination to rationalize the critical issues prior to using any of the management models to guide long-term operation and sustain business viability. Adaptation and agility to change seems to offer a viable solution.

9.2.2 Dependability Assurance Initiatives

Dependability as a management function should be integrated into the framework of an organization's infrastructure to facilitate dependability policy decisions and influence implementation of technical direction. Dependability engineering as a technical discipline should be closely associated with ongoing activities of engineering projects for design and process improvements. Dependability should work in collaboration with quality assurance to achieve a common objective for customer satisfaction. The organization's quality management system should provide the linkage for dependability policy coordination. A review of the quality management progress and evolution will provide practical insights for dependability management. Continuous learning for knowledge enhancement and progressive improvement should be targets to attain dependability excellence. It would be counterproductive to limit the scope of dependability management functions to barely meet conformance targets of minimum standard requirements.

Lessons learned from the evolution of quality management and quality assurance have shown that no single model is satisfactory as a management standard for all organizations and business operations. Management has to be adaptive to changing market needs and must make appropriate adjustments in an organization's policy, processes and resource deployment for alignment with its goals and mission. Dependability is greatly influenced by technology advancement and innovation. The skills and knowledge of dependability professionals contributing to an organization's endeavors in technological systems and projects should be recognized to foster performance excellence achievement and value creation. Dependability management leadership should enlighten the direction and envision proactive change management. Dependability assurance should ensure relevant management initiatives and practical dependability functions for value creation. The following presents some typical dependability assurance initiatives:

- dependability management systems integration
- dependability policy, vision and mission statement to reflect change;

- system life cycle management from a dependability application perspective;
- effective practices for management of dependability resources including technical, information, and human resources;
- outsourcing and supply chain management to facilitate system dependability realization and implementation;
- joint ventures in research and development of new technology platform for diverse applications;
- management process for coordinated design of hardware and software functions and human aspects incorporation;
- management and coordination of dependability-related information systems and data usage;
- technology and environmental influences on dependability applications;
- support services and customer relations influencing dependability applications;
- project tailoring for dependability performance optimization;
- dependability training and learning, and knowledge base capture and retention;
- dependability performance simulation, testing and measurement, assessment and review for achievement of system performance excellence;
- assurance focus for dependability value achievement.

It has been acknowledged that quality and reliability feedback loops do not always work in practice [18]. With the increasing complexity of products comes an increase in quality and reliability demands, with a shorter time period to accomplish them. It was found in a case study [14] that the information fed back to an organization was insufficient to have the desired impact on quality and reliability. This was mainly due to cost pressures on the after-service process to minimize cost. It had resulted in the lack of high-quality information to base improvements on quality and reliability. The quality of business processes must receive attention along with the functionality, quality and reliability of the products and services being provided.

9.3 LIFE CYCLE APPROACH TO DEPENDABILITY ASSURANCE IN SYSTEM PERFORMANCE

The life cycle approach involves a series of dependability activities to ensure achievement of the system performance objectives. The system life cycle stages include concept/definition, design/development, realization/integration, operation/maintenance, enhancement and retirement. The assurance objective at each life cycle stage is to ensure that system dependability requirements are met through judicious application of relevant methods for assurance of system dependability performance to achieve the required dependability of service. Dependability assurance strategies reflect the relevant technical domains in system-specific applications. They collaborate with system performance management and the routine maintenance and support activities by providing specific dependability engineering efforts. The dependability assurance strategies are focused from the system operation viewpoint to achieve performance excellence. They include:

- delivery of QoS and dependability of service to end users;
- ensuring integrity, security and safety in performance;
- enhancement of system performance functions and support processes.

Table 9.2 presents examples of dependability assurance processes for system life cycle stages.

Table 9.2 presents the system life cycle activities for assurance process implementation. The activities supporting the system assurance process and the dependability assurance process are addressed in Appendix B, where all of the life cycle activities are combined.

9.4 DEPENDABILITY ASSURANCE FROM A BUSINESS PERSPECTIVE

9.4.1 Assurance to Ascertain System Dependability Performance

From a system developer perspective, assurance is used as a management tool to ascertain system dependability performance achievement. The purpose of the assurance process is to build confidence that the required system performance can be achieved. Assurance is an enabling mechanism to support incorporation of dependability in engineering design and system development. This is achieved by implementing relevant assurance activities through the concept/definition, design/development and realization/implementation stages early in the system life cycle. The objective of the dependability assurance is to increase the probability that the specified system will meet its performance requirements when ownership is transferred to the organization that will use it for business operation. The dependability assurance objective can be achieved by progressive accomplishment of project milestones with auditable arguments to support objective evidence, such as an assurance case study.

The system developer assurance program should also address important business issues for possible consideration. The objective is aimed at business growth and enhancement to gain customer confidence and user trust. Some business related examples for assurance include ensuring:

(a) stakeholders' interests to reflect their investments objectives;
(b) adequate financing of project resources and funding of capital expenditures;
(c) long-term business goals in project commitments;
(d) compliance to contract agreements and legal obligations;
(e) appropriate deployment of availability technical resources;
(f) application of make-buy decision policy;
(g) selection of technology design platform;
(h) capability maturity of the system development process;
(i) implementation of supply chain management and outsourcing policy;
(j) application of confidentiality and security policy;
(k) proper disclosure of proprietary information in joint development projects;
(l) customers relation and suppliers liaison;
(m) brand value protection.

9.4.2 Assurance to Sustain System Dependability Performance

From a system service provider perspective, assurance is utilized as a management process to sustain system dependability performance. The purpose of the assurance process is to build customer confidence that the needed system performance will be achieved. In order to assure continuous satisfactory service to the customers, relevant activities are implemented during the long period of operation and maintenance. At times when service performance upgrades are required, appropriate enhancements will be incorporated based on market needs and sustainability of the business service commitments. When certain

TABLE 9.2 Dependability assurance processes for system life cycle stages

Concept/Definition Stage		
System Assurance Process	**Dependability Assurance Process**	**Typical Assurance Methods Implementation**
• Identify customer requirements • Define system performance objectives and constraints • Establish system assurance plan • Identify schedule delivery targets and resource requirements • Establish management responsibilities and project review process	• Define system dependability performance requirements • Identify system operational scenarios • Identify dependability application constraints and technical limitations • Identify functional performance specifications and applicable dependability standards	• Perform requirements analysis • Perform operational scenario analysis • Perform preliminary dependability assessment • Determine feasible architectural design configurations meeting functional performance requirements and dependability objectives
Design/Development Stage		
System Assurance Process	**Dependability Assurance Process**	**Typical Assurance Methods Implementation**
• Establish assurance plan for system design verification and requirements conformance • Establish system interface and integration plan for testing and verification of interoperability • Establish supplier assurance program • Establish assurance plan and procedures for full-scale system development	• Determine and allocate system and sub-system dependability requirements • Determine critical functions for mission-specific safety and security applications requiring specific dependability attention • Establish dependability database for information capture • Determine system failure criteria • Determine spares provisioning program • Define warranty conditions	• Perform detailed dependability assessment • Conduct system reliability prediction, FMEA, FTA and simulation where applicable • Perform design trade-off analysis for system performance optimization • Justify outsourcing requirements • Finalize system dependability specifications • Recommend make-buy decisions
Realization/Implementation Stage		
System Assurance Process	**Dependability Assurance Process**	**Typical Assurance Methods Implementation**
• Establish product assurance plan for realization of hardware/software functions and production • Establish system integration plan for system testing and acceptance validation • Establish system installation and transition plan	• Implement system dependability program • Implement quality assurance program • Implement supplier dependability programs • Implement system maintenance and logistics support program • Implement failure reporting, analysis, data collection and feedback system	• Perform product testing and evaluation of realized functions for product acceptance • Perform maintenance and logistic support analysis • Conduct fault insertion tests for risk evaluation of system fault exposures to determine and develop system recovery and restoration process • Analyze test data to identify and resolve critical design and procedural problems

(Continued)

TABLE 9.2 Dependability assurance processes for system life cycle stages (*Continued*)

Operation/Maintenance Stage

System Assurance Process	Dependability Assurance Process	Typical Assurance Methods Implementation
• Establish plan and procedures for system operation, maintenance and support • Identify service level agreements for provision of QoS • Define criteria for system performance value to achieve customer satisfaction • Establish fault management system and customer care services	• Implement monitoring and control system to sustain dependability performance in system operation • Implement reliability growth program as appropriate • Implement field data collection system and incident reporting • Implement fault management system and customer care services • Determine adequacy of QoS provision • Determine service operability, service supportability and dependability of service • Determine service integrity, security of service and safety of service where applicable	• Conduct customer satisfaction survey • Perform incident reporting and data collection • Analyze failure trends • Conduct root cause analysis for problem resolution • Recommend design or procedural changes for continual improvement

Enhancement Stage

System Assurance Process	Dependability Assurance Process	Typical Assurance Methods Implementation
• Identify new customer requirements • Establish enhancement strategy and plan relevant to the new customer requirements • Evaluate the need for change and resulting benefits	• Evaluate impact on dependability performance due to changes with added new features • Implement enhancement efforts	• Conduct life cycle cost impact study for change incorporation • Conduct risk and value assessments • Conduct customer satisfaction survey resulting from change reactions

Retirement Stage

System Assurance Process	Dependability Assurance Process	Typical Assurance Methods Implementation
• Identify status of aging system performance capability against market competition • Rationalize decisions on timing and extent for partial or complete system retirement • Initiate decommissioning plan as required • Identify reuse or salvage value of disposable items, including data	• Implement reuse and redeployment strategy • Implement waste treatment on disposal items • Notify the customers on service termination • Provide information on new or alternative service provision	• Evaluate constraints on system deactivation and impact on removal of system from operation • Evaluate impact on environments of disposal items • Conduct customer satisfaction survey due to termination of service

services are deemed obsolete and considered as candidates for retirement, new replacement services or renewal of the existing services should be considered for incorporation to sustain viable business operation. The dependability assurance objective is to ensure delivery of QoS, dependability of service and other value added system performance services to the customer.

The system service provider assurance program should also consider important business issues. The objective is aimed at sustaining ongoing business operation and maintaining service commitments. Some business related examples for assurance include ensuring:

(a) market relevance of service provisions with competitive advantage;
(b) business growth and investment strategies;
(c) efficiency and effectiveness of policies and operations;
(d) effective maintenance and logistic support strategies;
(e) effective incident reporting system for data collection, analysis and corrective action;
(f) integrity of data transfer, retrieval and recovery;
(g) compliance to service level agreements on delivery of QoS;
(h) compliance with statutory obligations such as occupational health, safety and security;
(i) customers care and support services;
(j) brand value protection.

9.5 ASSURANCE CASE

9.5.1 What Is an Assurance Case

An assurance case [19, 20] is a study of report findings on evidence gathered to support a claim or an argument. The assurance case study provides a means of achieving progressive assurance that dependability requirements are being met or will be satisfied throughout the life cycle of a technological system of interest. The framework for establishing an assurance case includes:

- a reasoned, auditable argument to support the contention that a defined system satisfies the dependability requirements;
- a summary of evidence and arguments to support the claims for dependability achievement;
- progressive assurance throughout the life cycle of the system as target of evaluation.

The assurance case provides a focal point for determining uncertainties and managing related risks. Therefore, assurance has become a key factor in risk assessment and risk management and in the life cycle activities that plan, design, achieve, demonstrate, sustain and monitor the dependability performance in system operation.

Management of dependability assurance activities should use existing performance monitoring systems to generate the needed information for process and service improvement. Typical examples include:

- failure reporting, analysis and corrective action system;
- customer care and feedback system;
- maintenance and logistic support system;

- incident reporting and fault management system;
- health monitoring system;
- quality management system.

Assurance technology has evolved with time from classic quality assurance principles to embrace new unknown challenges. Current assurance activities extend beyond the traditional product assurance boundaries of practice. New technologies and innovative approaches have been used to address issues such as climate change, energy conservation, health and safety of socio-economic systems, and vulnerability of security, industrial and financial operations, all under the general principles of risk management. In most cases, the assurance activities have direct or indirect dependability involvement as a fundamental valued attribute for inclusion. Dependability activities have placed emphasis on system survivability, recoverability and sustainability in performance to deliver dependability of service. The users' understanding of dependability is leaning towards the perception of trustworthiness of proven system performance records where the users are willing to entrust their valued assets.

The dependability assurance case study has been successfully used for independent audits for compliance assessment of system dependability projects. The case study can also be used for self-assessment to determine the extent of dependability risk exposures for possible risk mitigation. However, it should be noted that such case studies often involve a large amount of documentation and data handling that need to be created, compiled and analyzed for substantiation and presentation of objective evidence. It is prudent to justify the time and effort required for information gathering and dissemination that adds value to the outcome of the dependability assurance case study. Sometimes a simple functional test to demonstrate performance speaks much louder than words to convey a convincing message of trust and confidence.

9.5.2 Assurance Case Studies

A considerable number of case studies have been done where dependability has either been the main focus or has shared the stage with quality, service provision or safety. Some of these apply to products, others to networks that provide a service, and yet others situations that involve personal and public safety and environmental impact.

The electrical grid is such an integral service to society that it attracts a wide range of assurance studies. Some of these studies are concerned with the electrical grid. One of these looks at power system security [21], which is the ability of the power system to withstand disturbances resulting from faults of equipment failure as opposed to reliability, which is the probability of satisfactory operation. Event and fault tree models were developed of the substation protection and tripping functions, and then combined with dynamic simulations of the power system. Thus system performance could be analyzed.

Related to this are the issues of connecting distributed generation hardware to existing utilities [22]. These issues include protection, power quality and system operation. Simulations were run and cases were studied with respect to power quality, protection and reliability. The impact of faults in distributed generation equipment can be significant and becomes significant if many small units or a few large ones alter circuit levels enough to cause miscoordination between current protection devices. Both equipment and system solutions are necessary to maintain power system integrity.

Another approach taken recently is to analyze the electrical transmission network with the objective of identifying critical components with respect to vulnerability [23]. The analysis was done from four different perspectives: topological, reliability, electrical and electrical-reliability. Weighted indicators were calculated for the network nodes based on these characteristics to identify the most critical nodes. The reliability analyses identified areas with critical nodes that needed to be enhanced for network robustness. The electrical network analysis pointed out the bus where electricity flow was the highest and most vulnerable to faults. These analyses can then suggest where the most important network improvements can be made.

An interesting dependability situation is the one faced by Spain, with power generation provided by both hydro and thermal plants [24]. Network performance has to account for both load variations in hydro energy and maintenance outages in thermal plants. With an installed capacity of 25,000 MW from 80 thermal units and 16,500 MW from 200 hydro units, an efficient evaluation of reliability and availability was needed to assess the capacity of the system. A number of reliability indicators were established, along with load variations from the hydro plants and maintenance outages from the thermal plants. Scenarios were studied for normal years and dry years divided into monthly intervals. A nearly complete leveling of the reliability of the system was found achievable. This methodology was used to assess the reliability of the generating system for expansion planning purposes.

9.6 SOFTWARE ASSURANCE

9.6.1 Software Assurance Overview

Software assurance is the planned and systematic set of activities ensuring that software life cycle processes and products conform to requirements, standards and procedures. The capability maturity models [25] are common management tools recommended for implementation of software assurance programs in software development organizations. There are also extensive documented software assurance methodologies and procedures for software development and applications [26].

The software assurance process is to plan, develop, maintain and provide grounds for confidence and decision-making. The assurance life cycle [27] is conducted for conformity assessment purposes throughout the system life cycle on software products to meet applicable safety, security, dependability and other objectives. Assurance case [28] studies are claim records on process performance and the physical properties and functional characteristics of the software system audited for proof of conformance to system specifications. Software assurance engages in risk assessment, verification and validation testing, documentation and maintenance of audit records as objective evidence. Software assurance utilizes relevant, project-based measurement data to monitor the software product and relevant process for possible improvements.

Software dependability emphasizes software reliability as an intrinsic part of software assurance through implementation of the software reliability engineering process [29]. Software dependability and quality are prerequisites for achievement of safety and security in system operation.

9.6.2 Technology Influence on Software Assurance

Software technology has provided numerous advancements for efficient software development and versatility in software applications. However, software systems are often operating in an environment exposed to unknown viruses and furtive cyber intrusions. Recent hacking and intentional cyber attacks into well-established software system operations have become more frequent, more prominent and increasingly sophisticated. They affect the software developers, providers and users, causing various degrees of time-lost problems and in some cases resulting in massive data corruption requiring significant technical efforts for system restoration. Software vulnerability has become a major challenge to software security protection and dependability assurance efforts.

Software assurance is critical to organizations involved in safety, security and financial transactions in view of the vulnerability in software applications. Software assurance encompasses the development and implementation of methods and processes for ensuring that software functions as intended while mitigating the risks of vulnerabilities, malicious code, faults or errors that could bring harm to the end user. Software assurance is vital to ensuring the security of critical information technology resources. With the rapidly changing nature of the threat environment, even highest-quality software is not impervious from cyber intrusions if the software is improperly configured and maintained. Managing threats in cyberspace requires a layered approach on security prevention and collaboration: the developers build more secure and robust software, the system integrators ensure that the software is installed and configured correctly, the operators maintain the system properly, and the end users are using the software in a safe and secure manner.

Industrial control systems and networks that previously were deemed to be impervious to outside interference have been shown to be much more vulnerable than anticipated [30]. Hackers have found that it is "cool" to hack into industrial networks. For example, a malicious Trojan program inserted into a supervisory control and data acquisition (SCADA) system interfered with valve positions and compressor outputs on a natural gas pipeline and caused a massive explosion. This is leading industry groups to take up the cyber challenge and create security guidelines, standards and certifications for the protection of SCADA systems.

9.6.3 Software Assurance Challenges

Organizations involved with software are redefining their priorities in software assurance for their operations. The software assurance challenges can be interpreted as the "level of confidence that software is free from vulnerabilities, either intentionally or unintentionally designed into the software or accidentally inserted at any time during the software life cycle, and that the software functions in the intended manner" [31]. Software assurance should provide a reasonable level of justifiable confidence that the software will function correctly and predictably by a manner consistent with its documented requirements. The assurance objective is to ensure that the software function is not compromised either through direct attack or through sabotage by maliciously implanted code.

The level of confidence in software assurance addresses:

- *trustworthiness* – that no exploitable vulnerabilities exist, either maliciously or unintentionally inserted;
- *predictable execution* – that software functions, when executed as intended, will provide justifiable confidence;
- *conformance* – that the planned and systematic set of multi-disciplinary activities to ensure software processes and products conforms to requirements, standards and procedures.

The software assurance challenges include:

(1) accidental design mistakes or implementation errors that lead to exploitable code vulnerabilities;
(2) the changing technological environment, which exposes new vulnerabilities and provides the cyber attackers with new tools for exploitation;

(3) malicious insiders and outsiders who seek to do harm to the developers or the end users.

The first challenge is accidental and unintentional. The second and third challenges are intentional and deliberate by the cyber attackers. The countermeasure is to manage risks associated with these challenges through software assurance best practices.

9.6.4 Network Security Implications

(a) Network security service objectives
Information and communication networks such as SCADA systems are software-driven distributed network services. The network security service objectives are aimed at preventing unauthorized network access or intrusion and protection from security attacks without the loss or disclosure of secured information [32]. Software as the network data processing engine plays a pivotal role in the assurance of network security services.

The provision of network security services should take into consideration:

- the threats involved in network security attacks;
- the planning of the security service functions for network protection;
- the vulnerabilities of network security provisions;
- the implementation of network security layers.

(b) Threats to network security
In the provision of network services, the normal flow of information originating from a legitimate source should reach the designated destination. This normal flow of information could be disrupted when the network is being attacked, exposed to security breaches or experiencing other threats.

The network security threats consist of four main categories:

- *interruption*: an asset (software or software system) of the network is destroyed or becomes unavailable or unusable; this is an attack on network availability;
- *interception*: an unauthorized party gains access to an asset of the network; this is an attack on confidentiality;
- *modification*: an unauthorized party not only gains access to an asset of the network, but also tampers with the asset; this is an attack on integrity;
- *fabrication*: an unauthorized party inserts counterfeit objects into an asset of the network; this is an attack on authenticity.

(c) Network security service functions
Security services are intended to counter attacks or threats to the network. The network security services deploy one or more security mechanisms utilizing security algorithms (software programs) to provide the security service functions [33] for network protection purposes.

- *Confidentiality* is the protection of transmitted data from passive attacks such as the release of data contents. Confidentiality also includes protection of traffic flow through the network from unauthorized analysis where the attacker may gain access to observe the traffic characteristics, such as source and destination, on a communication facility.

- *Authentication* is to assure that the communication is authentic. The authentication function assures that the connections initiated between legitimate parties are authentic and not interfered with by unauthorized transmission or reception.
- *Integrity* is to assure that messages are received, as sent, with no duplication, insertion, modification, reordering, relays or destruction of data.
- *Non-repudiation* prevents either the sender or the receiver from denying a transmitted message.
- *Access control* is the ability to limit and control the access to host network and applications via communication links. This is to assure access rights to the network by means of identification and authentication for controlling purposes.
- *Availability* is the provision of continuity of service upon demand without denial of service due to security service reasons.

The incorporation of automated recovery mechanisms in network service functions and data protection is an important service feature that may prevent serious consequences in devastating situations such as loss of information, data corruption and server crashes.

(d) Vulnerability

Vulnerability is a flaw in the security system and an exposure to the network. When a specific vulnerability is exploited, the target of vulnerability may affect a person or an organization as the victim of the attack. The vulnerability may also affect the security system or the network. In some cases, the response time to an attack is instantaneous or within seconds, such as a worm entering into the security system via an email message to perform destructive tasks. In other cases, the response time to an attack may take months or years, such as breaking the encryption codes or circumventing the security protection policy of an organization.

There are four basic types of vulnerability causes [28] as shown in Table 9.3.

(1) *Social engineering*: an area of exposure to attack directly against the security policy of an organization, such as an internal worker committing sabotage or the deliberate planned destruction of the organization's intellectual property. "Social engineering" is a term related to the modern day cyber phenomenon. It is used in the security business to describe the practice of obtaining confidential information by manipulation of legitimate users.

(2) *Policy oversight*: a flaw in the planning of a security system, such as inadequate software backup and duplicate files for data protection.

(3) *Logic error*: a software fault inherent in the system design, such as a poorly written software code that allows unwanted or uncontrolled access.

(4) *Weakness*: a design flaw that could lead to a security breach, such as breaking the encryption irrespective of how well it is designed. When the attack occurs to a design weakness, the response time is instantaneous.

TABLE 9.3 Types of vulnerability causes

Target of vulnerability	Affecting a person or an organization	Affecting the security system or the network
Response time to an attack	Over a period of time	Instantaneous
Causes of vulnerability	(1) Social engineering (2) Policy oversight	(3) Logic error (4) Weakness

Software-driven computers are used extensively for control of security systems. The vulnerabilities to the computers and the associated software systems have the following characteristics:

- *fault*: describing how the vulnerability has become a mistake and created the problem;
- *severity*: describing the degree of compromise, such as gaining security administrator access to certain files not normally granted to the regular user;
- *authentication*: describing the intruder's successful registration with the host computer with proof of identity before exploiting the vulnerability of the system;
- *tactic*: describing the issue of who is exploiting whom in terms of location and access to a valid account for entry to the security system;
- *consequences*: describing the outcome and the mechanics behind the access promotion showing how a small amount of access can lead to far greater compromises.

Network security protects the network infrastructure from exposure to attacks by deploying appropriate defense at all possible entry points of attacks before they spread to other parts of the network. This is a preventive measure, which attempts to protect individual computers and other shared resources connected to the network.

Computer security is the means and measures of self-defense taken to protect individual computers connected to a network. A computer whose security is compromised is likely to infect other computers connected to a potentially unsecured network.

(e) Layers of network security

There are generally six layers of network security that can be used to manage security systems to prevent exposures and vulnerabilities [28]. The combination and application of these layers provides increased security for the network.

(1) The foundation of the first layer of network security is the work performed by the *network administrators*, who understand the operating systems and know how to lock down specific systems to allow access only the ports and processes that need to get through. Ongoing training of network administrators is essential to stay ahead of network knowledge from exploits and virus attacks.

(2) The second layer of network security is the *physical security* to protect the network assets from "social engineering" manipulation. A typical example is to gain access to a security system by calling the help desk on a forgotten password and requesting a change to a temporary or new password. Physical security includes access control and identification of legitimate users, and implementation of good organizational practices in security policies.

(3) The third layer is *monitoring*. Most attacks involve repeated attempts to access the security system. By monitoring the system logs on a regular basis, it is possible to identify certain patterns in the log files. There are software programs available for monitoring the log files to identify and alert suspicious patterns of attempts to gain access to the security system.

(4) The fourth layer is the *software* installed on the servers for applications. Every piece of software should be tested and evaluated with appropriate security procedures for assurance purposes before loading onto the servers.

(5) The fifth layer is the deployment of *security tools* for protection and preventive measures. Security tools include but are not limited to firewalls, intrusion detection software and proxies (i.e., an intermediary function within the network to ensure security, administrate control and caching service).

(6) The sixth layer is *security auditing*. This provides regular checking and examination of the security system against new exploits since network security is a continuous, ongoing process. A security audit should test every aspect of the network security system. This should be done on a regular basis due to new software installation, system upgrade, changing user base and revision of security policies.

9.6.5 Software Assurance Best Practices

There are software technology and software assurance forums [29] that involve government, industry, academia and user participation in implementation of software assurance best practices. The recommended software development practices are identified in section 5.4.1. The recommended software assurance best practices are presented here:

- establishment of a software assurance policy to guide software development and process implementation;
- training on software product-related technology applications and the use of reference resources;
- use of a common software architecture design platform to facilitate diverse software product development;
- implementation of software life cycle processes;
- initiation of software assurance case studies for risk assessment where warranted and appropriate;
- established common criteria for verification and validation for software qualification and conformance;
- configuration management control of software version releases;
- established software performance and fault tracking and data collection systems for software design and process improvement;
- established customer help centers to facilitate users service support and software product application.

REFERENCES

[1] ITU-T Recommendation E.800, *Definitions of terms related to quality of service.*

[2] Akao, Y., ed. (1990), *Quality Function Deployment*, Productivity Press, Cambridge MA.

[3] ISO 9001, *Quality management systems – Requirements.*

[4] ISO 14001, *Environmental management systems – Requirements with guidance for use.*

[5] TL 9000, *The Telecom Quality Management System Requirements Handbook, Release 5.0, and Measurements Handbook, Release 4.5,* QuEST Forum, http://tl9000.org .

[6] ISO/TS 16949, *Quality management systems – Particular requirements for the application of ISO 9000:2008 for automotive production and relevant service part organizations.*

[7] AS9100, *Quality Management Systems – Aerospace – Requirements,* SAE International, http://standards.sae.org .

[8] *CMMI for Development, Version 1.2*; Software Engineering Institute, Carnegie Mellon University, Pittsburgh, PA USA 2006.

[9] *System and Software Reliability Assurance Notebook*, Rome Laboratory, Peter B. Lakey and Ann Marie Neufelder, 1996.

[10] *EFQM Forum*, www.efqmforum.org.

[11] *Baldrige National Quality Program*, www.baldrige.nist.gov.
[12] *The Deming Prize*, www.juse.or.jp/e/deming/.
[13] *Six Sigma Fundamentals: A Complete Guide to the System, Methods, and Tools,* D. H. Stamatis, Productivity Press, 2004.
[14] Kuei, C.-H. and Madu, C.N., 2003. "Customer-centric six sigma quality and reliability management," The International Journal of Quality & Reliability Management; 2003; 20, 8/9; p. 954.
[15] Smart, P.K., Tranfield, D., Deasley, P., Levene, R., Rowe, A. and Corley, J. 2003, "Integrating 'Lean' and 'high reliability' thinking," Proceedings of the Institution of Mechanical Engineers, Vol. 217 No. 5, pp. 733-9.
[16] Sawhney, R., Subburam, K., Soontag, C., Rao, P.R.V. and Capizzi, C., 2009. "A modified FMEA approach to enhance reliability of lean systems," International Journal of Quality & Reliability Management, Vol. 27, No. 7, 2010, pp. 832-855.
[17] Aggogeri, F. and Mazzola, M., 2008. "Combining Six Sigma With Lean Production to Increase the Performance Level of a Manufacturing System," ASME 2008 International Mechanical Engineering Congress and Exposition (IMECE2008) October 31–November 6, 2008 , Boston, Massachusetts, USA, pp. 425-434.
[18] Molenaar, P.A., Huijben, A.J.M., Bouwhuis, D. and Brombacher, A.C., 2002. "Why do quality and reliability feedback loops not always work in practice: a case study," Reliability Engineering and System Safety 75 (2002) 295-302.
[19] ISO/IEC 15026-2, *Systems and software engineering – Systems and software assurance – Part 2: Assurance case.*
[20] BS 5760-18, *Reliability of systems, equipment and components. Guide to the demonstration of dependability requirements. The dependability case.*
[21] Haarla, L., Pulkkinen U., Koskinen, M. and Jyrinsalo, J., 2008. "A method for analysing the reliability of a transmission grid," Reliability Engineering and System Safety 93 (2008) 277-287.
[22] Kroposki, B. (technical monitor), 2003. "DG Power Quality, Protection and Reliability Case Studies Report," GE Corporate Reasearch and Development, Niskayuna, New York, National Renewable Energy Laboratory, NREL/SR-560-34635, August 2003.
[23] Zo, E., and Golea, L.R., 2012. "Analyzing the topological, electrical and reliability characteristics of a power transmission system for identifying its critical elements," Reliability Engineering and System Safety 101 (May 2012) 67–74.
[24] Gonzalez, C. and Juan, J., 1999. "Reliability evaluation for hydrothermal generating systems: Application to the Spanish case," Reliability Engineering and System Safety 64 (1999) 89-97.
[25] *CMMI-DEV (Version 1.3, November 2010)*, Carnegie Mellon University Software Engineering Institute, 2010.
[26] NASA-STD-8739.8 w/Change 1, *Software Assurance Standard*, May 2005.
[27] ISO/IEC 15026-4, *Systems and software engineering – System and software assurance – Part 4: Assurance in the life cycle.*
[28] ISO/IEC 15026-2, *Systems and software engineering – System and software assurance – Part 2: Assurance case.*
[29] Lyu, M. R. (Ed.): *The Handbook of Software Reliability Engineering*, IEEE Computer Society Press and McGraw-Hill Book Company, 1996.
[30] Anonymous 3, 2009, "Hacking the Industrial Network", White Paper by Innominate Security Technologies AG.
[31] *National Information Assurance (IA) Glossary*, (CNSS Instruction No. 4009), National Security Telecommunications and Information Systems Security Committee (NSTISSC), published by the United States federal government (unclassified), June 2006.
[32] IEC 61907, *Communication network dependability engineering.*
[33] *Software Assurance: An Overview of Current Industry Best Practices*, Software Assurance Forum for Excellence in Code, February 2008.

Chapter 10

DEPENDABILITY VALUE

10.1 THE VALUE OF DEPENDABILITY

Value is the relative worth of something desirable or significant. The worth of something may be expressed as monetary or material worth, or interpreted as its usefulness or importance to a process outcome. The economic value of an item reflects the value of its use or utilization in terms of goods and services. The exchange value of an item is determined by the market value and affected by supply and demand. In this respect, value is linked to price. In business, as Warren Buffett, the American investment entrepreneur, once said: "Price is what you pay. Value is what you get."

Dependability is the ability of a system to perform as and when required. Dependability value reflects the intrinsic worth of the system's ability in achieving its expected performance or service. A technological system is considered as a tangible asset, with a price that can be negotiated for its acquisition. System dependability performance, on the other hand, is an intangible asset. The value of performance or service achievement can only be appreciated by those involved in the delivery process. The value of dependability is either determined by means of assessment of system performance capability or experienced during actual system application. Dependability value can be achieved by application of the value chain process. A value chain is a sequence of activities that adds value through the process of achieving an operation. The value chain process is the delivery system in support of value creation.

10.2 CONCEPT OF VALUE CREATION

Business enterprises focus on value creation to sustain growth and enhance business operations. Creating value for customers enhances product sales. Creating value for shareholders increases their stock prices. Creating value for the organization insures return on investment. The annual financial statement generally reports the status of an organization's assets, liabilities, equity, and profit and loss. The organization's worth is expressed in terms of business revenue versus expenditure, reflecting primarily the organization's tangible assets. When the total revenue exceeds the operating expenses, the business is profitable. When revenue equals expenses, the business breaks even. When revenue is less than expenses, the business is running at a deficit. The traditional accounting methods of assessing the organizational performance on a balance sheet are no longer adequate in today's global economy. The value of stock prices is not solely dependent on business earnings and capital assets. The strict financial measures of performance by cutting cost to produce short-term results lack long-term business insights on competitiveness and growth that enhance investments. Value creation has become a sustainable business necessity. The value phenomenon is recognized as a progressive business management goal. Value creation in today's competitive enterprises is increasingly represented by intangible assets such as innovation, research and development, patents and brand to measure business success.

Figure 10.1 shows a strategic map on how an organization creates value.

The generic map reflects the generalization of Kaplan and Norton's work with a large number of organizations in developing strategic maps for corporate management [1]. The objective of the strategic map is aimed at converting intangible assets into tangible outcomes. The underlying principle of the strategic map is based on Kaplan and Norton's earlier work on the Balanced Scorecard [2]. The Balanced Scorecard is a strategic management tool for tracking managerial performance in task assignments and monitoring the consequences arising from their execution actions and results. The task performance activities are measured from four perspectives: financial, customer, internal business processes,

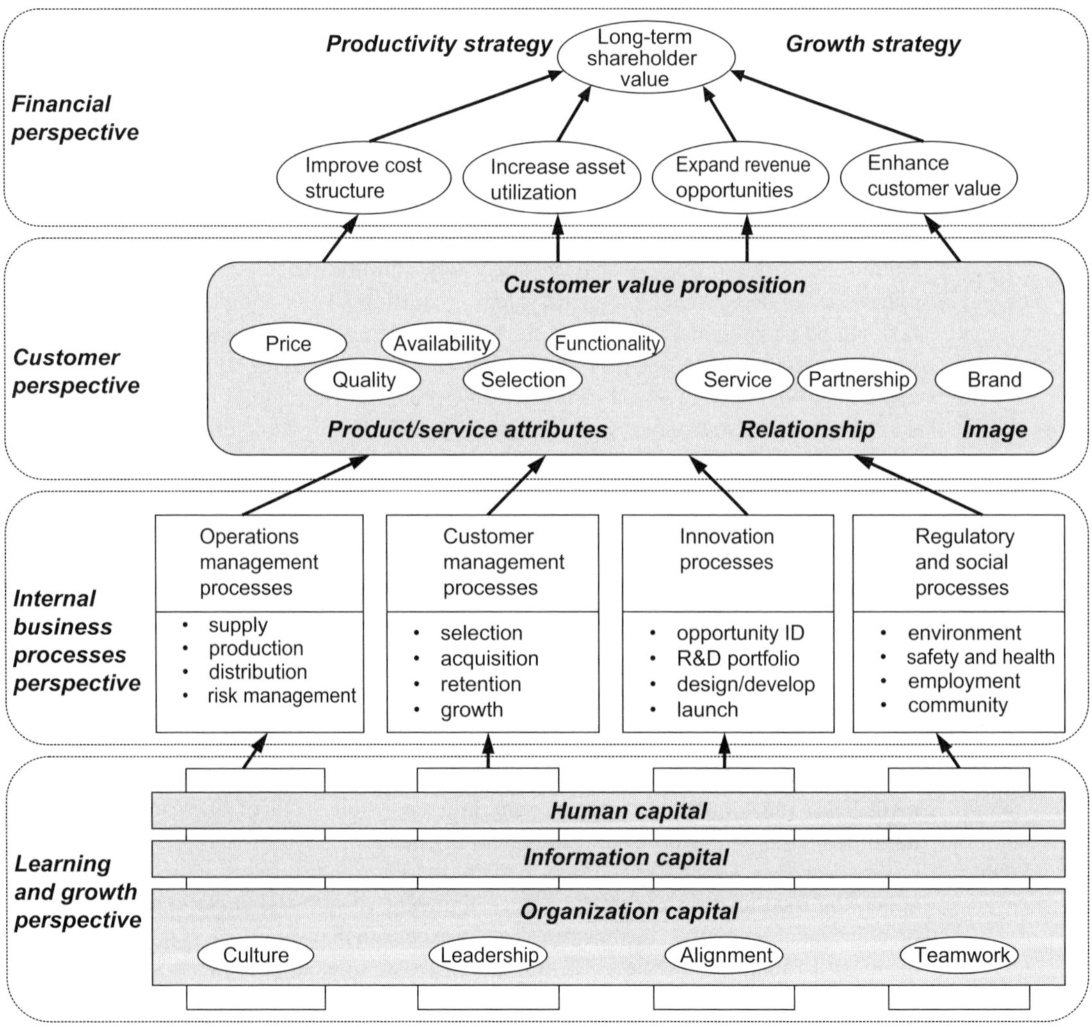

Figure 10.1 A strategic map on how an organization creates value

and learning and growth, with respect to established targets of the assigned tasks. The performance measurements permit objective setting and alignment of the organization's goals and strategic priorities by focusing on a balanced set of performance measures. The Balanced Scorecard does not replace the traditional financial statements, which address the organization's tangible assets, but it complements the development of intangible assets to enhance value for the organization. Common intangible assets recognized by organizations include technology, innovation, intellectual property, alliances, management capabilities, employee relations, customer relations, community relations and brand value. Most industry operations share similar intangible assets in doing business, although there may be some differences among industries requiring specific refinements of methodology for practical implementation. The link between these intangible assets and value creation is the corporate strategy, hence the strategic mapping exercise. The investments made to enhance intangible assets usually provide indirect rather than direct benefits to the organization. The focus on value creation forces an organization to adopt a strategy to align its resources for achievement of long-term goals.

In Figure 10.1, the fundamental resources of an organization are derived from ownership properties and capital, human and information assets, and are fostered by organizational infrastructure. These resources are nurtured by learning and growth of an adaptive corporate culture to promote leadership and teamwork for alignment of organizational activities that influence the internal business processes. Customer value propositions are derived from appropriate selection of relevant performance attributes provided by the internal business processes. Organizations focusing on customer value will lead to productivity improvement in business and growth for long-term shareholder value.

10.3 THE VALUE CHAIN PROCESS

The concept of the value chain was introduced by Michael Porter, Professor at Harvard Business School, in his 1985 book on *Competitive Advantage* [3]. The value chain describes a set of coordinated activities that run efficiently to add value to the organization's products and services. Porter highlighted the competitive advantage and distinct capabilities of value to improve profit margin and enhance customer satisfaction. The value chain allows alignment of processes with customers to generate quality advantage by focusing on cost management efforts supported with efficient processes to sustain and improve operations. The value chain thus helps managers identify the activities that are especially important for competitiveness attainment of the organization's overall strategy.

The value chain consists of two main groups of activities. The primary activities are directly concerned with the production or delivery of certain products or services. The support activities reinforce the efficiency and effectiveness of the primary activities. Porter's value chain framework is a model that helps analyze specific activities through which the firms or organizations can create value and gain competitive advantage. Figure 10.2 illustrates Porter's value chain framework.

(a) Primary activities

Inbound logistics: The raw materials are obtained from the organization's suppliers ready to be used for producing the end product. Typical activities include receiving, storing, inventory control, transportation and scheduling.

Operations: The raw materials received by the organization are processed into the final product. Value is added to the manufactured product at this stage as it moves through the production process. Typical activities include machining,

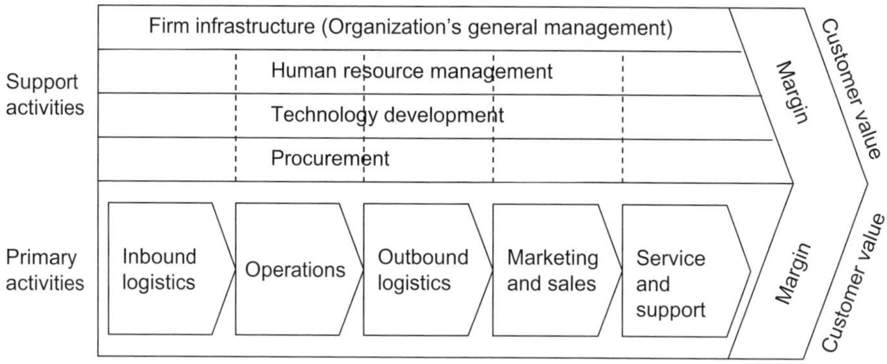

Figure 10.2 Porter's value chain framework

packaging, assembly, testing and verification to transform inputs into desirable outputs.

Outbound logistics: The finished end products are ready to be shipped to distribution centers, wholesalers, retailers or directly to the customers. Typical activities include packing and shipping, warehousing, order fulfillment, transportation and distribution management.

Marketing and sales: Marketing should ensure that the product is targeted toward the correct customer group. The marketing mix is used to establish an effective strategy. Any competitive advantage should be clearly communicated to the target group by promotional means. Typical activities include advertising, promotion, selling, pricing and retail management.

Service and support: After the product has been sold, the organization may offer additional support services such as user training, maintenance and service warranties. Typical activities include customer support, installation, training, repair services, upgrading and spare parts management.

(b) Support activities

Procurement: Purchasing should identify and source the required raw materials for the organization and obtain the best value, price and quality to meet procurement requirements. Typical activities include procurement of raw materials, servicing, spare parts and outsourcing.

Technology development: Technology should be utilized to enhance productivity within the organization. Competitive advantage should be recognized in a technological driven market environment. Typical activities include research and development, process automation and design optimization.

Human resource management: The organization should embrace a participative culture for employee motivation and career development. Staff training and update of the organization's products and services knowledge are essential to gain competitive advantage in the market place. Typical activities include recruiting, retention and compensation and knowledge development.

Firm (organization) infrastructure: The organization should ensure that the finances, legal obligations and management structure are established and well-coordinated to work efficiently in driving the organization forward. Typical activities include general management, planning and management of finance, accounting, pubic relations, legal council and quality management.

It should be noted that some of the examples provided herein are based on Porter's observations in the mid-1980s and might now be outdated due to technology advancement and the changing global business environment over the past two decades. However, Porter's value chain framework is still valid in today's business application. It lays the foundation for business model development for most organizations in the manufacturing and service sectors, with some industry-specific adjustments. It should also be noted that Porter's value chain framework process bears some resemblance to the process model (Chapter 3, Figure 3.3) and the system life cycle process (Chapter 1, Figure 1.5), which form the basis for further value chain model development.

The overall strategic goals of the organization should be oriented toward supporting the achievement of the value chain effort for sustainability of business operation. Value chain analysis may be used for developing organizational infrastructure, identifying the essential core operating functions, determining customer requirements and outsourcing needs, formulating joint ventures, alliances, and marketing strategies, and allocation of appropriate resources to achieve competitive advantage for value creation.

10.4 DEPENDABILITY VALUE FRAMEWORK

10.4.1 Framework Overview

Since its introduction, the value chain has taken various forms and is extensively used by various industries for value engineering, supply chain management, and value-added service applications. In the context of technological systems, the value chain can be represented by the sequence of life cycle stages, which constitutes the primary process for value creation. Each life cycle stage from concept initiation to in-service operation adds value to the process. In this respect, the value chain process has become the delivery system in support of value creation. Figure 10.3 illustrates the dependability value framework for value delivery in system life cycle application.

The dependability value framework is based on the value chain derived from the system life cycle process that adds value in system development and implementation. The system support infrastructure will likely involve multiple organizations in control at different stages and time periods over the system life cycle. For example, the system can be designed by one organization as system developer, manufactured by different organizations as system suppliers, and operated by another organization as system service provider to the customers. In an open-system application, the system is connected to other systems to form a network, such as the Internet or an energy distribution network. In such cases, there are multiple system service providers that need to cooperate with one another to provide a common set of system performance values for customer services. The formula that binds the cooperation of various parties is usually addressed in a contract agreement to meet established quality of service requirements, in which dependability specifications are stipulated.

Most technological systems are complex. It is highly unlikely that a complete new system is developed and built from scratch for any application. This is due to practicality in project implementation, business economics rationale and time and cost constraints, and this is often influenced by regulations in diverse regional jurisdictions when the service crosses distribution boundaries. Hence, new systems are normally built as add-ons to or renewals of existing legacy systems already in service. The new system may also be an

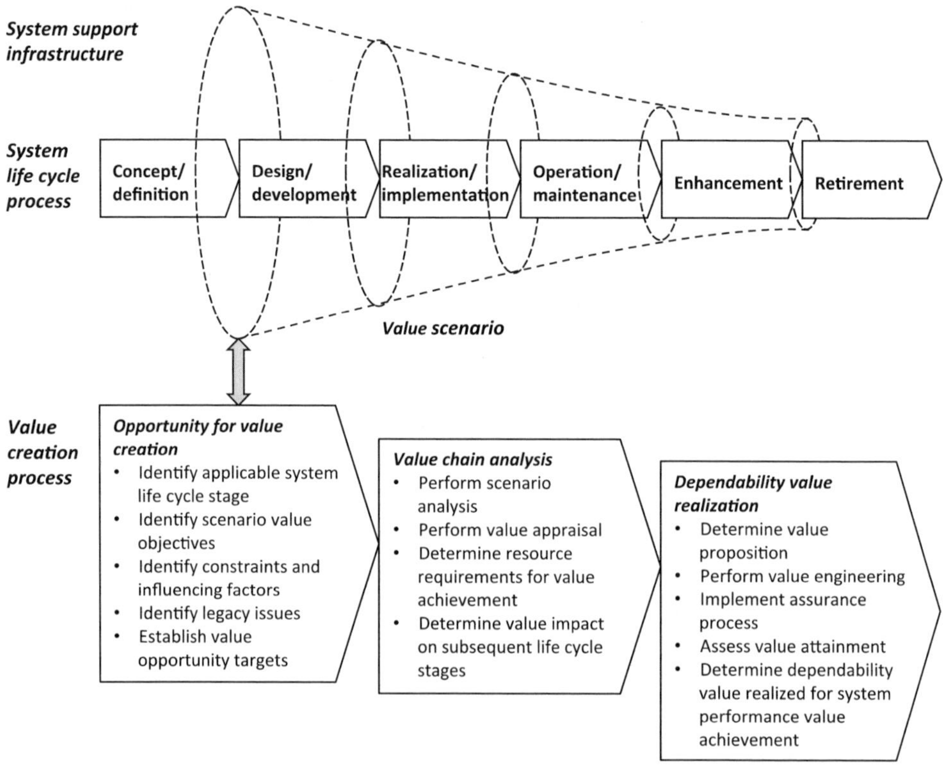

Figure 10.3 Dependability value framework for system life cycle application

integrated set of sub-systems consisting of commercially available equipment assembled to provide the new service operation. Typical examples include automated systems where new application software is developed to provide new features in system operation and add-on service extensions to an existing distribution delivery system. In this respect, dependability value can be appraised at any stage of the system life cycle based on the operational service needs, resource availability conditions and value targets for the new development project. The applicable value scenario has to be clearly defined and addressed to facilitate initiation of the dependability value creation process.

10.4.2 Value Scenario

The value scenario describes the value of a set of system performance attributes appraised at a certain point in time in the system life cycle. The appraised value represents the potential value of system performance capability that can be created at this point. The logical timing for appraisal is at the transition points of the system life cycle stages prior to decisions made for resource commitments. From a dependability performance viewpoint, the appraisal process provides crucial information on the effects of influencing factors affecting the outcomes of subsequent dependability value achievements. The influencing factors include, where applicable, management policy, support infrastructure, design approaches, outsourcing agreements, supply chain, manufacturing process, assurance practices, operating procedures, maintenance and logistic support strategies and customer relations. At the early system life cycle stages in concept/definition, design/development and realization/implementation, the value scenario provides a broader scope and opportunities for potential value creation than at later operation/maintenance and enhancement stages. During the long in-service duration of the operation/maintenance stage, there can be more than one enhancement initiative. In practice, multiple enhancements may be required to adapt to changing market needs with time. Continuous system improvements may be necessary to sustain viable business operations to meet changes in customer requirements. At the point of entering the retirement stage, the potential value creation opportunities will be limited to reuse or salvage value.

The value scenarios are shown in the framework overview (Figure 10.3) as circles encircling the sequence of system life cycle stages. Each circle portrays a view of the value creation opportunities reflecting the potential system performance value status at the time of appraisal. It should be noted that, for dependability value application, these circles decrease in size to indicate the narrowing of the scope of value creation possibilities as they approach the end of the system life cycle process.

10.4.3 Value Creation Process

The value creation process provides a methodology for envisioning the value of system performance, which is its worth as perceived by customers or experienced by end users. This is achieved by appraisal of the opportunity for value creation to maximize value at the applicable system life cycle stage. A major focus for realizing value is on dependability in delivering system performance. System performance characteristics are application specific associated with the system operation, and thus dependability as a system performance attribute varies with the specific system application.

The value creation process engages activities to:

(a) identify the opportunity for value creation;
(b) perform the necessary value chain analysis;
(c) ensure realization of dependability value.

The application of dependability processes is described in Chapter 2, including relevant information on process inputs and outputs, influencing factors and enabling mechanisms for process applications within each system life cycle stage in Appendix B. Tables 10.1 through 10.5 describe typical outputs of the value creation process to realize dependability value at each system life cycle stage.

10.5 DEPENDABILITY VALUE REALIZATION

Figure 10.4 illustrates the value concept relating to a system's worth and investments.

A system's worth reflects the results of the value creation process. The system's worth prior to system realization only represents the appraised value targets. The actual value exists only after the system's tangible and intangible assets have been realized and implemented; they are verified and assured of their value achievements. The tangible asset is the physical structure of the system configuration that contributes to the system operational value. The intangible assets consist of the contribution of system performance value, which includes the dependability performance value. Dependability is highlighted in value realization because of its intended system design and implementation focus. Dependability

TABLE 10.1 Transition from concept/definition stage to design/development stage

a) Opportunity for Value Creation

Activities	Descriptions
• Identify applicable system life cycle stage	At the end of the concept/definition stage prior to decision for resource commitments to initiate the design/development stage.
• Identify scenario value objectives	Optimize achievable dependability value through technology selection and system architecture design approach to meet established system performance requirements.
• Identify constraints and influencing factors	Competition, budget constraints, technology limitations, available resource capability, design and development timeline, outsourcing design feasibility, project risk management.
• Identify legacy issues	Reusable designs, interface with existing systems and interoperability of equipment, knowledge and experience with previous system design issues.
• Establish value opportunity targets	Value comparisons of system design approaches with alternative technology applications; recommendation of design options, selection of best value achievable within established time constraints and budgetary limits to meet delivery targets.

b) Value Chain Analysis

Activities	Descriptions
• Perform scenario analysis	Analysis of achievable dependability value and possible performance outcomes of the identified value objectives in technology selection and system design approach.
• Perform value appraisal	The value added for the selected technology for design application and the design approach for recommending cost-effective implementation. Examples include technology platform designs and material standardization that would benefit multiple product development for expediency and time to market.
• Determine resource requirements for value achievements	The extent of resources needed and allocated for value achievements in the recommended design implementation, including outsourcing requirements.
• Determine value impact on subsequent life cycle stages	Possible value impact related to time and cost savings for implementing the selected technology and design approach on design/development and operation/maintenance effort. Recommendation of make-buy decisions to maximize value for cost-effective solutions. Potential problem areas of value impact such as special design capability and manufacturing capacity should be identified and resolved.

c) Dependability Value Realization

Activities	Descriptions
• Determine value proposition	The results of value chain analysis and possible impact resolution should be presented as a value proposition for recommending dependability value realization.
• Perform value engineering	Where applicable, engineering for value improvement of system functions or reduction in cost of implementation should be identified as part of the value proposition. Examples include process standardization, modularization, and functional design simplification.
• Implement assurance process	Appropriate assurance procedures should be implemented at this stage to ensure dependability value attainment leading to subsequent stages in dependability value achievement.
• Assess value attainment	The extent of value attained at this stage should be verified for assurance purposes.
• Determine dependability value realized for system performance value achievement	Potential dependability value targets in delivering system performance value as perceived by the customer.

TABLE 10.2 Transition from design/development stage to realization/implementation stage

a) Opportunity for Value Creation

Activities	Descriptions
• Identify applicable system life cycle stage	At the end of the design/development stage prior to decision for resource commitments to initiate the realization/implementation stage.
• Identify scenario value objectives	Optimize achievable dependability value through implementation of design process and full-scale system development incorporating recommended outsourcing and make-buy decisions.
• Identify constraints and influencing factors	Design capability, resource commitments, development timeline, outsourcing agreements, supply chain, production capacity, verification and validation capability, special test facility requirements, hardware and software assurance programs, project risk management.
• Identify legacy issues	Applicability of reusable designs, interface and protocol with existing systems for interoperability, knowledge and experience with previous system development and production issues.
• Establish value opportunity targets	Maximizing the value-added process in design and development to achieve seamless transition for system realization and implementation within scheduled project completion targets for production.

b) Value Chain Analysis

Activities	Descriptions
• Perform scenario analysis	Analysis of achievable dependability value objectives identified in the design process and full-scale system development, including value benefits in outsourcing of alternatives for make-buy decisions.
• Perform value appraisal	The value added for design process and full-scale system development and recommendations for cost-effective implementation. Examples include preferred suppliers, shared inspection schemes and supply chain management.
• Determine resource requirements for value achievements	The extent of resource needed and allocated for value achievements in design process implementation and full-scale system development, including outsourcing, test facilities and enabling support requirements.
• Determine value impact on subsequent life cycle stages	Possible value impact related to design process improvement and standardization of development procedures, and verification of outsourcing requirements that may impact production and system realization and implementation.

c) Dependability Value Realization

Activities	Descriptions
• Determine value proposition	The results of value chain analysis and possible impact resolution should be presented as a value proposition for recommending system realization for cost-effective implementation.
• Perform value engineering	Where applicable, engineering for value improvement of system design and cost-effective implementation for product and system realization. Examples include common design feature functions and interchangeable assemblies to facilitate production process, inventory control and spares provisioning.
• Implement assurance process	Appropriate assurance procedures should be implemented at this stage to ensure dependability value attainment leading to realization in dependability value achievement.
• Assess value attainment	The extent of value attained at this stage should be verified for assurance purposes.
• Determine dependability value realized for system performance value achievement	Potential dependability value targets in delivering system performance value as perceived by the customer.

TABLE 10.3 Transition from realization/implementation stage to operation/maintenance stage

a) Opportunity for Value Creation

Activities	Descriptions
• Identify applicable system life cycle stage	At the end of the realization/implementation stage prior to customer delivery and acceptance for system operation.
• Identify scenario value objectives	Optimize achievable dependability value through verification and validation process for customer acceptance for system in-service operation. Dependability performance requirements and incentives, as well as customer training value, should be considered to avoid costly maintenance support services, especially under warranty.
• Identify constraints and influencing factors	System installation, delivery and commissioning schedule, customer training, warranty requirements and incentives, customer support, transition management.
• Identify legacy issues	System interoperability with existing systems and equipment, knowledge and experience with system commissioning process. Previous customer relations and established cooperative effort.
• Establish value opportunity targets	Value perceived and/or experienced during cooperation in system acceptance and transition management. Established criteria for response time for customer support services.

b) Value Chain Analysis

Activities	Descriptions
• Perform scenario analysis	Analysis of achievable dependability value to validate system performance outcomes in the system acceptance and transition process.
• Perform value appraisal	The value added for cooperative effort to achieve system acceptance and support system performance objectives. Examples include simplification of customer training needs and maintenance support procedures to expedite system acceptance process.
• Determine resource requirements for value achievements	The extent of resources needed in customer support services to sustain system operation and maintain cooperative customer relations.
• Determine value impact on subsequent life cycle stages	Possible value impact related to failure-free operation and timely response to support system operation and maintenance effort on customer premises.

c) Dependability Value Realization

Activities	Descriptions
• Determine value proposition	The results of value chain analysis and possible impact resolution should be presented as a value proposition for efficient customer support services to minimize warranty cost and improve customer relations.
• Perform value engineering	Where applicable, engineering for value improvement of system operation and maintenance procedures and logistic support strategies to optimize inventory holdings and cost-effective spares provisioning.
• Implement assurance process	Appropriate assurance procedures should be implemented to ensure operational dependability value achievement and effective maintenance support process.
• Assess value attainment	The extent of value attained in system operation and maintenance should be verified for assurance purposes.
• Determine dependability value realized for system performance value achievement	Potential dependability value in delivering system performance value as perceived by the customer.

TABLE 10.4 Transition from operation/maintenance stage to enhancement stage

a) Opportunity for Value Creation

Activities	Descriptions
• Identify applicable system life cycle stage	During the operation/maintenance stage prior to decision for resource commitments to initiate the enhancement effort.
• Identify scenario value objectives	Extent of value added with design upgrade or incorporation of new service features to sustain market competitiveness and customer demands.
• Identify constraints and influencing factors	Customer demands for enhanced features and performance improvements. Market competitiveness of new service features incorporation.
• Identify legacy issues	New service features incorporation in existing system infrastructure to achieve system performance enhancement.
• Establish value opportunity targets	Value of new service features for system performance enhancement to meet competitive performance leverage and value targets.

b) Value Chain Analysis

Activities	Descriptions
• Perform scenario analysis	Analysis of achievable value objectives in the enhancement effort.
• Perform value appraisal	Determination of the value added for the enhancement effort. Examples include improvement of system enhancement process and service update procedures to minimize downtime frequency and duration affecting continuity of customer services.
• Determine resource requirements for value achievements	The resource needed to justify the enhancement effort for return on investment.
• Determine value impact on subsequent life cycle stages	Possible return on investment for the extended duration of system in-service operation.

c) Dependability Value Realization

Activities	Descriptions
• Determine value proposition	Value chain analysis results to justify the need for enhancement effort for return on investment within reasonable time period. Timing for implementing enhancement should be rationalized.
• Perform value engineering	Not applicable for minor enhancement effort due to short-term improvement project that may not justify need for value engineering effort.
• Implement assurance process	Appropriate assurance procedures should be implemented for any enhancement effort.
• Assess value attainment	The extent of value attained in enhancement should be verified for assurance purposes.
• Determine dependability value realized for system performance value achievement	Potential dependability value in delivering system performance value as perceived by the customer.

TABLE 10.5 Transition from enhancement stage to retirement stage

a) Opportunity for Value Creation

Activities	Descriptions
• Identify applicable system life cycle stage	At the start of the retirement stage.
• Identify scenario value objectives	Timing for termination of service.
• Identify constraints and influencing factors	Impact on continuity of service provision, replacement of terminated service, disposal cost of obsolete equipment.
• Identify legacy issues	Disposability of obsolete equipment.
• Establish value opportunity targets	Reuse or salvage value of obsolete equipment.

b) Value Chain Analysis

Activities	Descriptions
• Perform scenario analysis	Not applicable.
• Perform value appraisal	Determining the reuse or salvage value of obsolete equipment and the cost of disposal and impact to the environment.
• Determine resource requirements for value achievements	Not applicable.
• Determine value impact on subsequent life cycle stages	Not applicable.

c) Dependability Value Realization

Activities	Descriptions
• Determine value proposition	Not applicable.
• Perform value engineering	Not applicable.
• Implement assurance process	Appropriate assurance procedures should be implemented for disposition of obsolete equipment to minimize impact to the environment.
• Assess value attainment	Not applicable.
• Determine dependability value realized for system performance value achievement	Not applicable.

value is an intangible asset reflecting the intrinsic worth of a system's ability to perform on demand. The dependability performance is dependent on the functional capability of the operating system in delivering value outcomes. Dependability provides a measure of valuation of the system performance achievement. The customer perceived value is the combined system in-service operational value and the system performance value.

The investments needed for system support reflect the realization of the infrastructure value. The investments are expressed in terms of costs at the appropriate system life cycle stages. Investments are essential to support system design and development where the system architecture and configuration are specified to meet dependability value targets for full-scale development. Investments are required for subsequent resource commitments for system realization and implementation when the system is built, verified and tested for in-service operation. The system developer is normally accountable for resource commitments and funding of the system infrastructure support up to this point prior to transfer of ownership responsibility to the system service provider upon system acceptance. Although the system developer still has responsibility for system warranty, the extent of warranty incentives is subject to contract agreements with the system service provider during the warranty period.

The system service provider is the owner of the system after system acceptance. Investments are needed to sustain continual system operation and maintenance support.

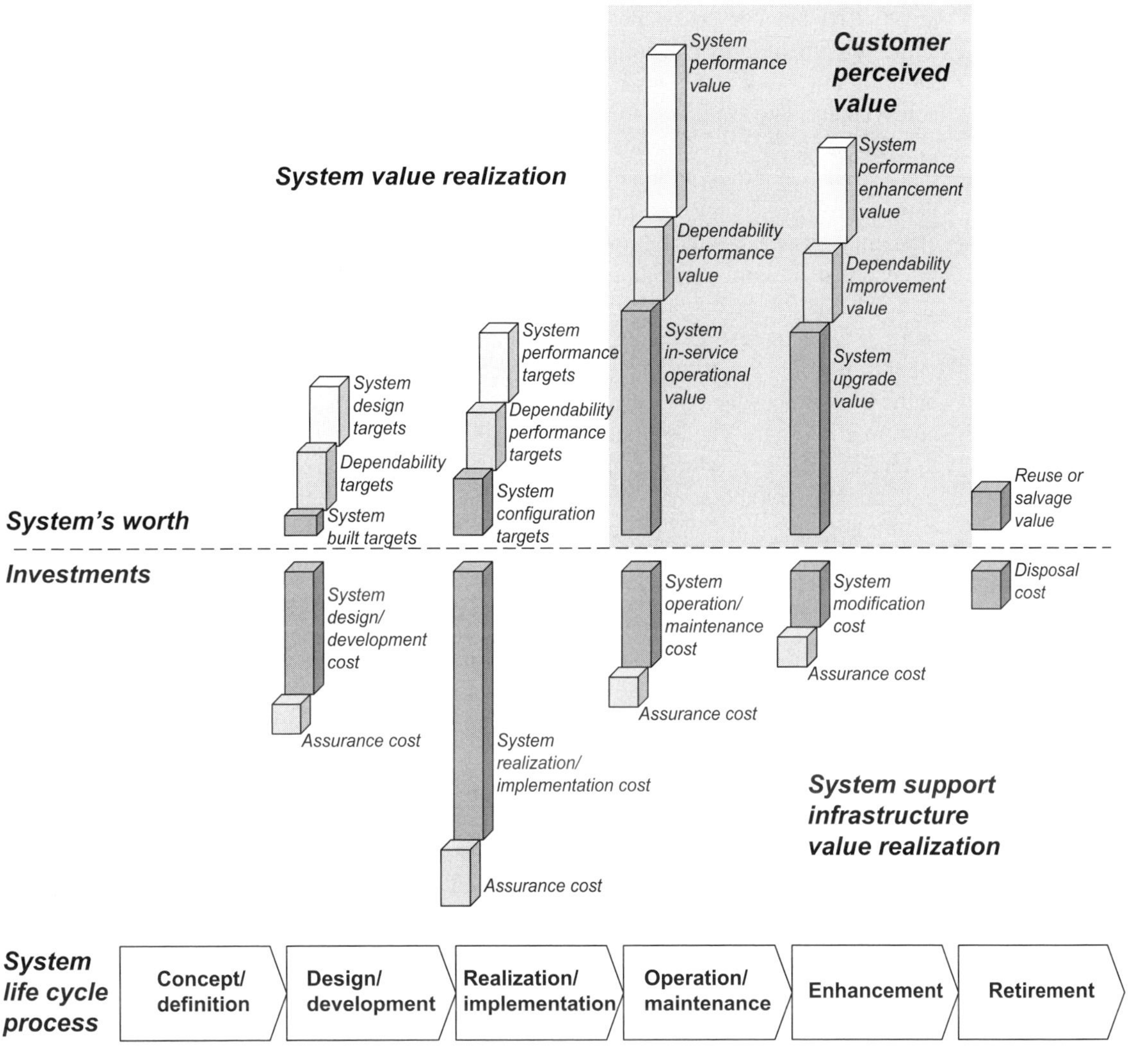

Figure 10.4 Value concept relating to a system's worth and investment

Subsequent and periodic enhancement needs for system upgrade and modification would be required to ensure market competitiveness of system performance value as perceived by the customers or experienced by the end users. At the retirement stage, the system residual value is only represented by its tangible value for reuse or salvage. It should be noted that there is no significant system residual value if the disposal cost equals or exceeds the tangible reuse or salvage value.

10.6 ASSURANCE IN VALUE ACHIEVEMENT

Assurance plays an important role in fostering value achievement at various system life cycle stages. The assurance process ensures confidence that value objectives have been or can be fulfilled. Dependability assurance ensures that the capability of system performance targets have been or can be met. In this respect, the dependability activities are focused on value attainment in system realization throughout the system life cycle stages. The collaboration of assurance attributes contributing to system performance is associated with value achievement. The assurance process is addressed in Chapter 9 (Figure 9.2).

The assurance of dependability value is closely linked to the organization's investment strategies in strengthening its business operation. The investments should focus on new capabilities, innovative processes and experienced know-how in value creation and efficiency in performance improvement. Assurance procedures should be implemented to ensure successful value attainment through the process in value realization. Priorities should be set for the opportunities identified at various system life cycle stages for achievement of value targets. Dependability assurance activities should not be limited to enhancement of tangible assets or solely for product improvement, but also to increase proficiency in the work process and training of people, simplify administrative procedures, and improve customer relations. The assurance process should engage up-to-date methods and IT tools to leverage the key value drivers to enable achievement of significant value on return on investments. Collaboration with customers and suppliers should be in the business plan to ensure alignment of their value propositions and adapting to changing market needs and business environments.

Technology and resource investments should not be viewed as short-term ongoing business survival tactics. Long-term impact of such investments should be assured for business development and growth. There are advance indicators or early warnings to signal that the organization's products and services are approaching technological obsolescence. This is often associated with the deployment of a non-versatile workforce incapable of timely transition to new or renewed operation. The assurance objectives should engage in all aspects of the business process to enable realistic business transformation where technology advancements and knowledgeable workforce go hand in hand in progress for business enhancement. Successful organizations are constantly looking for innovative ideas that create competitive advantages and differentiated capabilities for leverage in the market. Business survival hinges on the strategic emphasis on continuous improvement and organizational capability to sustain viable operation in times of changing business environment. The success formula for dependability value creation links the strategic objectives to process improvement, technology innovation, people empowerment, stakeholder interests, and cultivates a creative and cooperative organizational working environment. The assurance process is essential to ascertain value attainment. It ensures management capable of overseeing and tracking the strategic, operational and tactical operations of the organization.

10.7 VALUE INFRASTRUCTURES

10.7.1 Dependability Value Expressions

The value of dependability is related to the ability of functional and non-functional requirements to be satisfied from a time perspective. The value of dependable operation is in enhancing availability and reliability as well as in avoiding consequences caused by cessation of required functions that are not protected by a backup system.

In general, dependability value can be expressed in the following ways.

1. *Safety is enhanced.*
   In many industries, such as transportation, safe execution of the service is of paramount importance. Businesses go to great lengths to ensure that no injuries or

deaths occur. Hopefully no person is under the illusion that all risks can be eliminated. Moreover, there may be different acceptable safety levels for the public as opposed to industry employees.

2. *Customer or user satisfaction is achieved.*
   In particular for customer products and services, satisfaction is the measure of success, even though it is unlikely that everyone will be equally satisfied. This satisfaction will be linked to the performance of the product or the service and whether any product failures or service interruptions are experienced. Availability upon demand is important to the user or customer.
3. *Life cycle cost is optimized.*
   The cost of new system development and equipment acquisition for new design should seek an appropriate balance with the cost of system operation and service provision. The dependability value is reflected in the optimization process of life cycle costing. The cost of operation of the existing system and equipment should seek an appropriate cost-effective approach to reduce preventable failure occurrences by good operating and maintenance practices. The dependability value is to sustain service continuity with minimum cost and avoid adverse consequences and potential risk exposures.
4. *Maximum asset life can be attained.*
   Dependable products and systems are much more likely to have been designed for long life, something that is most important for infrastructure and involves expensive assets. As long as the failure rate and associated costs do not increase dramatically, sustaining longer service operation maximizes asset life and reduces life cycle costs.
5. *Environmental impact is minimized.*
   Failures can seriously impact emissions and environmental damage due to loss of containment of hazardous substances or increased emissions. Environmental sustainability is enhanced by incorporating dependability in the system design and its implementation, and is further realized by proper operation and maintenance efforts.
6. *Reputation is maintained or enhanced.*
   Although this is challenging to quantify, a loss of reputation can impact business value in many ways, such as the stock price, and may result in a loss of market for products that could even lead to the end of an organization and its business survival.

10.7.2 Common Value Infrastructures

There are common value infrastructures among industries that share similar dependability value objectives relating to customer-perceived value of performance services. The following examples in Table 10.6 illustrate the common value infrastructures in delivering dependability value.

10.7.3 Asset Management

Asset management is also closely linked to obtaining the maximum value for any organization that is heavily dependent on its assets. Asset management is defined in PAS 55 [4] as "systematic and coordinated activities and practices through which an organization optimally and sustainably manages its assets and asset systems, their associated performance, risks and expenditures over their life cycles for the purpose of achieving its organizational strategic plan." In its focus on performance, risk management and expenditures over the

TABLE 10.6 Common system infrastructures

Common Value Infrastructures	Value Creation Sources	Value Chain and Realization Process	Dependability Value as Perceived by Customers
Telecommunications	Voice/data/video	Wire-line and wireless transmission and distribution in delivering voice/data/video information, availability and reliability of end-to-end process in transmission, distribution and reception	Subscriber satisfaction, security and integrity in dependability of service
Information technology	Intellectual property, proprietary information and secured data, resources required to acquire, process, store and disseminate information	Application of computerized systems for routine processing of both data and analysis to provide useful information, utilization of multimedia information access, facilities for electronic data storage retention and retrieval, security protection schemes for sensitive data	End user confidence and trust, security and integrity of processed information, efficiency of electronic data access, retrieval and storage
E-commerce	Monetary exchange and transfer, financial transactions, bank accounts, credit notes	Auditable monetary account transactions, owner account identity verification and control	Customer confidence and trust, credibility and expediency of financial transaction services
Electric power	Various energy sources and conversion methods for hydro, nuclear, fossil, fuel, wind, solar, bio energy generation	Energy conversion, high voltage transmission, distribution stations and substations, power outlets	End user satisfaction of utility services, availability and efficiency of energy supply, safety applications
Oil and gas	Energy extraction methods for crude oil and natural gas	Wells, processing plants, refineries, transportation, pipelines, service depots, distribution to industry, homes and offices	Delivery of efficient oil/gas production, public/employee safety, structural integrity and secure supply in transportation methods, recovery process, environmental impact, availability and dependability of services
Transportation	People, merchandise	Transport and shipment via land/sea/air, depots, warehouses, distribution networks to wholesalers and retailers, insurance, traceability and recovery process	Customer satisfaction in delivery of services, safety and security concerns of people and merchandise in transportation
Sales and marketing	Products and services for market promotion	Multimedia, radio, TV, Internet, Web services, ordering and delivery process, cancellation, return and refund policy	Customer satisfaction, credibility of advertisement, value for money spent on product/service ordered

life cycle, asset management mirrors the basic principles of dependability and shares the same benefits.

In fact, it can be said that asset management is not possible without due attention to dependability of performance and service. In this sense, dependability is a key enabler assisting asset management in delivering the value of assets. Some of the tools and methodologies mentioned in PAS 55-2, such as life cycle costing, reliability-centered maintenance and risk-based inspection, point directly to dependability as the means of reducing asset risk [5]. Among the asset performance targets that can be measured, all of the characteristics of dependability and associated factors are noted, such as levels of service/supply, functionality, survivability, capacity, customer satisfaction, safety and/or environmental impact. All of these combine to create value for an organization and its stakeholders.

10.7.4 A Pipeline Example

The actual value proposition adopted by an organization operating a pipeline will vary with its stakeholders. Starting with its customers, value is created by dependable delivery of the product being transported. The target will be 100% availability of the design capacity with no interruptions in service. By its very nature, a pipeline needs to consider the public as a major stakeholder. The value proposition of the public is focused on safe and environmentally acceptable operation and the prevention of incidents that can lead to personal, property or environmental damage. The value for shareholders is not only financial gain but also reputation and avoidance of major consequences that could threaten the ability to continue operation. Finally, employees have a stake in personal safety and work satisfaction. Figure 10.5 combines these value propositions into a balanced approach that points to the necessity for compromise between them. This then leads to risk management as a fundamental means of ensuring that they can be met. It is clear that dependability is a critical factor in most of these value propositions.

Satisfying dependability performance for a pipeline happens at several levels, starting with the pipeline system as a whole and supported by specific and different approaches for the pipe portion and the compression or pumping facilities. The pipeline system level is essentially a network consisting of pipe and facilities with input and delivery points. The delivery is naturally also dependent on adequate supply volumes, but this has to be assumed and so will not be considered further.

Delivery from a pipeline system is measured by availability as a function of flow with expectation by the customer that contracted volumes will be met. For a gas pipeline, even with loss of compression, expected volumes can often be made up due to changeable linepack and delivery requirements satisfied unless downtime is extensive. For oil pipelines, unless it is operating well below capacity, this may not be possible. For this reason, redundancy for pumping is more crucial than for compression.

Compared to pipe, compressors are less reliable and require service downtime for maintenance. This would argue for less compression and larger pipe diameters except for the fact that installing pipe is considerably more costly than compression. Determining the most effective tradeoff options can be conducted by dependability analyses. Compressor unit and station availability studies play a fundamental role in providing information for options that will support decision making, such as installing stand-by units.

The value of dependability is centered on ensuring safety of the public, employees and contractors as well as environmental protection while minimizing total cost over the long

<u>CUSTOMERS</u> 100% availability of design capacity No service interruptions Product quality	<u>PUBLIC/SOCIETY</u> No personal incidents/accidents No damage/loss of property No environmental damage
<u>SHAREHOLDERS</u> Financial gain No loss of reputation No loss of capacity	<u>EMPLOYEES</u> No personal incidents/accidents Work satisfaction

Figure 10.5 The value proposition for pipelines according to the balanced approach

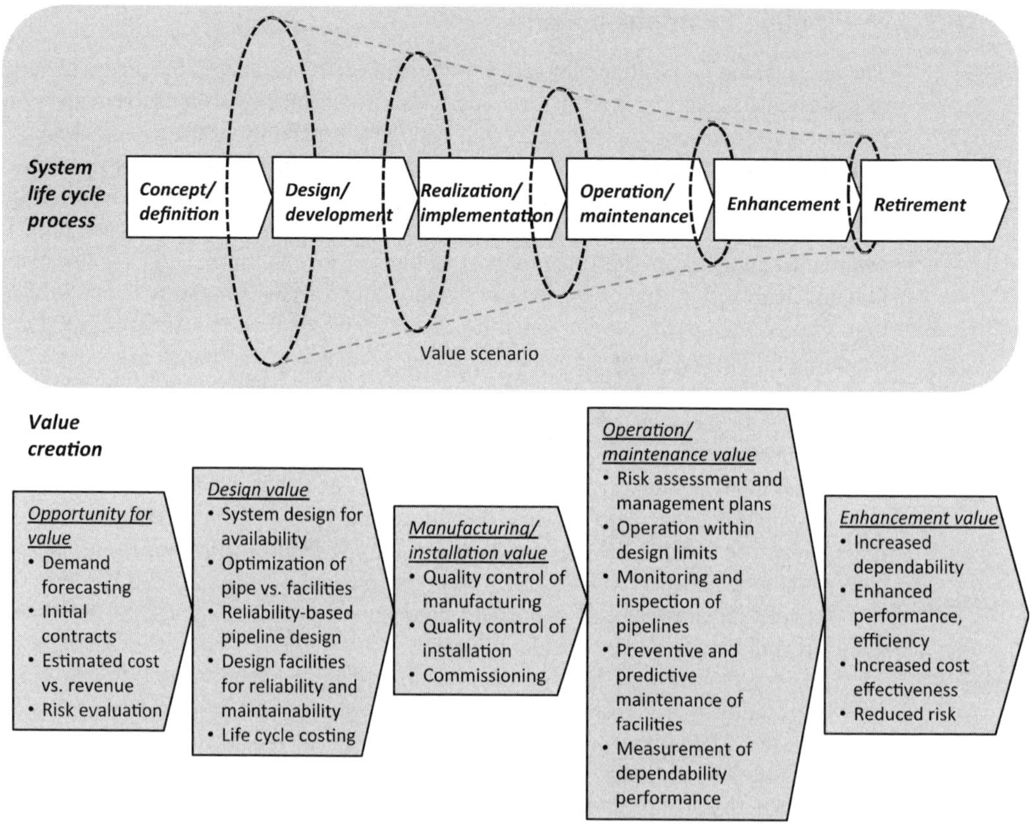

Figure 10.6 Dependability value framework for system life cycle application

term. Being proactive in mitigating risk is especially critical toward the end of a pipeline's life. The risk for pipelines increases substantially as corrosion advances and begins to show in more sections of the pipeline. At some point, replacement of major sections is needed, and it is desirable from a cash flow point of view to spread these over a longer time period. The pipeline may even face being shut down by the regulator. For oil pipelines, environmental damage from major leaks or ruptures may cause extended shutdowns of the line due to public pressure and cleanup can be extremely costly financially and in terms of corporate reputation, as recent experience has shown.

Creating value from dependability for compression or pumping facilities leads primarily to these benefits: improved safety, high availability and reduced costs. Improved safety applies mainly to employees and contractors and not so much to the public, so it is less critical for compressor and pump facilities than for pipelines. Availability is linked to meeting delivery contract requirements and providing customer satisfaction. The impact of compressor/pump downtime depends heavily on the number of stand-by units installed and, especially for gas pipelines, the flexibility of the pipeline itself in handling short-term downtime, which will guide the tradeoff between availability and capital and operating costs.

The value chain in the framework of the life cycle can be interpreted for pipelines and facilities as shown in Figure 10.6.

It has been shown that dependability value is evident in three distinct but interrelated aspects of pipeline operation: the pipeline system, the pipeline and compression/pumping facilities. The key factors in ensuring that dependability value is realized include the following:

- be proactive and not reactive;
- use long-term thinking and planning;
- consider all relevant aspects of dependability during the different stages of the life cycle;
- satisfy stakeholders in a balanced way;
- understand the nature of the assets and apply the right dependability techniques;
- apply necessary resources;
- measure and analyze dependability data and use the analysis results to support decision-making.

The importance of dependability in creating value for pipeline organizations is clear and can be summarized as shown below:

- Dependability has far-reaching implications and influences all aspects of achieving successful pipeline operation.
- Dependability sustains viable business ventures.
- Dependability ascertains value investments for long-term stakeholder benefits.
- Dependability implies trustworthiness in delivery of service performance.

REFERENCES

[1] Robert S. Kaplan and David P. Norton, *Strategy Maps: Converting Intangible Assets into Tangible Outcomes,* Harvard Business Press, 2004.
[2] Robert S. Kaplan and David P. Norton, *The Balanced Scorecard: Translating Strategy into Action,* Harvard Business Press, 1996.
[3] Michael E. Porter, *Competitive Advantage: Creating and Sustaining Superior Performance*, The Free Press, (1st edition) 1985.
[4] PAS 55-1:2008, "Asset Management, Part 1: Specification for the optimized management of physical assets," The Institute of Asset Management, British Standards Institute.
[5] PAS 55-2:2008, "Asset Management, Part 1: Guidelines for the application of PAS 55-1," The Institute of Asset Management, British Standards Institute.

Appendix A

GLOSSARY OF TERMS

A.1 INTRODUCTION

A.1.1 Terms and Definitions

As with most fields of endeavor, definitions of standard terms are important to common understanding and transfer of knowledge. Not surprisingly, there are various sources of definitions for the areas associated with dependability with the usual range of variations in definitions. The definitions in this appendix are intended to support the chapters of this book and are thus not complete as a dictionary of dependability engineering terms.

A.1.2 Concept Diagrams

In order to illustrate some of the relationships between the terms, several concept diagrams have been developed based on the understanding of the authors. Many types of concept diagrams exist. The method used in this appendix is to define three primary forms of concept relationships: generic, partitive and associative.

Generic relations are shown as solid lines. They depict a hierarchy where the subordinate term inherits all of the characteristics of the higher or superordinate term. An example is the generic term system, which breaks down into the more specific terms of subsystem or equipment (Figure A.1).

Where subordinate terms form parts of another term, the relationship is called partitive. These are shown as a rake. For example, hardware is either repairable or non-repairable (Figure A.1).

Associative relationships are more general in that they describe an association between two terms such as cause and effect, location and activity. These are shown as dotted lines. For example, the operation of software is sustained by means of maintenance support, whereas the security of software performance is influenced by the integrity of data (Figure A.1). A similar argument exists where the failure mode can be identified as a fault associated with hardware or software. A failure symptom is discovered that can be categorized as the failure cause due to hardware defect or software error (Figure A.2).

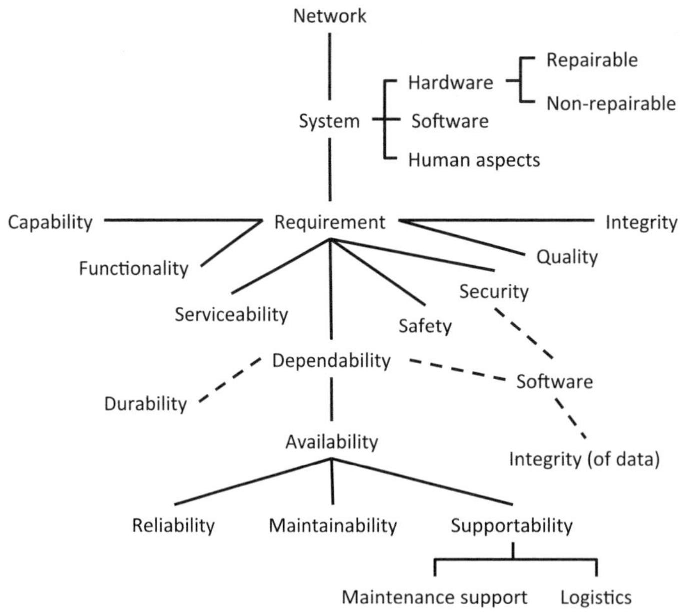

Figure A.1 Concept Diagram for Terms Related to System and Dependability

A.2 CONCEPT DIAGRAM FOR SYSTEM AND DEPENDABILITY-RELATED TERMS

The dependability of systems entails both physical assets depicted by terms such as subsystems, equipment, devices, components for hardware and virtual assets such as software and human aspects. The major relationship between a system and dependability is that dependability is one of the requirement characteristics in the performance expectations of a system.

A.3 CONCEPT DIAGRAM FOR RELIABILITY-RELATED TERMS

Reliability addresses system failure. Hardware and software failures relate to fault, defect and error. A fault is usually the result of an error made in design, operation, maintenance or a defect in material. Ageing will cause system failure. Wearout will result in hardware failure.

A.4 CONCEPT DIAGRAM FOR MAINTENANCE-RELATED TERMS

Maintenance strategy drives the design features in maintainability as a design characteristic. The associated maintenance program directs the supportability of the system and defines the logistic support policy. Maintainability influences maintenance actions and system supportability. Maintenance actions can be preventive or corrective. Scheduled maintenance is planned maintenance activities, including perfective maintenance for software updates. Supportability is associated with maintenance support and logistics to facilitate all maintenance actions (Figure A.3).

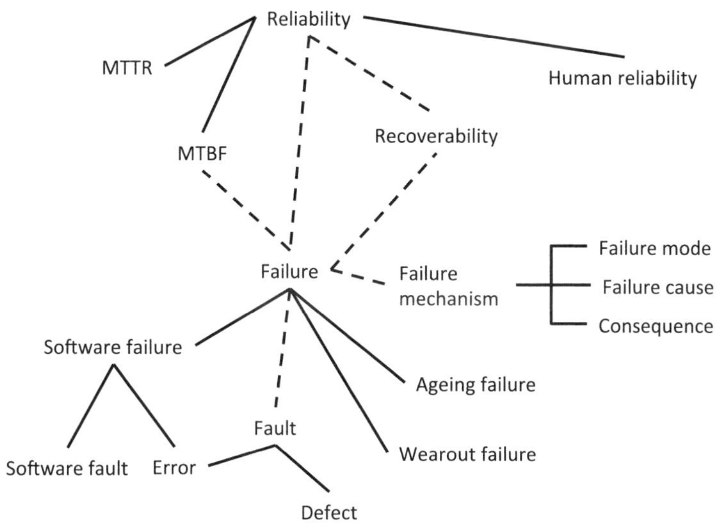

Figure A.2 Concept Diagram for Terms Related to Reliability

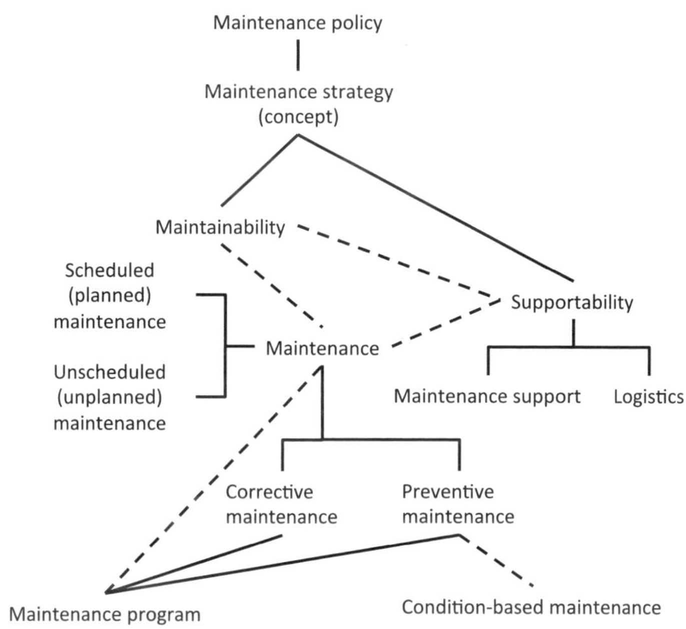

Figure A.3 Concept Diagram for Terms Related to Maintainability and Supportability

A.5 TERMS AND DEFINITIONS

ageing failure – failure whose probability increases with time due to cumulative deterioration; a physical or chemical phenomenon that involves changes in characteristics of the material with time, in some circumstance related to interaction with its environment

architecture – fundamental concepts or properties of a system in its environment embodied in its elements, relationships and principles of its design and evolution

assurance – grounds for justified confidence that a claim has been or will be achieved

assurance case – a reasoned, auditable argument created to support the contention that its claims are satisfied

audit – systematic, independent and documented process for obtaining audit evidence and evaluating it objectively to determine the extent to which agreed criteria are fulfilled

availability – degree to which a system is in a state where it can perform as required

capability – ability to meet a service demand of given quantitative characteristics under given internal conditions

code – character or bit pattern that is assigned a particular meaning to express a computer program in a programming language

condition-based maintenance – preventive maintenance based on the assessment of physical condition by operator observation, conducted according to a schedule, or by condition monitoring of system parameters

conformity – fulfillment of a requirement

consequence – quantitative or qualitative outcome of an event that affects objectives

constraint – externally imposed limitation on system requirements, design or implementation, or on the process used to develop or modify a system

continual improvement – recurring activity to increase the ability to fulfill requirements

corrective maintenance – maintenance carried out after detecting a fault or failure to restore functions

criteria – set of policies, procedures or requirements determined as a reference

customer satisfaction – customer's perception of the degree to which the customer's requirements have been fulfilled

data – collection of values assigned to base measures, derived measures and/or indicators

data analysis – systematic investigation of the data and their use in a real or planned system

defect – non-fulfillment of a requirement related to an intended or specified use

degradation – detrimental change in ability to meet requirements

dependability – ability to perform as and when required depending upon its application; also used as a collective term for the time-related quality characteristics of a product or service

dependability of service – effect of providing the required dependability performance for user services

dependability risk factor – potential problem or issue associated with dependability risk exposures that would have negative effects to a planned dependability activity, characterized by the probability of the dependability risk occurrences and a potential loss should the dependability risk occur

durability – ability to perform as required, under given conditions of use and maintenance, until a limiting state of degradation is reached such as when it is no longer considered as a viable asset, for reasons that may include obsolescence, economics, technology and legislation

element – combination of hardware, software, information and/or human components that form the basic building block to perform a distinct function

embedded software – software within a system whose primary purpose is not computational

enabling system – system that complements a system of interest during its life cycle stages but does not necessarily contribute directly to its function during operation

ergonomics – study of scientific information concerning humans to the design of objects, systems and environment for human use incorporating elements from many subjects including human anatomy, physiology and psychology in the design; ergonomics is sometimes used interchangeably with human factors engineering, although there are minor differences in approach

error – discrepancy between a computed, observed or measured value or condition and the true, specified or theoretically correct value or condition

event – occurrence of a particular set of circumstances

failure – loss of ability to achieve required performance

failure cause – set of conditions that lead to failure originating during specification, design, manufacture, installation, operation or maintenance

failure mechanism – process that leads to failure that may be physical, chemical, logical or a combination thereof and described by failure mode, cause and consequence

failure mode – manner in which failure occurs

failure modes, effects and criticality analysis (FMECA) – quantitative or qualitative method of analysis that involves failure modes and effects analysis together with a consideration of the probability of the failure mode occurrence and severity of the consequences

fault – inability of the system to perform as required, due to an internal state

fault tree analysis (FTA) – analysis using a logic diagram that shows the faults of system functions, external events or combinations thereof, which causes a predefined, undesired system event

function – elementary operation performed by the system which, when combined with other elementary operations (system functions), enables the system to perform a task

human factors – human abilities, limitations and other human characteristics that are relevant to the design and application of products affecting human-system performance

human factors engineering – application of human factors knowledge to the design of tools, machines, systems, tasks, jobs and environment for safe, comfortable, efficient and effective human use

human reliability – study of human performance in terms of probability that a person will correctly perform some system-required activity during a given time period (if time is a limiting factor) without performing any extraneous activity that can degrade the system

information – facts, data or instructions in any medium or form

information management – information processing system that functions for controlling the acquisition, analysis, retention, retrieval and distribution of information

information processing – systematic performance of operations upon information, which includes data processing and may include operations such as data communication and office automation

inherent or intrinsic availability – availability value provided by the design, under ideal conditions of operation and maintenance but excluding delays associated with maintenance, such as logistic and administrative delays

inspection – conformity evaluation by observation and judgment accompanied as appropriate by measurement, testing or gauging

integrated logistic support – process during design and development to manage the provision of all materials and resources required to meet the needs for operation and maintenance

integrity – ability of a system to sustain its form, stability and robustness, and maintain its consistency in performance and use

integrity (of data or information) – ability to ensure that the data contents are not contaminated, corrupted, lost or altered between transmission and reception

knowledge – understanding and comprehension of information gained through experience, education, observation or investigation

knowledge management – strategies and practices used in an organization to identify, create, represent, distribute and enable adoption of insights and experiences that comprise processes or practices

level of risk – magnitude of a risk expressed in terms of the combination of consequences and their likelihood

life cycle – series of identifiable stages through which a system progresses from its conception to disposal or retirement

life cycle cost – total cost of acquisition, ownership and disposal over the full life of a system

life cycle costing – process of economic analysis to assess the cost of a system over all or part of its life cycle

logistic support policy – criteria established for organizing resource provisions and establishing procedural guidelines for logistic support operation, including outsourcing and training of certified personnel to perform maintenance support tasks

logistics – management of services and goods flowing from the point of origin to the point of consumption to fulfill customer requirements normally consisting of the combination of packaging, material handling, inventory, warehousing, transportation, security and information

maintainability – probability that an item will be retained in or restored to a specific condition within a given period of time, when the maintenance is performed in accordance with prescribed procedures and resources ability to be retained in, or restored to, a state to perform as required, under given conditions of use and maintenance, including location for maintenance, accessibility, maintenance procedures and maintenance resources

maintenance – combination of all technical, supervision and management actions intended to retain an item in, or restore it to, a state in which it can achieve required performance

maintenance concept or strategy – combination of the maintenance tasks and maintenance support resources that provide the basis for maintenance planning and determining supportability requirements

maintenance policy – general approach by an organization to the provision of maintenance and maintenance support resources as high level guidance to the maintenance strategy

maintenance program – list of all the maintenance tasks developed for a system for a given operating context and maintenance concept or strategy

maintenance support – resources to maintain a system under a given maintenance concept or strategy including human resources, support equipment, materials and spare parts, maintenance facilities, documentation, information and maintenance information systems

management – system of controls and processes required to achieve the strategic objectives set by the organization's governing body

management information system – system that provides information needed to manage organizations efficiently and effectively involving primary resources: technology, information and people

Markov analysis – statistical technique used in forecasting the future behavior of a variable or system whose current state or behavior does not depend on its state or behavior at any time in the past

mean time between failures (MTBF) – average operating time between failures; this should only be applied to repairable items

mean time to failure (MTTF) – average operating time to failure for non-repairable items

mean time to restoration (MTTR) – average time to restore a system malfunction back to normal operation

network – set of interconnected and interacting systems such as those that deliver mail (e.g., postal service), transport people and goods (e.g., railroad or airline services), distribute energy sources (e.g., pipeline fuel, electricity) or exchange information (e.g., telecommunications, Internet)

nonconformity – non-fulfillment of a specified requirement

non-repairable item – item that cannot, under given conditions, after a failure, be returned to a state in which it can perform as required, where "given conditions" may include technical, economic and other considerations

objective evidence – data supporting the existence or verification of something that may be obtained through observation, measurement, test or other means

operability – ability of a system function to be easily controlled and successfully operated by the end user or the system operator

operational availability – availability value experienced under actual conditions of operation and maintenance as determined by consideration of downtime due to failures, outages and associated delays, but excluding external causes

operational scenario – description of a sequence of events of system operation and interaction of the system elements with its environment and users for analysis and evaluation of system requirements

perfective maintenance – maintenance actions carried out to improve software performance, maintainability, or other attributes of a computer program during software in-service operation

preventive maintenance – maintenance actions carried out prior to failure to mitigate degradation and reduce the probability of failure

procedure – specified way to carry out an activity or a process

process – set of interrelated or interacting activities that transforms inputs into outputs

product – result of a process

program (computer) – set of coded instructions executed to perform specified logical and mathematical operations on data consisting of a combination of coded instructions and data definitions that enable computer hardware to perform computational or control functions

project – unique process, consisting of a set of coordinated and controlled activities with start and finish dates, undertaken to achieve an objective conforming to specific requirements, including the constraints of time, cost and resources

quality – degree to which a set of inherent characteristics fulfills requirements

quality assurance – part of quality management, focused on providing confidence that quality requirements will be fulfilled

quality control – part of quality management, focused on fulfilling quality requirements

quality management – coordinated activities to direct and control an organization with regard to quality

quality planning – part of quality management focused on setting quality objectives and specifying necessary operational processes and related resources to fulfill the quality objectives

quality of service – collective effect of service performance that determines the degree of satisfaction of a user of the service

recoverability – ability to recover from a failure, with or without corrective maintenance

reliability – probability than an item can perform its intended function for a specified interval under stated conditions

reliability block diagram (RBD) – logical, graphical representation of a system showing how reliability of the system functions, represented by blocks, and combinations thereof, affect reliability of the system

reliability centered maintenance (RCM) – systematic method for determining the respective maintenance tasks and associated frequencies, based on the probability and consequences of failure using data from analysis or experience, which may also result in recommendations for improvements such as redesign, modifications or operational and maintenance procedures

reliability growth – iterative process for reliability improvement by addressing design and manufacturing weaknesses

repairable item – item that can, under given conditions, after a failure, be returned to a state in which it can perform as required, where the "given conditions" may include technical, economic and other considerations

required function – function or combination of functions, stated or implied, that may be considered necessary to provide a given service

requirement – statement or specification that translates or expresses a need and its associated constraints and conditions

requirements engineering – interdisciplinary function that mediates between the domains of the acquirer and supplier to establish and maintain the requirements to be met by the system, product or service of interest

risk – degree to which objectives are affected by uncertainty, often expressed in terms of a combination of the consequences of an event or a change in the circumstances, and the associated likelihood of occurrence

risk analysis – systematic use of information to identify the risk situation and to estimate the level of risk

risk assessment – overall process of risk identification, risk analysis and risk evaluation

risk avoidance – decision not to become involved in, or action to withdraw from, an activity based on the level of risk determined by the result of risk evaluation

risk criteria – terms of reference by which the significance of risk is assessed

risk evaluation – process of comparing the estimated risk against given risk criteria to determine the significance of the risk

risk exposure – potential loss presented to a project or an organization by a risk, commonly defined as the product of a probability and the magnitude of a consequence, to express an expected value or exposure

risk factor – potential problem that would have negative effects to a planned activity, characterized by the probability of problem occurrences and a potential loss should the problem occur

risk identification – process to find, list and characterize elements of risk; in the context of dependability application, risk identification is a systematic approach to determining the dependability risk factors associated with a planned dependability activity

risk management – coordinated activities to direct and control an organization with regard to risk

risk management system – set of elements of an organization's management system concerned with managing risk, including strategic planning, decision making and other processes for dealing with risk

risk mitigation – limitation of any negative consequence of a particular event concerning risk and presenting a course of action to reduce the probability of a potential loss from a risk factor

risk retention – acceptance of the burden of loss, or benefit of gain, from a particular risk, including the acceptance of risks that have not been identified

risk scenario – description of an event that can lead to negative impact if and when it should occur as characterized by an internal or external threat to the organization or project

risk transfer – sharing with another party the burden of loss or benefit of gain, for a risk

risk treatment – process of developing, selecting and implementing controls to avoid, remove, share or retain a risk

safety – relative freedom from danger, risk or threat of harm, injury or loss to personnel and/or property, whether caused deliberately or by accident

scheduled (or planned) maintenance – maintenance carried out in accordance with an established time schedule, usually to perform preventive maintenance although it may also identify the need for some corrective maintenance action

security – state of freedom from danger or threat

security of service – effect of providing the required security for user services

service – set of functions offered to a user

serviceability – ability of user access to system service functions; once obtained, the service functions continue to be provided by the system under given conditions for a requested duration

software – programs, procedures, rules, documentation and data of an information processing system

software configuration item – software item that has been configured and treated as a single item in the configuration management process

software dependability – ability of the software item to perform as and when required when integrated in system operation

software failure – failure that is the result of a software fault that will continue to manifest itself as a failure until it is removed

software fault (bug) – state of a software item that may prevent it from performing as required due to specification faults, design faults, programming faults, compiler-inserted faults or faults introduced during software maintenance

software function – elementary operation performed by the software module or unit as specified or defined as per stated requirements

software module (or unit) – software element that can be separately compiled in programming codes to perform a task or activity to achieve a desired outcome of a software function or functions

specification – document stating requirements

system – defined set of components or functions within a defined boundary that cooperate collectively to satisfy a requirement

system requirements specification – structured collection of the requirements (functions, performance, design constraints and attributes of the system and its operational environments and external interfaces requirement)

supportability – ability to provide specified logistic and maintenance resources to support system operation and sustain defined service life

tailoring (process) – process to adapt, adjust or alter an organization's set of established processes and activities to fulfill, satisfy or meet requirements as they apply to dependability

test – determination of one or more performance characteristics according to a procedure

unscheduled (unplanned) maintenance – corrective maintenance that has not been anticipated and cannot be deferred

validation – confirmation, through the provision of objective evidence, that the requirements for a specific intended use or application have been fulfilled

verification – confirmation, through the provision of objective evidence, that specified requirements have been fulfilled

wearout failure – failure due to deterioration caused by the accumulated stresses imposed during use usually affecting material dimensions and properties with increasing probability of failure

Weibull analysis – statistical analysis of failure and repair data based on the Weibull distribution, which is a probability density function consisting of a shape parameter and a scale parameter that is most suitable for failure rate analysis

Appendix B

SYSTEM LIFE CYCLE PROCESS FOR DEPENDABILITY APPLICATION

B.1 INTRODUCTION

The system life cycle process provides a sequence of process stages to facilitate review and transition where specific time-related activities are identified for dependability application. Typical system life cycle process activities are presented for each stage. Relevant dependability engineering activities are identified for incorporation into the system process activities to achieve the desirable dependability performance outcome prior to advancing to the next stage. Implementation of dependability in technological systems is time sensitive for effective attainment of expected performance results. Specific project application of the life cycle process should be tailored for appropriate dependability implementation.

The data requirements of each system life cycle stage include:

(a) inputs necessary to initiate the process activities of each stage;
(b) relevant influencing factors for consideration;
(c) enabling mechanisms for consideration;
(d) outputs resulting from the process generated at the end of each stage;
(e) description of key process activities to be performed at each stage.

The application of the system life cycle process for specific project requirements can start at any stage and continue forward, or stop at the completion of any stage for practical reasons. Typical examples requiring specific implementation of dependability activities include:

- defining the scope of the dependability project;
- system design using mainly OEM (original equipment manufacturer) products;
- realization of specific system functions by outsourced manufacturing;
- engaging third party maintenance support during system operation;
- enhancement of an existing system performance with added service features.

The system life cycle process identifies the key process activities and the relevant dependability engineering activities applicable to each stage. The process activities presented herein are intended to provide guidance that should be tailored for specific project application.

B.2 PROCESS FOR THE SYSTEM CONCEPT/DEFINITION STAGE

B.2.1 Data Requirements for Concept/Definition

(a) Inputs:
- customer requirements, needs and wants;
- established standards and regulatory requirements related to health, safety, security and environmental concerns;
- company policy and business decisions;
- market intelligence and competition.

(b) Influencing factors for consideration:
- competition;
- economic issues;
- technology issues;
- capability issues;
- environmental issues;
- legal issues;
- timing for investments issues.

(c) Enabling mechanisms for process applications:
- human resources;
- financial resources;
- assets and facilities;
- integrated design and implementation processes;
- assurance process.

(d) Outputs:
- system specifications;
- systems design knowledge.

(e) Description of key process activities to be performed is shown in Table B.1 and B.2.

B.2.2 Description of Key Process Activities for Concept/Definition

TABLE B.1 Requirements definition

System Life Cycle Process Activities	Dependability Engineering Activities
• Identify the customer • Identify system application and operating environment • Identify system constraints and use conditions • Identify system interface and legacy issues • Identify system performance requirements, applicable standards and regulations • Identify human interface requirements • Identify system support requirements • Identify system development resource requirements • Identify project schedules and deliverable targets	• Identify dependability requirements • Identify system performance limits and allowable outage downtimes • Identify technology constraints and limitations for dependability application • Identify dependability resource requirements • Identify feasible approach to achieve dependability performance objectives

TABLE B.2 Requirements analysis

System Life Cycle Process Activities	Dependability Engineering Activities
• Determine system boundaries, operating functions and performance characteristics from the set of defined system requirements • Evaluate the constraints affecting architectural design options • Determine technical approaches and feasibility for system realization • Determine feasible system configurations and functional partition • Determine critical functions required for safety and security application • Determine technical support requirements • Determine quality requirements • Determine documentation requirements • Determine capability of undertaking the system design and development work • Identify potential partnership, outsourcing needs and suppliers requirements • Establish architectural design requirements • Conduct requirements review • Develop system specifications	• Determine operating scenarios • Determine probable system failures and performance degradation limits • Determine probable risk exposures and criticality of system failure impact • Determine system structure and breakdown of system and sub-system functions • Determine means for dependability assessments • Determine human interface conditions with the system • Determine system maintenance and logistic support requirements • Determine quality and dependability assurance requirements

B.3 PROCESS FOR THE SYSTEM DESIGN/DEVELOPMENT STAGE

B.3.1 Data Requirements for Design/Development

(a) Inputs:
- system specifications;
- architectural design requirements.

(b) Influencing factors for consideration:
- availability and access to relevant skills resources;
- commitment targets for development schedules;
- project risks.

(c) Enabling mechanisms for process applications:
- availability of specific tools required for development;
- training needs.

(d) Outputs:
- system prototype;
- system and sub-system support requirements.

(e) Description of key process activities to be performed is shown in Table B.3, B.4, B.5 and B.6.

B.3.2 Description of Key Process Activities for Design/Development

TABLE B.3 Architectural design

System Life Cycle Process Activities	Dependability Engineering Activities
• Establish system configuration • Partition system functions • Establish design rules and criteria for system functions • Formulate make-buy decisions of system functions • Identify outsourced OEM products and design options for system configuration • Select technologies for functional design application • Establish criteria for realization of system functions utilizing hardware/software elements • Formulate solution to meet system requirements and design options • Establish means for verification and integration of system functions • Conduct preliminary design review • Document architecture design specification	• Establish dependability plan • Establish criteria for dependability evaluation • Perform system availability/reliability functional model analysis • Perform preliminary system reliability prediction as baseline reference • Perform reliability allocation of system functions to meet performance objectives • Determine failure criteria and criticality of system functions • Evaluate reliability of each partitioned function and recommend alternative design options if required • Establish maintainability criteria for self-test and diagnostic functions • Establish criteria for maintenance and logistic support functions • Determine levels of maintenance actions • Establish database for information capture of data records and incident reporting • Perform system availability and reliability prediction based on feasible system configurations and functional partition • Perform fault tree analysis to determine critical areas requiring design attention • Conduct system level failure modes, effects and criticality analysis to support design alternatives and justifications • Evaluate system availability and cost trade-off affecting design options

TABLE B.4 Functional design/evaluation

System Life Cycle Process Activities	Dependability Engineering Activities
• Formalize the functional design process • Identify design composition of hardware/software elements for each function • Incorporate test functions for performance verification • Establish human factors design criteria • Establish environmental design criteria • Establish ergonomics design criteria • Establish electromagnetic compatibility design criteria • Establish safety, security and reliability design criteria • Establish hardware design rules • Establish software maturity design schemes • Determine fault coverage and system recovery strategy • Verify performance limits and interoperability of the functional design to meet architectural design requirements • Conduct detailed design review	• Conduct reliability assessment • Conduct maintainability evaluation • Conduct functional level failure modes, effects and criticality analysis • Conduct functional level design trade-off, fault tolerance and risk evaluation • Establish maintenance and logistics support plan • Establish process for supplier evaluation for quality assurance and reliability conformance • Establish process for off-the-shelf product evaluation and acceptance

TABLE B.5 Design specifications and documentation

System Life Cycle Process Activities	Dependability Engineering Activities
• Develop system specifications and document recommended design approaches • Develop OEM product specifications for procurement requirements • Develop system and sub-system development plan • Develop system and sub-system maintenance support plans • Establish a database for information capture and data records maintenance • Incorporate all process activities in a master project plan for ease of update, scheduling coordination and project management where appropriate	• Incorporate dependability requirements in system specifications • Document quality and dependability plan for system and sub-system development • Document quality and dependability specifications for sub-system functions • Document quality and dependability acceptance criteria for outsourced OEM products and maintenance support requirements • Establish failure reporting, analysis and corrective action system (FRACAS)

TABLE B.6 System and sub-system development

System Life Cycle Process Activities	Dependability Engineering Activities
• Implement system and sub-system development plan • Initiate in-house development for sub-systems and functions • Initiate interface development for interoperability of system and sub-system functions and human interactions • Initiate procurement of outsourced products and contract agreements of external development efforts • Develop software sub-systems and identify software configuration items • Develop system and sub-system test plan • Develop in-house production plan • Develop system operation plan • Develop packaging, handling, storage and transportation plan • Develop system integration and installation plan • Develop configuration management plan and design change procedures • Develop system operation and maintenance training programs • Initiate prototype construction of sub-systems and functions for system integration • Conduct prototype testing for system functional performance evaluation • Conduct system and sub-system development reviews	• Implement system dependability program • Implement quality assurance program • Establish suppliers' dependability programs • Establish dependability acceptance criteria and reliability growth programs for system and sub-systems • Establish system maintenance and logistics support program • Define warranty conditions • Implement failure reporting, analysis, data collection and feedback system

B.4 PROCESS FOR THE SYSTEM REALIZATION/ IMPLEMENTATION STAGE

B.4.1 Data Requirements for Realization/Implementation

(a) Inputs:
- system prototype.

(b) Influencing factors for consideration:
- transition management;
- commitment targets for system delivery schedule;
- warranty requirements and incentives.

(c) Enabling mechanisms for process applications:
- project management;
- customer or user training.

(d) Outputs:
- system performance demonstration for service operation;
- system acceptance by customer.

(e) Description of key process activities to be performed is shown in Table B.7, B.8, B.9, B.10 and B.11.

B.4.2 Description of Key Process Activities for Realization/Implementation

TABLE B.7 Realization

System Life Cycle Process Activities	Dependability Engineering Activities
• Implement system production plan • Fabricate sub-system assemblies • Construct hardware/software functions • Acquire OEM products according to outsourced procurement specifications • Initiate coding of software units according to design specifications • Configure applicable software units and labelling as software configuration items • Develop training program for system operators and maintainers • Establish test equipment and test facility requirements • Develop packaging, handling, storage and transportation instructions • Conduct supply management review • Conduct configuration management review	• Implement system dependability program • Implement quality assurance program • Implement suppliers' dependability programs

TABLE B.8 Verification

System Life Cycle Process Activities	Dependability Engineering Activities
• Implement verification plan • Implement measurement standards and metric evaluation criteria • Perform verification tests on system functions • Document verification test results • Analyze verification test results to determine achievement of specified design requirements • Conduct system verification review • Develop system integration plan and procedures	• Conduct system dependability assessments • Conduct qualification tests where applicable • Document failure reports from verification tests • Perform failure analysis and determine failure cause for classification • Resolve anomalies found during verification tests by corrective actions • Determine code coverage of software units • Determine software test completeness • Determine software fault coverage

TABLE B.9 Integration

System Life Cycle Process Activities	Dependability Engineering Activities
• Implement system integration plan • Involve OEM suppliers during system integration if required • Integrate system entity to demonstrate assembly procedures and training effectiveness • Document effectiveness of integration procedures • Develop procedures for system installation • Develop system validation/acceptance plan	• Implement integration-related system dependability program • Implement integration-related quality assurance program

TABLE B.10 Installation/transition

System Life Cycle Process Activities	Dependability Engineering Activities
• Implement installation plan • Document installation records and procedures • Evaluate transition strategy for improvement • Engage customer participation in system installation process	• Establish shared maintenance support and reporting schemes with customer maintainers on system installed on customer premises • Monitor turn-around time for system restoration and replenishment of spares • Maintain adequate spares inventory on maintainer's/customer's site

TABLE B.11 Validation/acceptance

System Life Cycle Process Activities	Dependability Engineering Activities
• Implement validation/acceptance plan • Demonstrate system performance met and fulfilled customer specified requirements • Document validation test results • Implement warranty schemes if applicable • Conduct acceptance review • Document customer sign-off for system acceptance	• Validate that system performance fulfills the dependability requirements such as reliability growth and accelerated test programs where applicable • Document failure reports from validation tests • Generate non-conformance reports for recommended corrective/preventive actions • Resolve anomalies found during validation • Resolve warranty issues with customers

B.5 PROCESS FOR THE SYSTEM OPERATION/MAINTENANCE STAGE

B.5.1 Data Requirements for Operation/Maintenance

(a) Inputs:
- system in full service operation.

(b) Influencing factors for consideration:
- system service capacity;
- supply chain for spares provisioning;
- responsive maintenance actions.

(c) Enabling mechanisms for process applications:
- project/operations/maintenance management;
- operator and maintainer training.

(d) Outputs:
- dependable system performance;
- customer satisfaction results.

(e) Description of key process activities to be performed is shown in Table B.12.

B.5.2 Description of Key Process Activities for Operation/Maintenance

TABLE B.12 Operation/Maintenance

System Life Cycle Process Activities	Dependability Engineering Activities
• Implement system operation strategy • Monitor system performance • Implement system maintenance support strategy • Monitor system maintenance support efforts • Provide customer care service and training where applicable • Conduct system performance review • Conduct maintenance and logistic support review	• Implement field data collection system or computerized maintenance management system • Conduct customer satisfaction survey • Analyze failure trends • Conduct root-cause analysis of problem areas • Review/update maintenance program • Recommend design or procedural changes for continual improvement • Determine quality of service

B.6 PROCESS FOR THE SYSTEM ENHANCEMENT STAGE

B.6.1 Data Requirements for Enhancement

(a) Inputs:
- new customer requirements;
- enhanced features.

(b) Influencing factors for consideration:
- timing for change;
- return on investment.

(c) Enabling mechanisms for process applications:
- change management;
- obsolescence management;
- customer "buy-in" or reaction to incorporation of new service features.

(d) Outputs:
- enhanced system performance;
- comparison of customer satisfaction results before and after enhancement efforts.

(e) Description of key process activities to be performed is shown in Table B.13.

B.6.2 Description of Key Process Activities for Enhancement

TABLE B.13 Enhancement

System Life Cycle Process Activities	Dependability Engineering Activities
• Identify new system service features for incorporation • Establish enhancement strategy and plan • Evaluate the need for change and resulting benefits • Identify feasible evolution/renewal opportunity • Implement enhancement efforts • Implement software update and perfective maintenance • Implement design change and configuration management • Assess new service introduction impact	• Evaluate dependability performance impact due to changes with added new features • Conduct life cycle cost impact study for change incorporation • Conduct risk and value assessments • Conduct customer satisfaction survey resulting from change reactions • Manage software version release

B.7 PROCESS FOR THE SYSTEM RETIREMENT STAGE

B.7.1 Data Requirements for Retirement

(a) Inputs:
- status of ageing system performance capability;
- obsolescence of major service functions;
- competitiveness and marketability of existing operational services;
- increased operation and maintenance and support costs.

(b) Influencing factors for consideration:
- timing for retirement;
- technology obsolescence;

- regulatory constraints;
- social impact due to termination of service.

(c) Enabling mechanisms for process applications:
- project management.

(d) Outputs:
- termination of service.

(e) Description of key process activities to be performed is shown in Table B.14.

B.7.2 Description of Key Process Activities for Retirement

TABLE B.14 Retirement

System Life Cycle Process Activities	Dependability Engineering Activities
• Implement system retirement/decommissioning plan • Implement reuse and redeployment strategy • Implement waste treatment on disposal items • Notify the customers on service termination • Provide information on new or alternative service provision	• Evaluate constraints on system deactivation and impact on removal of system from service operation • Evaluate impact on environments of disposal items • Conduct customer satisfaction survey due to termination of service

Appendix C

EXAMPLE OF A SYSTEM DEPENDABILITY SPECIFICATION

C.1 INTRODUCTION

A machinery protection system for a rotating equipment application is used as an example to illustrate the process of developing a system dependability specification.

The system configuration and operating procedures are typical for application in diverse locations such as pump and compressor stations, processing plants, petrochemical facilities and power generation. The system hardware and software elements are identified at the functional level for simplicity of explanation. Machinery protection consists of monitoring machinery conditions, such as vibration, bearing temperatures, process temperatures and pressures, oil pressure and so on. Human interactions with the machinery protection system functions are representative of plant operation and may happen at the individual unit level or at a higher plant control room or centralized control center. The dependability information and operational data in this example are for illustration purposes. They are not intended to reflect the performance capabilities of any specific manufacturer's product or service operation.

For the purpose of this example, the application is a machinery protection system for a gas turbine-driven centrifugal compressor in a process plant. The compressor unit is remotely operated and monitored by a central control room at the plant site.

Cyber security control, vulnerability protection and information privacy are not included in this example. Information security requires additional information technology (IT) infrastructure with a different technology skills set. This is normally dealt with by the IT function of the individual organization working at the business premises.

C.2 IDENTIFY THE SYSTEM

The machinery protection system is intended to protect machinery against excessive operating conditions such as high vibration, which could cause mechanical damage and ultimately a failure that will trip the machine. The machinery protection system consists of two separate but interacting systems:

- A local unit control panel connected to instrumentation on the gas turbine and compressor unit (see Figure C.1);
- An on-site control room operated full-time 24/7 by control room operators.

Figure C.1 Example of a unit control panel for a gas turbine-driven compressor

C.3 DESCRIBE THE SYSTEM OBJECTIVES

The objectives of the machinery protection system are:

- to prevent damage to the gas turbine and compressor and its components such as the bearings;
- to provide an alarm to operating personnel that a potentially damaging condition has been reached;
- to shut down the equipment when a potentially dangerous situation exists;
- to record when an alarm or shutdown has occurred and the source of the condition.

C.4 IDENTIFY THE FUNCTIONS TO MEET THE SYSTEM OBJECTIVES

The machinery protection system has five primary functions:

- detection function to sense a machine condition;
- control function for data processing and information distribution;
- alarm function to alert operators at the control center;
- shutdown function to prevent further damage and catastrophic machine failure;
- data recording function.

C.5 DESCRIBE THE FUNCTIONS

The detection, control, alarm, shutdown and data recording functions employ essential sensing and control equipment installed at the compressor unit location. They are integrated into a plant-wide system for access and control such as a distributed control system (DCS) and coordinated by the control center staff. This group of collaborating equipment functions meets the first objective of the protection system. The following provides a description of each function.

1. Detection function
   - Sensing the vibration of the gas turbine using accelerometers or velocity transducers and of the compressor, normally with the use of displacement probes.
   - Sensing pressure and temperature with transducers installed in locations such as the oil system, bearings and combustion section.
   - Detecting abnormal conditions such as high/low pressures and temperatures using dedicated switches.

2. Control function
   - Processing the detected signals from the sensors in order to activate an alarm. A programmable logic controller (PLC) industrial computer will compare the sensor reading against an alarm level and possibly use a voting system if several inputs are available to be compared. The alarm will be displayed on a computer screen.
   - Recording the date and time of the event.
   - Passing the alarm information to the remote control center for display.
   - The computer system is programmed to understand the status of the unit. For example, alarms will not be activated when the unit is not running and may be overridden on startup. A manual override control for false alarm deactivation is incorporated into the computer system control function.

3. Alarm function
   - Displaying an alarm indication on the computer screen both in the unit control room and the remote control center by turning red and flashing in order to attract the attention of the operators. There may also be a flashing light on the unit control room panel.
   - Activating an audible alarm by the computer system when triggered by the detection of an alarm situation.

4. Shutdown function
   - Producing a shutdown when that level is reached. Depending on the source of the shutdown, either a regular or emergency shutdown is initiated.
   - Notifying the operators and control center staff via the computer screen or audible alarms.

5. Data recording function
   - Recording the alarm or shutdown in a database with the date and time.
   - Providing a history of data recordings to facilitate data retrieval of the alarms and shutdowns for analysis purposes.

The combined effects of the interacting systems at the unit level and the remote control room permit a comprehensive protection system in meeting the specified system objectives.

C.6 IDENTIFY THE INFLUENCING CONDITIONS AFFECTING THE FUNCTIONS

The identification process focuses on factors having significant impact on the system functions, hence affecting system performance. The following key influencing factors are identified for evaluation of dependability characteristics associated with the design and acquisition of the necessary functions.

1. Detection function
   - Type and cost of various sensors, transducers and switches suitable for detection purposes
   - Number of sensors, transducers and switches and their location needed for full or partial coverage to minimize risk exposures
   - Technology and reliability of the sensors, transducers and switches
   - Ease of installation and maintenance of the sensors, transducers and switches
   - Expected life of sensors, transducers and switches
   - Means of transmitting the signal to the control room using either hard wiring or newer fieldbus or wireless technologies

2. Control function
   - Specific control system design using commercially available products
   - Flexibility of computer program modification
   - Ease of use and programming for arming and disarming the system
   - Automatic system bypass
   - Wired or wireless connections to sensors, transducers and switches
   - Trouble display for control system diagnosis
   - Communication to remote control room

3. Alarm function
   - Ability of computer system to display the alarm condition
   - Loudness of audible sounds when alarm is activated in meeting regulatory requirements
   - Response time when alerted

4. Shutdown function
   - Ability of the control system to execute the unit shutdown
   - Surveillance and monitoring service by the control center
   - False alarm/shutdown consequences and follow-up by the control center

5. Data recording function
   - Processing capacity of the computer system and network
   - Storage capacity of the computer database
   - Data backup and storage

C.7 EVALUATE THE TECHNICAL APPROACH TO ACHIEVE THE NEEDED FUNCTIONS

The technical approach deals with the acquisition, design, implementation and maintenance of the needed functions for sustained operation. The evaluation process determines that the performance capability of the needed functions can be attained or developed within the project's time frame and cost constraints. The complexity and interfaces of the functions are identified. The means of interconnection to achieve interoperability of collaborating functions can be determined. Key dependability performance characteristics associated with the functions are considered for functional application and capability of performance.

1. Detection function
   The level of technology sophistication of the sensor, transducers and switches is the main driver for the cost and coverage of the detection function. The technical approach and decision process starts with searching for available commercial-off-the-shelf products that could satisfy the functional requirements of the equipment. The equipment performance history in terms of reliability, durability and service life for indoor and outdoor applications and the cost of equipment replacement should be examined. Multiple supply sources should be investigated. Safety requirements must be met. Where necessary, the supplier's willingness to modify their existing products to accommodate specific requirements to suit the functions should be considered. Maintenance serviceability of the equipment installed on the gas turbine and compressor in various locations should be determined. The warranty scheme of the purchased equipment and spares provisioning and maintenance support services should be considered.

2. Control function
   The main control element is the PLC and issues similar to those for the detection equipment should be considered.

3. Alarm function
   The alarm function is mainly executed by the code in the PLC using industry standard ladder logic. The design of this software and the coding standards used are of paramount importance.

4. Shutdown function
   The shutdown function operates in software in a manner similar to that of the alarming function.

5. Data recording function
   The data recording function is dependent on both computer hardware and software, usually at the control center.

C.8 DESCRIBE THE HARDWARE, SOFTWARE ELEMENTS AND HUMAN INTERACTIONS INVOLVED IN SYSTEM OPERATION

The system functions are realized by design and construction. The functions consist of hardware components and software program instructions to realize the system performance capability for machinery protection. Human interactions with the system are kept simple with minimum intervention to facilitate user-friendly operation and working environment. The following describes the hardware and software elements and the human interactions with the operation of the machinery protection system.

1. Hardware
   All system hardware for the protection system is integrated for application. Sensors, transducers and switches are installed in the gas turbine and compressor to meet protection criteria. All sensing and detection devices are mounted at various strategic locations to achieve maximum protection coverage. Some of the devices can be bypassed temporarily so maintenance can be performed without interruption of unit operation. The unit control system is automated and uses highly reliable PLC-based computers for processing. It requires minimal attention to its normal operating functions. The control center utilizes more standard computers and a network, but it incorporates redundancy in these computers so a single failure will not jeopardize the operation.

2. Software
   The system control is driven by software to process the data from the sensors and decide whether an alarm level has been reached or a shutdown should be initiated. The software is based on industry standard ladder logic. The system control, monitoring and information display functions are automated with minimum human intervention. Authorization is required to permit access or change of the pre-set software program instructions for operation. A dedicated remote server is used to support data storage backup capability.

3. Human interactions
   The machinery protection system is designed to ensure compliance with industry standards and safety regulations for system operation. The remote control center is operated by trained operators, and access to the control room is restricted. Self-prompting instructions for human-computer interface are designed to facilitate diagnosis, communications and enquiry responses during control system operation.

   When an alarm is activated, standard procedures are followed to respond according to the nature of the alarm. An emergency button is incorporated in the unit control panel for easy access. This is used for manual override in case of an emergency situation.

C.9 DETERMINE THE OPERATIONAL SCENARIO

The protection system operational scenario describes the task requirements of operational usage and event sequence in system application. There are five primary functions involved in the protection system operation:

- detection function to sense a machine condition;
- control function for data processing and information distribution;
- alarm function to alert operators at the control center;
- shutdown function to prevent further damage and catastrophic machine failure;
- data recording and storage function.

There are three modes of operation in the protection system operating profile.

1. Normal mode of operation
   The normal mode of operation of the protection system requires all functions to be in readiness mode. The data recording function operates continuously with time indication and registration of system operating status.

2. Alarm mode of operation
   The alarm mode of operation of the protection system requires all functions except for the shutdown function. The alarm is acknowledged by an operator. The normal mode of operation will resume when the alarm situation is cleared.

3. Shutdown mode of operation
   The shutdown mode of operation of the protection system requires all of the functions except for the alarm function, which is no longer required during a shutdown condition. The unit will be restarted when the shutdown cause is determined and rectified.

C.10 DESCRIBE THE SYSTEM CONFIGURATIONS TO MEET SYSTEM OBJECTIVES

(a) System configuration for normal mode of operation
For the normal mode of operation, the protection system requires the availability of all the functions for system dependability performance. The alarm or shutdown signals are silent and are not activated. No alarm or shutdown indication is being displayed or recorded. Figure C.2 shows the system configuration for normal mode of operation.

(b) System configuration for alarm mode of operation
The unit control system, upon detection and confirmation of reaching the alarm limit, immediately activates an alarm signal to alert control center operators. At this point, all of the functions are required with the exception of the shutdown function. The alarm will be displayed, possibly along with an audible signal at the unit control panel as well as at the remote control center. Alerting the control center operators is more important since it is staffed at all times. The control center operators will decide if they need to obtain assistance from the compressor operator. Figure C.3 shows the system configuration for the alarm mode of operation.

(c) System configuration for the shutdown mode of operation
When a shutdown level is reached, the control system will immediately initiate either a normal shutdown or an emergency trip. The information is displayed and recorded as for an alarm. The control center operator may be able to restart the compressor unit under certain circumstances but more commonly will advise operations and maintenance staff to investigate the cause and rectify the situation. The system configuration for the shutdown mode of operation is shown in Figure C.4.

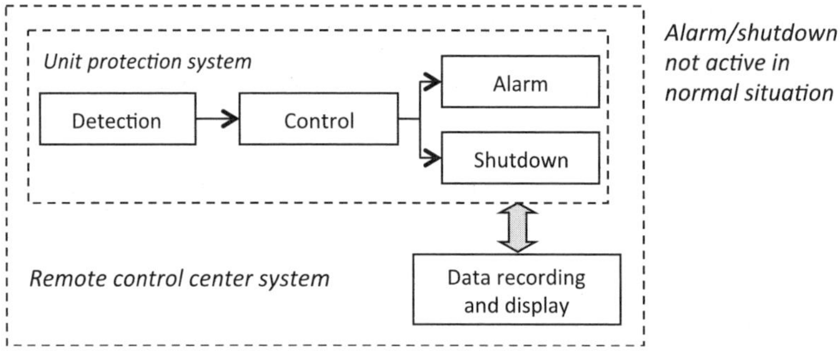

Figure C.2 System configuration for the normal mode of operation

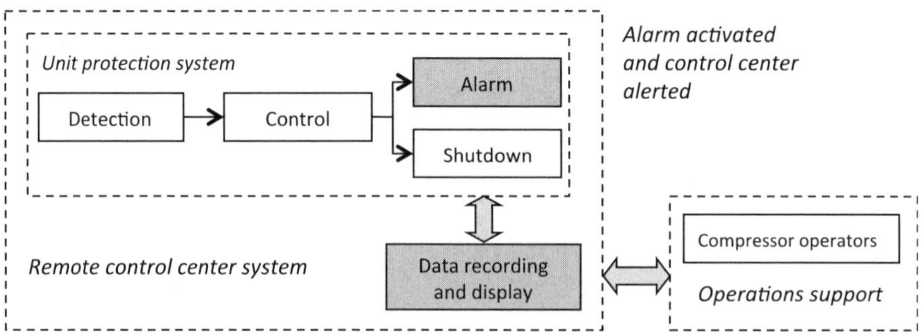

Figure C.3 System configuration for the alarm mode of operation

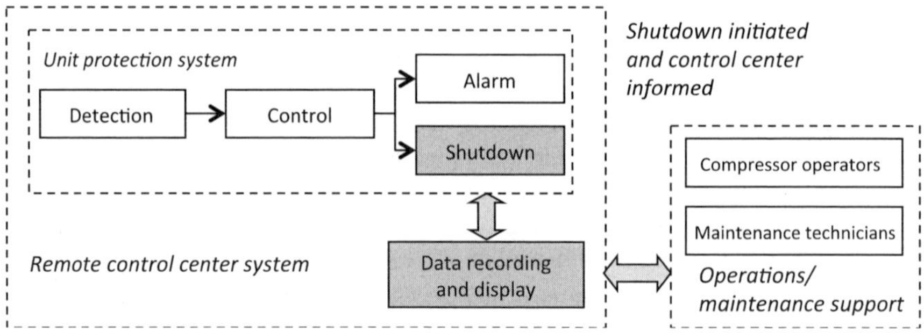

Figure C.4 System configuration for emergency mode of operation

C.11 DETERMINE THE DEPENDABILITY REQUIREMENTS

The dependability requirements of the machinery protection system functions are determined from information obtained from the system configurations for different modes of operation and the operational usage of the relevant system functions.

The following summarizes the dependability requirements of each function of the machinery protection system. The numerical figures are derived from the relevant dependability characteristics of the functions provided by equipment suppliers, industry data sources and actual experience. Dependability performance characteristics include availability and reliability. Availability is used in this example to represent the operational state of the equipment functions with the ability to perform as required. In practice, availability is determined by the ratio of uptime to the total operating time of equipment functions. The combined availability of collaborating equipment functions is calculated by application of probability principles. Reliability is measured by MTBF. Other relevant dependability characteristics such as expected life, response time and maintenance support conditions are obtained from field performance experience data. Note that these numbers are for illustration purposes only.

(a) Detection function
- Sensors: 99.9% availability; MTBF 50,000 hours; annual calibration check; expected life 15 years.
- Wiring connectivity to unit control room: 99.9% availability; MTBF 100,000 hours; annual loop check; expected life 20 years.
- Detection function availability is 99.9% × 99.9% = 99.8%.

(b) Control function
- PLC: 99.8% availability; MTBF 150,000 hours; hardware self-diagnostics; expected life 20 years.
- Software: 99.99% availability; MTBF 150,000 hours; software maintenance checks.
- Control function availability is 99.8% × 99.99% = 99.8%.

(c) Alarm function
- Software: 99.99% availability; MTBF 150,000 hours; software maintenance checks.
- Control center network to unit control: 99.7% availability; MTBF 25,000 hours; network self-diagnostics; expected life 20 years.
- Operator response: 99% availability; response time 5 minutes.
- Alarm function availability is 99.99% × 99.7% × 99% = 98.7%.

(d) Shutdown function
- Software: 99.99% availability; MTBF 150,000 hours; software maintenance checks.
- Control center network to unit control: 99.7% availability; MTBF 25,000 hours; network self-diagnostics; expected life 20 years.
- Operator response: 99% availability; response time 1 minute.
- Shutdown function availability is 99.99% × 99.7% × 99% = 98.7%.

(e) Data recording function
- Control center computers and data storage: 99.8% availability; MTBF 150,000 hours; hardware self-diagnostics; expected life 10 years.
- Control center network: 99.8% availability; MTBF 150,000 hours; network self-diagnostics; expected life 10 years.
- Software: 99.99% availability; MTBF 50,000 hours; software maintenance checks.
- Data recording function availability is 99.8% × 99.8 × 99.99% = 99.6%.

Probably the most problematic for overall protection system dependability is the human factor. Control center operators need to respond quickly to both an alarm and a shutdown. The shutdown will be more important because it will have consequences for the overall process plant. Other actions to mitigate the effects of the shutdown will have to be taken. There is the possibility that the alarm or shutdown will be ignored or that the response will be delayed or will be inappropriate for the alarm or shutdown condition.

For normal operation, the availability is the combined detection, control and data recording functions and is 99.8% x 99.8% x 99.6% = 99.2%.

For the alarm scenario, the availability is the combined detection, control, data recording and alarm functions and is 99.8% x 99.8% x 99.6% x 98.7% = 97.9%.

For the shutdown scenario, the availability is the combined detection, control, data recording and shutdown functions and is 99.8% x 99.8% x 99.6% x 98.7% = 97.9%.

C.12 DOCUMENTATION OF SYSTEM DEPENDABILITY SPECIFICATION

The system dependability requirements form part of the overall system requirements. The following summarizes the data inputs for inclusion in documenting the system dependability specification. Note that the human factors are not included in the specification.

1. System identification
   - The protection system is used to protect the compressor unit from abnormal conditions and prevent a major failure.
   - The remote control center monitors all alarms and shutdowns and takes appropriate action based on the nature of the alarm or shutdown.

2. System objectives
   - To prevent damage to the gas turbine and compressor and its components such as the bearings;
   - To provide an alarm to operating personnel that a potentially damaging condition has been reached;
   - To shut down the equipment when a potentially dangerous situation exists;
   - To record when an alarm or shutdown has occurred and the source of the condition.

3. System functions
   - Detection function to sense a machine condition;
   - Control function for data processing and information distribution;
   - Alarm function to alert operators at the control center;
   - Shutdown function to prevent further damage and catastrophic machine failure;
   - Data recording function.

4. System operating profile
   - Normal mode of operation
   - Alarm mode of operation
   - Shutdown mode of operation

5. System configurations
   - Figures C.2, C3 and C.4 illustrate the system configurations for the modes of operation.

6. Dependability requirements for each protection function
   - Detection function: 99.8% availability; annual maintenance check, expected life 15 years.
   - Control function: 99.8% availability; self-diagnostics; expected life 20 years.
   - Alarm function: 99.7% availability; self-diagnostics; expected life 20 years.
   - Shutdown function: 99.7% availability; self-diagnostics; expected life 10 years.
   - Data recording function: 99.6% availability; self-diagnostics; expected life 10 years.

7. A statement on the security system dependability
   The following specifies the availability performance measures of the security system in various modes of operation of assigned tasks:

   - Protection equipment performance availability for normal mode of operation shall be >99.2%. The alarm operation is a transitional operation that forms part of the normal operation process.
   - Protection equipment performance availability for alarm or shutdown modes of operation shall be >97.9%.

   The following specifies the life expectancy of equipment and maintenance requirements:

   - 15 years expected life for the detection function;
   - 20 years expected life for the control function;
   - 20 years expected life for the alarm and shutdown functions;
   - 10 years expected life for the data recording function;
   - Data backup and software update is automatic without service interruption;
   - Annual maintenance checks shall be performed on all equipment identified.

Appendix D

DEPENDABILITY ENGINEERING CHECKLISTS

D.1 CHECKLISTS FOR SYSTEM LIFE CYCLE PROJECT APPLICATIONS

(a) Requirements checklist

(1) The nature and applications of system dependability performance are defined, and the intent to replace an existing system or to enhance its performance is known.

(2) The timing for new system introduction with specific dependability features has been established.

(3) The system operating environments, specific dependability influencing factors and related regulatory issues have been identified.

(4) The technical capabilities for dependability engineering of system development have been identified.

(5) The resources required to support the dependability project have been identified and estimated.

(6) The capital investments and acquisition of specific dependability tools and enabling mechanisms for system development have been identified.

(7) The potential customers and probable competitors interested in system development with dependability performance focus have been identified.

(8) The expected system performance requirements and special system features are defined, including the identification of unique dependability issues and customer expectations (e.g., software robustness).

(9) The system dependability operating scenarios, interoperability with other systems, technological design preference and legacy issues involved have been identified.

(10) The system maintenance and logistics support requirements for dependable operation have been identified.

(11) The marketing strategy and plan for leveraging dependability in system performance have been established.

(12) The project team for proposal and technical work with dependability expertise support has been established.

(13) The go/no go decision for system development can be justified with a strategic dependability performance focus.

(b) System design and development checklist
   (1) A project development plan for implementation of dependability tasks has been established.
   (2) System requirements have been analyzed, and dependability characteristics have been assessed.
   (3) Design strategy, technology selection and dependability activities for system development have been determined.
   (4) Quality plan and dependability assurance process are established and implemented.
   (5) Standardization process and dependability design rules have been implemented.
   (6) System architecture and physical configuration in meeting system performance requirements have been determined.
   (7) System and sub-system integration plan has been established.
   (8) Hardware partitioning, software interfaces and human factors designs to meet system performance requirements have been determined.
   (9) System dependability performance requirements and operating conditions are specified.
   (10) System test strategy, test coverage and functional evaluation are completed.
   (11) System functions to meet dependability performance needs have been evaluated.
   (12) System designs and dependability of system functions have been validated.
   (13) Outsourced project work, development partnerships and preferred suppliers have been identified, coordinated and established.
   (14) Second sources have been identified and coordinated to support alternative project needs.
   (15) Applicable enabling systems and support strategy are deployed for system dependability achievements.
   (16) Manufacturability for product realization and related dependability issues are determined.
   (17) Design documentations, training instructions and test procedures are completed.
   (18) System operation and support plan have been established.
   (19) Logistics support plan is established.
   (20) Maintenance policy and the levels of repair of lowest level assembly are established.
   (21) Go/no go decision for product realization can be justified.

(c) Product realization checklist
   (1) A product implementation plan has been established.
   (2) Product quality and dependability assurance tasks are implemented.
   (3) Supplier product coordination and control for dependability assessment are completed.
   (4) Commercial-off-the-shelf products required for incorporation in system functions have been evaluated.
   (5) Product and sub-system evaluation for dependability verification are completed.
   (6) System and sub-system test and performance evaluation have been completed.
   (7) System integration and sub-system incorporation have been achieved.
   (8) Design freeze and configuration control plan is established.

(9) System performance requirements have been validated.
(10) System acceptance strategy has been established.
(11) Failure reporting analysis and corrective action system have been established and implemented.
(12) Go/no go decision for system transfer and customer acceptance can be justified.

(d) System acceptance checklist
(1) System acceptance plan is established with customer consultation.
(2) System dependability performance demonstration plan and applicable warranty period have been established and accepted by the customer.
(3) Incident reporting system is implemented and criteria for reporting established.
(4) System operation and support plan to achieve dependability performance are implemented.
(5) Training for system operators and maintainers is conducted, and the trainees are certified, where applicable.
(6) System support for third-party participation such as calibration services has been identified, coordinated and approved.
(7) System hand-over procedures are established for transfer to customer operation.
(8) Legal transfer of system ownership to the customer under contract is completed.
(9) Go/no go decision for system in-service operation can be justified.

(e) System in-service operation checklist
(1) System operation and support plan are implemented.
(2) System performance monitoring and control procedures are implemented.
(3) Incident reporting system is implemented to track dependability performance, service continuity operation, maintenance support activities and corrective and preventive actions.
(4) Maintenance actions are tracked.
(5) Design change procedures and configuration control plan are activated.
(6) Logistics support plan is implemented.
(7) System operational analysis is implemented.
(8) Operational anomalies and areas for improvement are identified.
(9) System dependability performance trend is established.
(10) End user satisfaction surveys are conducted.
(11) Go/no go decision for retention of existing system in-service operation can be justified.

(f) Enhancement checklist
(1) Market needs for system enhancement have been established.
(2) Risk and value assessments are conducted to justify the enhancement effort.
(3) Impact on dependability performance due to enhancement changes is verified.
(4) Impact on the environment and other influencing factors including regulatory, safety and security issues concerning the enhancement changes are investigated and validated.
(5) The cost and time schedules for the enhancement work estimated.
(6) The resources needed for the enhancement work are determined.
(7) Go/no go decision for system enhancement can be justified.

(g) Retirement checklist
   (1) The need and timing for system retirement have been established.
   (2) The causes for retirement, such as technology obsolescence, economic and regulatory constraints, have been determined.
   (3) Replacement system to provide continued system service is determined.
   (4) Social implications due to termination of service are assessed.
   (5) Go/no go decision for system retirement can be justified.
   (6) Plan for seamless transition from old to new system has been established and assured.

D.2 CHECKLISTS FOR TECHNOLOGY DESIGN APPLICATIONS

(a) System hardware design checklist
   (1) System hardware requirements have been established.
   (2) The hardware elements selected for designing system functions have been identified.
   (3) The hardware technology and reliability history are known and have been assessed.
   (4) The system hardware configuration is determined.
   (5) Hardware design specifications have been established.
   (6) The hardware packaging concept and modularization scheme have been determined.
   (7) The thermal budget in an operating profile has been analyzed to determine hot spots and cooling schemes with respect to the module ambient conditions and the system operation environment.
   (8) The electromagnetic compatibility budget in an operating profile has been established to identify the extent for shielding, filtering, partitioning and placement requirements.
   (9) Functional module interface and connectivity has been established.
   (10) Power feed and supply plan and voltage standardization for the system have been determined.
   (11) System reliability performance modeling has been evaluated for redundancy and design options.
   (12) Functional analysis and reliability allocation for each system function have been determined.
   (13) System and sub-system integration plan has been developed.
   (14) System maintainability and testability have been analyzed and test coverage determined.
   (15) Built-in test capability and self-checking features where applicable are incorporated into the module design to facilitate fault identification and fault isolation.
   (16) Fault-tolerant and fault-avoidant designs are incorporated into critical system functions.
   (17) System maintenance concept and the levels of maintenance have been established.
   (18) Spares provisioning of the lowest level assemblies has been determined.
   (19) Turn-around time for spares replenishment has been determined.

(20) System simulation where needed to demonstrate availability performance has been conducted.
(21) System test cases for fault detection, isolation and repair, and restoration time have been verified.
(22) Commercial-off-the-shelf hardware products are evaluated for incorporation into system functions.
(23) System, sub-system and functional module test plans and procedures are developed.
(24) Design documentation is complete for hardware product and assembly production.

(b) System software design checklist
(1) System software requirements have been established.
(2) The system architecture is determined.
(3) Software standards are implemented for software design and development.
(4) Software tools and services are acquired to support software development.
(5) Software partitioning and allocation of functions have been established.
(6) Software functions interface and protocol have been established.
(7) Software design specifications have been established.
(8) Criteria for data integrity have been established.
(9) Software delivery schedules and plans for preliminary design and detailed design are established.
(10) Software module functions are tested and verified to meet design specification.
(11) Commercial-off-the-shelf software products are evaluated for incorporation into system functions.
(12) Acceptance criteria of software product and subsystem have been established.
(13) Acceptance testing has been conducted to determine the software product and subsystem in meeting acceptance criteria.
(14) Software system test and evaluation are validated to meet performance specification.
(15) Software tools for system operation and maintenance support have been identified.
(16) Design documentation is complete for software product replications.

(c) Human factors engineering design checklist
(1) The objective for human factors engineering design is defined.
(2) The human factors engineering plan has been established for design applications.
(3) The human factors engineering design concepts are established for usability, operational suitability, function allocation and level of automation, recognition of the human capabilities and limitations in system operation and maintenance.
(4) The human system interfaces have been evaluated in terms of design simplicity, identical functions for consistency in operation, compatibility with other existing systems of that kind and user awareness of the information displays and communications.
(5) The human-computer interfaces have been evaluated in terms of screen design to facilitate user-friendly interaction, input controls and control

mechanisms, ease of data entry and editing, graphic information and display, update and interrupt features, file management functions, message windows and help services. It is important that system messages are correct, complete, not misleading and easy to understand.

(6) The system designs have incorporated fail-safe features, error resistance and tolerance, ease of handing critical situations and emergencies, ease of enabling and disabling automated functions, simple diagnostic routines for fault management and ease of navigation through a degraded mode of system operation for corrective action.

(7) The system designs have incorporated ease of access to replace removable units and lowest level assemblies, adequate labelling for safety warning and operation, and access to technical manuals and support documents for maintenance, installation and repair instructions.

(8) The system designs for operation have determined the level of automation and the skills and training needs for the operators and maintainers.

(9) Design documentation is complete for development of system operation and maintenance manuals.

(d) Environmental compatibility design checklist

(1) The objective of designing for the environment is defined.

(2) The environmental design requirements have been established for design applications.

(3) The environmental standards and regulations have been reviewed and incorporated in the environmental design concepts and implementation plan focusing on reducing the number of hardware assemblies and parts and aiming at reuse or recycling.

(4) The number of parts used in an assembly has been minimized to reduce assembly and disassembly time to improve recycling process efficiency.

(5) Modular design for the lowest level replaceable unit with a single function has been considered to permit service options, functional upgrade and recycling of parts.

(6) Grouping non-recycled parts in one location has been considered to facilitate disassembly and quick removal for disposal.

(7) Placement of high-value parts in a location for easy access has been considered to enable partial disassembly for optimum return and salvage.

(8) Designing parts for ruggedness and stability has been considered to enhance manual disassembly.

(9) Avoidance of moulded-in metal inserts and reinforcement in plastic parts in assembly has been considered to enhance separation and recycling of plastic parts.

(10) Making access and break points obvious in logical sequence has been considered to enhance disassembly and maintenance service training.

(11) Power down or standby condition whenever possible to save energy and reduce pollution.

(12) The number of fasteners has been minimized to reduce assembly and disassembly time.

(13) Standardizing the use of tools for assembly and disassembly has been considered to save tooling cost and time.

(14) Easy access to fastening points has been considered to enhance maintainability and servicing.

(15) Using snap-fits where applicable and practical has been considered to enhance disassembly and ease of parts removal.
(16) Using compatible fastening materials with connecting parts has been considered to enhance recycling of parts.
(17) Making incompatible parts easily separable when joining together has been considered to enhance parts separation for recycling.
(18) Use of adhesives is generally not recommended due to difficulty of disassembling the parts, especially when the two joined materials are incompatible in recycling. Furthermore, even for compatible materials the adhesives can contaminate the materials, making recycling difficult.
(19) The number and length of interconnecting wires and cables should be minimized to reduce assembly and disassembly time and avoid potential electromagnetic interference.
(20) Designing breakable connections for throwaway parts should be considered to enhance disassembly.

D.3 CHECKLISTS FOR OUTSOURCED PRODUCTS USED IN SYSTEM APPLICATIONS

(a) Outsourced product identification checklist
(1) The outsourced or commercial-off-the-shelf product is commercially available with a unique identification for purchase and has sufficient product information and functional description for evaluation of its fitness for use for the intended application.
(2) There are multiple suppliers of similar products in the commercial market to choose from.
(3) The product identification is designated by name, model or version, and serial number or date on the manufacturer's product label.
(4) The product description contains product specification, instructions for installation and operation of the product, procedures for product connections and interface requirements for applications, and the need and extent for product maintenance and support services.
(5) Warning labels and procedures for safety-related operations where applicable are provided.
(6) Product warranty information is provided.
(7) Product reliability and maintainability information, performance history, and supporting test data are available for verification.
(8) A statement of product quality attestation is provided.

(b) Outsourced product evaluation checklist
(1) Product performance records containing relevant documentations to substantiate conformance to product specification are available for verification.
(2) The relevant documentation, including test plan, test procedures, test environment and conditions, and test records, are used to demonstrate product conformance to specification.
(3) Test cases designed to evaluate fault-tolerant conditions where applicable to the product claims are available for verification.

(c) Outsourced product assurance checklist
   (1) Product quality information and quality records are available for verification.
   (2) Product conformity assessment data are available for verification.
   (3) Product field data is available to support reliability performance claims.
   (4) Product return rates and failure trends are available for verification.
   (5) Product maintenance records are available for verification.
   (6) Product risk assessment and evaluation of product features and related process attributes are completed for critical system application. Specific evaluation includes but not limited to fault detection, redundancy needs, and establishing the integrity level of the commercial-off-the-shelf product appropriate for critical system operation. Integrity level denotes a range of values of the product property necessary to maintain system risks within tolerable limits.

Appendix E

RELIABILITY IMPROVEMENT OF RELIQUEFACTION SYSTEM FOR LNG CARRIERS

E.1 OBJECTIVES OF THE STUDY

The objective of this study was to quantify the reliability of a boil-off gas (BOG) reliquefaction system for liquefied natural gas (LNG) carriers and to verify design improvements based on the reliability analysis [1]. In contrast with the conventional use of steam turbines for propulsion, new LNG carriers are beginning to use diesel-electric engines with dual fuel that utilize BOG, with the cargo LNG as one of the fuel sources. Reliquefaction of BOG in LNG carriers presents formidable challenges, although natural gas reliquefaction is a well-proven technology onshore. Safety should be given top priority since the BOG reliquefaction process contains both fuel and ignition sources within itself. The fuel sources in the BOG reliquefaction process correspond to its separators, exchangers and vessels, while the ignition sources are, for example, its pumps and compressors. Handling of the two sources together has prevented the BOG reliquefaction process from entering into commercial LNG carriers. However, recent development in the safety design of BOG reliquefaction technology makes the application acceptable. For example, Hamworthy BOG reliquefaction is a combination of the conventional BOG handling system and a nitrogen (N2) cooling system. The inertness of the nitrogen cooling system does not add any significant hazard to the conventional system.

Another critical issue for the BOG reliquefaction process is reliability and availability. Reliability is the probability of failure-free operation for a specified time interval, and availability is the ratio of uptime to the sum of uptime and downtime. In the conventional steam turbine LNG carriers, the failure of BOG compression system requires special consideration to treat the BOG. The same is true for the new LNG carriers with the reliquefaction process. If the BOG reliquefaction process fails to operate properly, the BOG should be vented to the atmosphere or burnt in combustion units. This may lead to hazardous situations as well as a significant financial loss, thus erasing the merits of BOG reliquefaction. Accordingly, rigorous investigation of the BOG reliquefaction process in terms of reliability and availability is of critical importance.

A reliability study may provide guidelines for system design and maintenance [2]. System reliability analyses may suggest design improvements, such as using more reliable components, introducing redundancy and mitigating loads on specific components during operation. The reliability analyses may also assist in setting up a proper maintenance strategy to assign inspection and maintenance resources to more important components.

The objectives of the study were two-fold: to verify whether the configuration of the current Hamworthy reliquefaction system is designed reasonably in the sense of reliability and availability requirements, and to show what the maintenance policy should be like to improve or sustain its availability. For the first objective, a bare or non-redundancy system is conceived, and its development is proposed in order to maximize stepwise the reliability and availability of the improved system. The second objective of the optimum maintenance policy is mainly described rather qualitatively, stating which parts should receive concentrated maintenance efforts.

E.2 SYSTEM DESCRIPTION

In this study, the general liquefaction system consisted of the sub-systems of BOG preparation, BOG compression, BOG liquefaction, N2 cooling, buffer N2 reservoir, and N2 generation. BOG from the cargo tank enters the sub-system "BOG preparation system," where mist droplets are removed for compression and BOG is cooled, if necessary. Sub-system "BOG compression system" compresses BOG and sends it to sub-system "BOG liquefaction system," where BOG liquefies after indirect contact with cooled nitrogen gas in cryogenic exchangers. Sub-system "N2 cooling system" provides a cooling source through the usual compression/expansion gas cooling cycle. The sub-system is relatively complicated and bulky because it contains compressors, aftercoolers, expanders and control devices. The cooling capacity is controlled by the flow rate circulating the cooling cycle. Sub-system "Buffer N2 reservoir system" withdraws or provides nitrogen to sub-system "N2 cooling system." When nitrogen storage is not adequate, sub-system "Buffer N2 reservoir system" should take delivery of nitrogen from sub-system "Buffer N2 generation system."

E.3 RELIABILITY AND AVAILABILITY ESTIMATION

The system availability, or the ratio of uptime to the sum of uptime and downtime, is affected by the maintenance efforts as well as the sub-system reliability. In other words, the system availability is increased by either enhancing the system reliability or shortening the time for repair. The availability assessment in this study is based on usual repair times, mean-time-to-repair (MTTR).

The following assumptions are made in estimating the reliability and availability.

(a) All the components obey the exponential failure model.
(b) Only critical failures are considered, with degradable failures neglected. Exceptionally, degradable failures are considered for pressure sensors.
(c) Repair time is active repair calendar time.
(d) An instrument consisting of transmitters, switches and indicator control is considered an assembly, denoted by a "sensor" with one failure rate and repair time.
(e) Any instrumentation and equipment for emergency shutdown do not affect the reliability and availability, equivalently the production efficiency, of the system. Meanwhile, control systems working in normal operation such as control valves and transmitters are taken into account.
(f) Unless expressed explicitly, normally required components for a unit are included in the unit itself. In other words, the unit failure rate comprises the failure rates of the required components.

E.4 RELIABILITY AND MAINTENANCE DATA

Estimation of reliability and availability intrinsically requires the failure rate, mean-time-between-failure (MTBF) and MTTR for all components. Since the BOG reliquefaction has not been in long-time operation, a database was not available. Alternatively, this study extracted the data mainly from the OREDA handbook [3]. Table E.1 exemplifies a list of failure rate and MTTR for sub-system "BOG liquefaction system."

It is worthy noting some points here. A unit listed in the OREDA handbook operates under harsher conditions than the corresponding one for BOG reliquefaction. As a consequence, it is reasonable to assume that the latter has a failure rate far less than the former and to consider the results of this study as a worst case of the actual process. The actual reliability is expected to be higher than the estimate of this study. Another point is that the OREDA handbook does not contain all the units used in BOG reliquefaction. For such units, the data from a similar one are taken from the OREDA handbook. For example, cryogenic heat exchangers have not been in use offshore and are thus not registered in the OREDA handbook. Their failure rate was taken from that of plate heat exchangers in OREDA, and their MTTR was based on the information from the manufacturer.

E.5 ANALYSIS USING RELIABILITY BLOCK DIAGRAMS

E.5.1 Results Without Redundancy

One way to see whether a system is well designed is to check its development from a bare or non-redundancy system, which contains its components and sub-systems with no standby substitutes. Too much redundancy may result in considerable cost and maintenance efforts with an insignificant increase in availability. Which sub-system or unit should have redundancy? Strictly speaking, the decision should be made on life cycle cost, a formidable effort in the development stage due to various uncertainties. Figure E.1 is the Reliability Block Diagram (RBD) of the bare system after one year of operation. The system is intrinsically a series system: a failure of any sub-system implies the failure of the whole system. Some of the sub-systems have a reliability factor less than 0.1 although their availability is above 0.98. Note that the reliability is the probability of being failure-free for a specified time interval, and any failure of a component or a sub-system causes the system reliability to be zero.

TABLE E.1 Failure rate and repair time for sub-system "BOG liquefaction system" [1]

Unit	Component	Failure Rate per 10^6 hr	MTBF, 10^6 hr	MTTR, hr
Cryogenic Heat Exchanger	Cryogenic Heat Exchanger	21.75	0.05	18.70
	Temperature Sensor	8.56	0.12	2.30
BOG Liquid Separator	BOG Liquid Separator	28.83	0.03	2.10
	Level Sensor	4.16	0.24	3.70
	Level Control Valve	3.88	0.26	2.00
	Pressure Sensor	5.48	0.18	2.30
	Pressure Control Valve	3.88	0.26	2.00
BOG Return Pump	BOG Return Pump	43.01	0.02	11.40
	BOG Return Motor	22.75	0.04	7.80

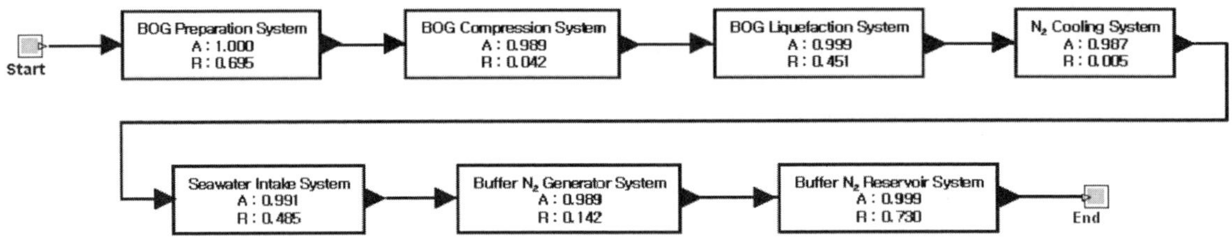

Figure E.1 Reliability block diagram of a system without any redundancy at 8,760 hours [1]

The reliability of the bare system decreases rapidly with time, as shown in Figure E.2. After 1,000 hours of operation, the system reliability falls to around 0.2. In order to keep the system available or operating properly, a repair action for the failed component or sub-system is taken. In steady-state operation, the repair action balances the failure rate to give a constant availability. The steady-state availability of the bare system is less than 0.96. Improvement of the availability may be attained by either accelerating the repair (equivalently, decreasing MTTR) or adding redundancy for a component or a sub-system that has the lowest availability. In the design phase, the latter consideration is critical for the reliability improvement. Which sub-system should have redundancy may be determined by checking the reliability of each sub-system.

Figure E.3 shows the reliability of each sub-system at each 1,000 hours. The sub-systems are classified into three ranks according to their reliability: the high reliability rank comprises the sub-systems "BOG preparation system," "Seawater intake system," and "Buffer N2 reservoir system." The medium reliability rank includes the sub-system "BOG liquefaction system," and the low reliability rank includes the sub-systems "BOG compression system," "N2 cooling system," and "Buffer N2 generation system." In view of a redundancy policy, the sub-systems belonging to the high reliability rank have the least requirement to have a redundant sub-system, whereas the sub-systems of the low reliability rank are recommended to have redundancies. Whether to add a redundant to the sub-systems of the medium reliability rank depends on the cost-and-benefit consideration.

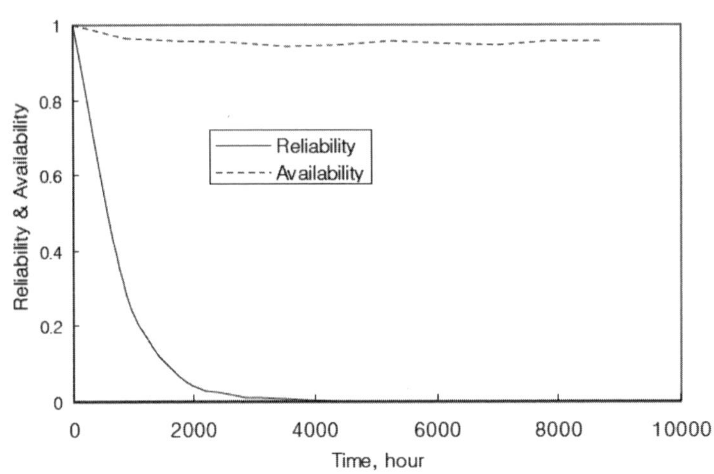

Figure E.2 Reliability and availability of a system without any redundancy [1]

E.5.2 Results With Redundancy

The sub-system "N2 cooling system" has the lowest reliability, and then sub-systems "BOG compression system" and "Buffer N2 generation system." This means that sub-system "N2 cooling system" should be considered the first candidate for redundancy, and consecutively sub-systems "BOG compression system" and "Buffer N2 generation system." It is worth checking the sensitivity of adding redundancy to each of the three systems. Before proceeding further, the system configurations are defined as:

System 0: The bare system where no sub-system has a standby.
System 1: System where sub-system "N2 cooling system" has a standby.
System 2: System where sub-system "Buffer N2 generation system" has a standby.
System 3: System where sub-system "BOG compression system" has a standby.
System 4: System where all three of the sub-systems have its own redundancy.

Figure E.4 shows the reliability block diagram of System 4 after one year of operation. The parallel sub-system configuration means redundancy.

Figure E.5 depicts the reliability of the various process configurations with operation time. Systems 1 to 3 show a significant increase in the reliability compared with System 0. However, they do not show a notable difference in reliability between each other. In other words, it is hard to tell which of Systems 1, 2 and 3 is superior to the others in terms of reliability. System 4 has a prominent increase in reliability compared with Systems 1 to 3.

As shown in Figure E.6, System 4 shows a definite improvement in availability compared with Systems 1 to 3, which in turn shows a significant increase from System 0. Systems 1 to 3 have an availability around 0.97, a remarkable increase from the availability of System 0, while System 4 has an availability around 0.99.

The common feature of the three least reliable sub-systems is that they have rotating machines, such as pumps and compressors. More specifically, compressors and pumps

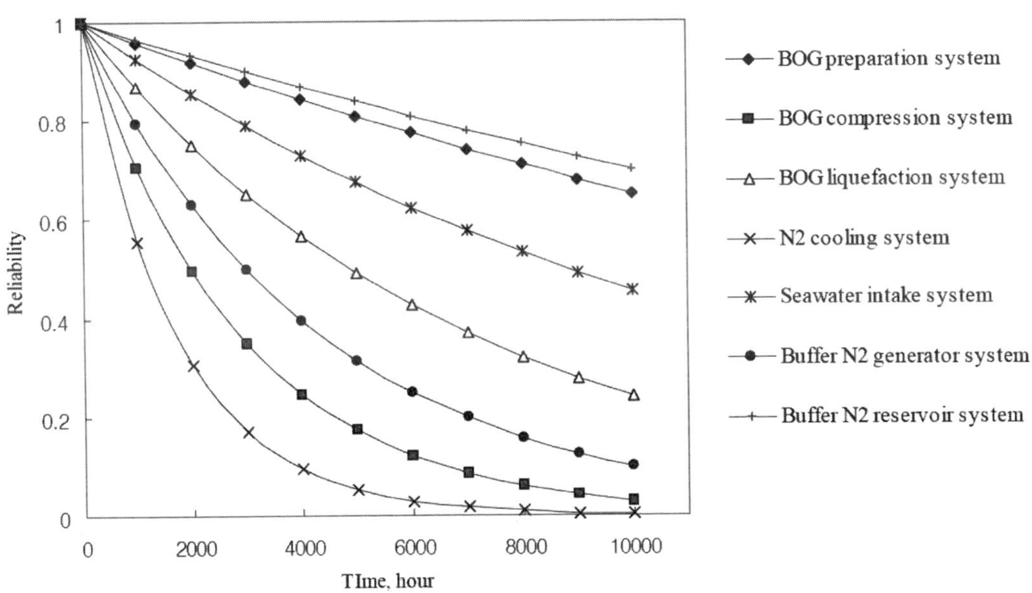

Figure E.3 Reliability of sub-systems with time [1]

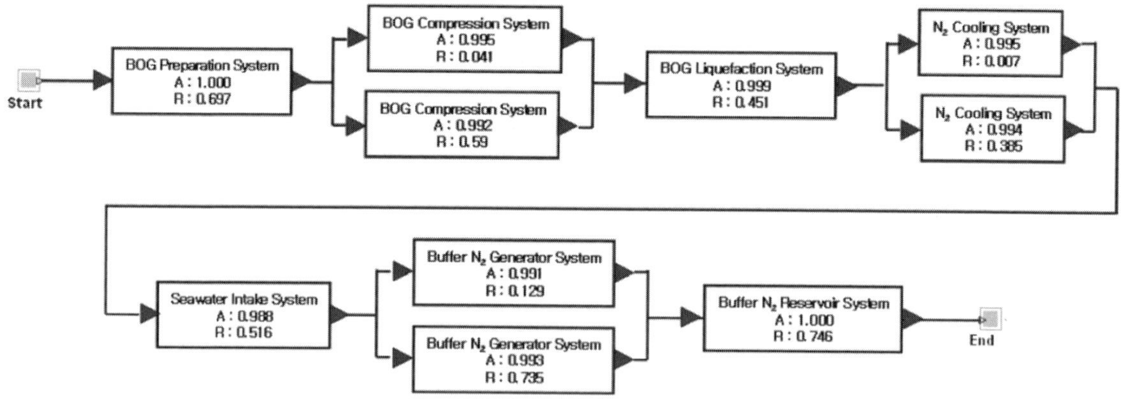

Figure E.4 Reliability block diagram of System 4 at 8,760 hours [1]

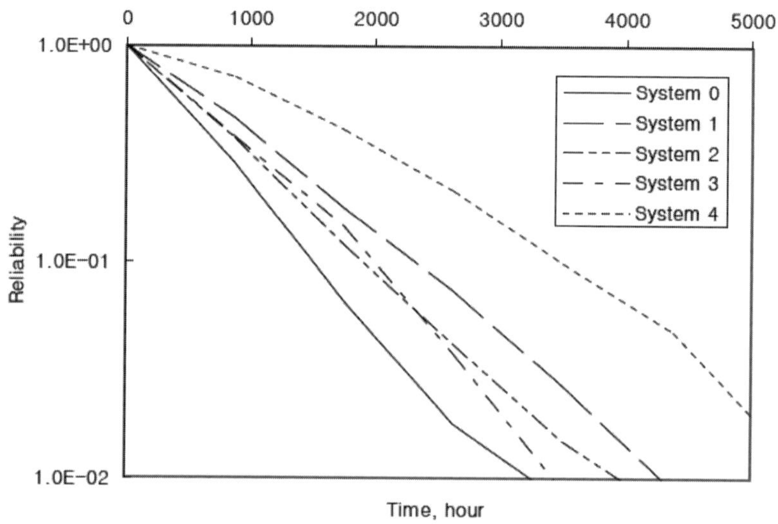

Figure E.5 Reliability of Systems 1 to 4 with time [1]

have a failure rate of around 250 and 60 per 10^6 hours, respectively, whereas separators and sensors have a failure rate of less than 10 per 10^6 hours. As a consequence, most of the maintenance efforts should be directed to rotating machinery. Another operational point to sustain high availability is to reduce operational and environmental loads on the least reliable units. As a result, rotating machinery should be placed in a favorable circumstance and operated in an alternating way.

The current BOG reliquefaction system has standbys for the least reliable three subsystems and is expected to have an availability factor of around 0.99. Considering the basis of the current analysis (i.e., that the failure rates of offshore equipment and components have been used for the reliability estimation), the actual BOG reliquefaction process should attain a higher reliability. Collection of operation data, including failure rate and maintenance efforts, is the prerequisite for optimal design and maintenance strategy. The BOG

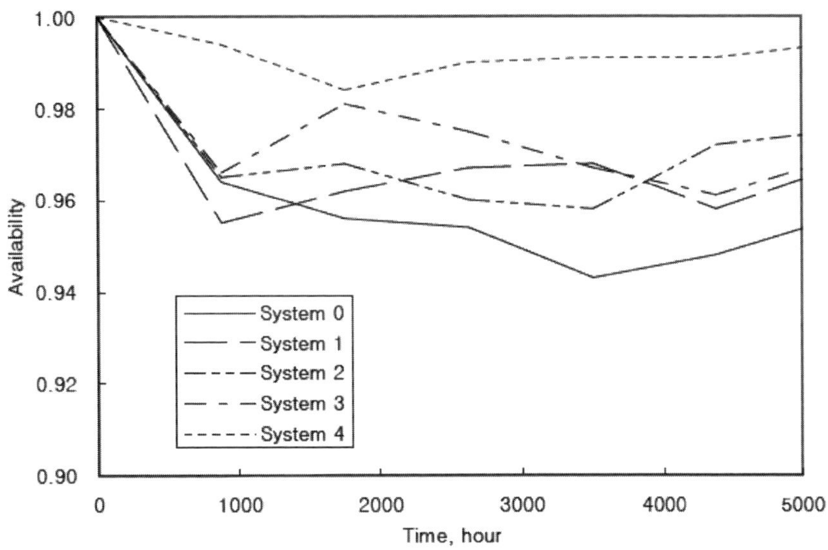

Figure E.6 Availability of Systems 1 to 4 with time [1]

reliquefaction process is on the verge of commercial service, and using redundancy provides the opportunity to optimize its design and maintenance. The collection of operation data should play a critical role in future verification and improvements.

REFERENCES

[1] Chang, D. et al, 2005. A Study on Reliability-Based Improvement of Reliquefaction System for LNG Carriers. OMAE2005-67023, pp. 17-23
[2] Marvin Rausand, System Reliability Theory, 2nd Ed., John Wiley & Sons, New Jersey, 2004.
[3] OREDA Participants, Offshore Reliability Handbook, 4th Ed., 2002.

Appendix F

COMPRESSOR STATION AVAILABILITY

F.1 INTRODUCTION

As the competitive market for gas transmission becomes more and more dynamic and there is an increasing influence from regulatory agencies, the pressure on reducing cost of service without affecting reliability and safety is a consequence. Notwithstanding this trend, transportation companies must act in a way that guarantees a fair return on investment and optimizes assets and operation costs. Contractual obligations play an important role since most of the contractual capacity is on a firm basis and subject to liabilities related to capacity shortage or interruption. Compressor station availability studies play a fundamental role in providing information that will support decision making in terms of defining a criterion for installing stand-by units.

This study [1] presented two methods adopted to evaluate the availability of the Bolivia-Brazil Gas Pipeline Project (GASBOL), a gas transmission system that has four compressor stations in Bolivia and ten compressor stations in Brazil: (1) scheduled and unscheduled maintenance, and (2) Monte Carlo simulation. In addition, compressor unit unavailability was calculated with a binomial distribution for the purpose of comparing its results with the other two methods. The objective of the study was to quantify the availability of the transmission system and to identify the required number of stand-by compressor units to be installed to fulfill contractual obligations for firm capacity. An optimum number of stand-by compressor units were defined, taking into account contractual liabilities (as a result of failure to provide total required firm capacity) and the total investment and operating cost for the new stand-by compressor units.

The study focused on the Brazil section of the GASBOL as presented below:

Compressor stations in Brazil:	10
Compressor units per station:	2
Compressor units installed power:	15,000 HP ISO
Maximum compressor ratio:	1.8
Mean spacing between compressor stations:	125 km

The availability values for the compressor station units were defined based on the following criteria:

(a) Obtained from the EPRI Report No. RP 4CH2983 as 0.971 for installed compressor stations with centrifugal compressor and gas turbine driver.

(b) Obtained by the following equation and without stand-by units:
Reliability = 1 - FOF

Availability = 1 – (FOF + SOF)

FOF = FOH/PH

SOF = SOH/PH

FOF = Forced Outage Factor

FOH = Forced Outage Hours

PH = Period Hours

SOF = Scheduled Outage Factor

SOH = Scheduled Outage Hours

The values of FOF and SOF presented below were taken from the North American Electric Reliability Council's NERC Report of January 2005 for gas turbine drivers.

FOF = 0.0282

SOF = 0.0424

Reliability = 1 – FOF = 0.9718

Availability = 1 – (FOF + SOF) = 0.9294

(c) Obtained from a binomial distribution based on the availability value taken from the NERC report for the gas turbine driver.
(d) Obtained from the scheduled maintenance as recommended by the gas turbine manufacturer.
(e) Obtained from Monte Carlo simulation, considering the availability number for the gas turbine drivers taken from the NERC report. The simulation considered the compressor stations operating initially without stand-by units and then defined the number of stand-by units to be installed to guarantee an adequate level of availability for the pipeline to cope with contractual obligations related to firm transportation capacity and to mitigate liabilities.

F.2 BINOMIAL DISTRIBUTION ANALYSIS

For the purpose of comparing the two adopted approaches, the model was simplified, as shown below, to assume two compressor units per station for all ten stations in Brazil. Stations #8 and #10 with four smaller units of 7,000 HP were replaced in the model by two bigger units of 15,000 HP. For this scenario of ten compressor stations with two units each, the most appropriate scenarios involved the compressor station availability with two units.

The binomial distribution can be calculated by the following equation:

$$BC(X, n) = \frac{n!}{X!(n-X)!} \qquad \text{Equation F.1}$$

where:

n, number of units = 20
X = simultaneous units unavailable
$BC(X, n)$ = binomial coefficient

The probability of any number of unavailable units is:

$$P(X) = \frac{n!}{X!(n-X)!} P^X (1-P)^{n-X} \qquad \text{Equation F.2}$$

where:

$p = 0.0706$ (unavailability value)
$1 - p = 0.9294$ (availability value)
$n = 20$ (compressor station total units)
X = unavailability of 0, 1, 2, 3… n units

The resultant table with the probability of 0 to 20 units being unavailable is:

TABLE F.1 Unavailability results for 20 units using a binomial distribution [1]

Units Unavailability

X	BC (X,n)	P(X)	Unavailability (days/year)
0	1	0.231235	84.40
1	20	0.351306	128.23
2	190	0.253519	92.53
3	1140	0.115549	42.18
4	4845	0.037304	13.62
5	15504	0.009068	3.31
6	38760	0.001722	0.63
7	77520	0.000262	0.10
8	125970	0.000032	0.01
9	167960	0.000003	0.00
10	184756	0.000000	0.00
11	167960	0.000000	0.00
12	125970	0.000000	0.00
13	77520	0.000000	0.00
14	38760	0.000000	0.00
15	15504	0.000000	0.00
16	4845	0.000000	0.00
17	1140	0.000000	0.00
18	190	0.000000	0.00
19	20	0.000000	0.00
20	1	0.000000	0.00
		1.000000	365.00

These results are compared to an equivalent Monte Carlo simulation in Figure F.1. It can be seen that, where a large enough number of iterations is used for the Monte Carlo simulation, the frequency values from the binomial distribution and the Monte Carlo simulation are essentially identical.

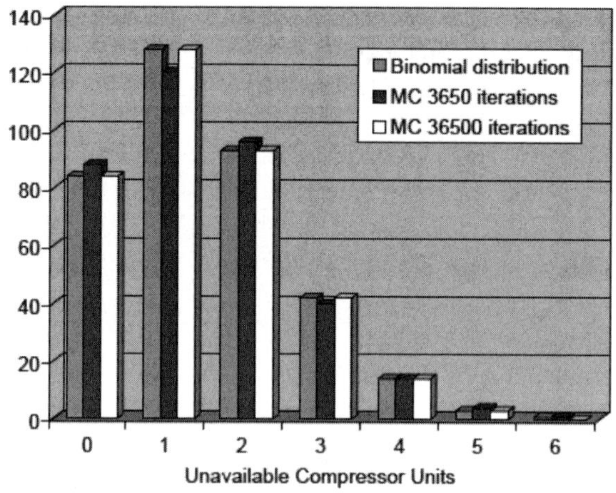

Figure F.1 Frequency comparison between binomial distribution and Monte Carlo simulation [1]

F.3 MONTE CARLO SIMULATION

Monte Carlo simulation is a very practical, simple and straightforward tool for predicting compressor station unit unavailability that would expose the gas transmission company's capacity shortage and consequently subject the company to risk of revenue losses and contractual liabilities as a result of failure to provide the firm capacity to the shippers. The software @Risk® from Palisade was used in conjunction with Excel® for Windows® to run all the Monte Carlo simulations. An availability value of 0.9294 has been adopted for each compressor station unit, and then the Monte Carlo simulation was applied to the gas pipeline compressor station model for three cases: (a) no stand-by units, (b) five stand-by units and (c) ten stand-by units. The criteria to evaluate the gas pipeline availability considered the available capacity taken from the thermo-hydraulic simulation software of the gas pipeline under different unavailability scenarios. Pipeline Studio® from Energy Solutions was used to run the scenarios. The maximum capacity taken from the simulation software for each scenario divided by the contractual firm transmission capacity provides the availability of the transmission system. Stand-by compressor units were defined based on feasibility analysis and also on transportation company experience in handling market demand and operation risk.

The Monte Carlo simulation was run for a configuration of a total of ten compressor stations, with two compressor units per station, and the failure outcomes were determined for (a) no stand-by units, (b) five stand-by units for the first five compressor stations (one at each compressor station) and also (c) ten stand-by units (one at each compressor station), respectively, for comparison purpose only. The feasibility study was performed to identify the best number of stand-by units to mitigate the exposure to contractual liabilities. The Monte Carlo simulation was performed with 36,500 iterations, with the results for configuration (b) shown in Table F.2.

From the average firm transportation capacity values of 27.77, 29.36 and 29.99 MMm^3/d calculated for each configuration, respectively, the availability of the gas pipeline

TABLE F.2 Failure simulation results of compressor station units with five stand-by units (one at each of the first five stations) using Pipeline Studio® [1]

Unavailable Compressor Units	Frequency days/year	Capacity, MMm3/d	Compressor Station									
			#5	#6	#7	#8	#9	#10	#11	#12	#13	#14
			unavailable days per year									
0	58	30.08										
1		30.08	11.2									
	67.7	30.08		14.1								
	101.6	30.08			13.1							
	169	30.08				15.3						
		30.08					14.0					
		27.50						9.4				
		28.70							8.8			
	112	29.80								7.9		
		29.30									7.6	
		25.80										10.5
2		25.90	1.6									
	5	26.10		1.1								
		26.70			0.6							
		26.70				0.5						
		27.20					1.1					
		27.50						0.6				
	2	28.70							0.7			
		29.80								0.5		
		29.30									0.2	
		25.80										0.0
3	0		0	0	0	0	0					
			Compressor Station									
			5&6	6&7	7&8	8&9	9&10	10&11	11&12	12&13	13&14	
1+1 (at two contiguous stations)		30.08	3.1									
	11	30.08		2.9								
		30.08			2.1							
		30.08				2.7						
		27.50					1.5					
		23.70						1.9				
	9	24.40							1.9			
		24.90								2.1		
		21.90									1.3	
	365.5	29.36										
System Availability		0.9761										

Note: For the case of *one unit unavailable per station* the quantity of 101.6 days are related to others non contiguous stations that also have one unit unavailable that do not reduce the pipeline simulated capacity.

transmission system was evaluated by simply dividing this capacity value by the firm contractual capacity of 30.08 MMm^3/d with the following results:

- No stand-by compressor units: 0.9231
- Five stand-by units (for the first five stations): 0.9761
- ten stand-by units (one at each station): 0.9971

F.4 SCHEDULED MAINTENANCE

Unavailability can also be estimated from the outage time associated with the quarterly, semi-annual and annual inspections and the turbine overhauls that occur at intervals of 30,000 running hours. Based on a schedule as recommended by the equipment manufacturer, the available capacity can be determined as shown in Figure F.2, which shows a

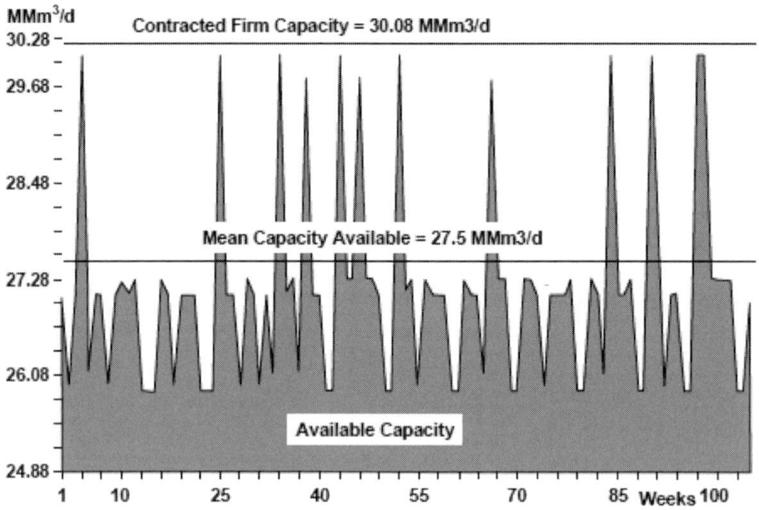

Figure F.2 Capacity loss due to maintenance services with no stand-by compressor units [1]

significant reduction of 2.48 MMm^3/d due to maintenance outages. This is well below the contracted firm capacity.

However, as additional turbines are installed as stand-by units, the overall pipeline availability increases, as the stand-by units override the unavailable ones without causing capacity shortage on the stations where they are installed. Figure F.3 presents the situation

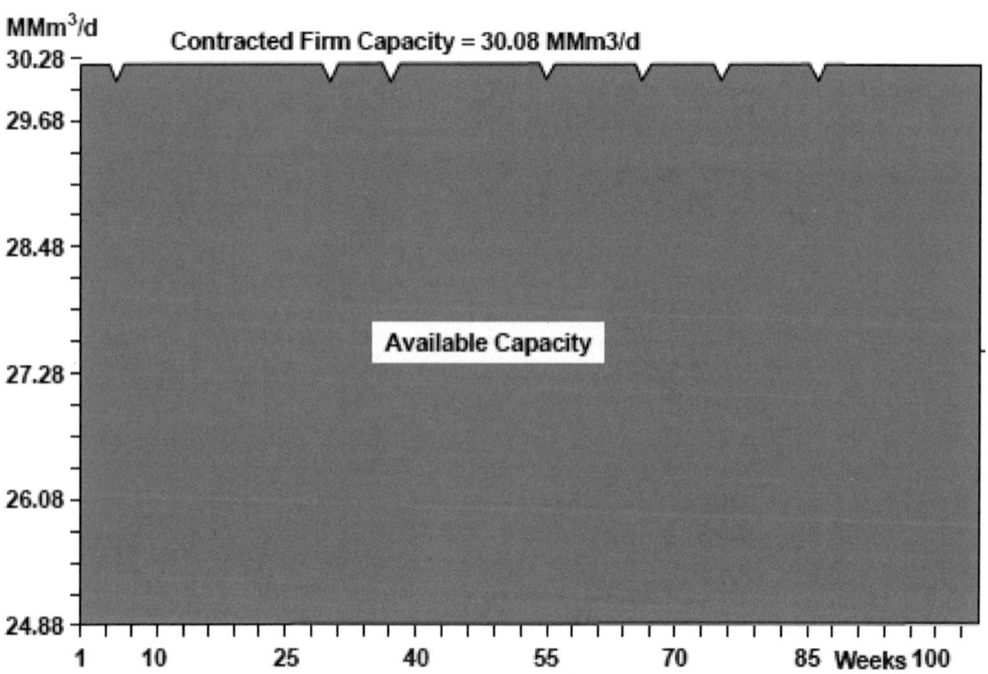

Figure F.3 Capacity recovery with the installation of five stand-by compressor units [1]

when five additional units are installed in five compressor stations. The gray area below graphic line now extends upward, covering almost 100% of the required firm gas transmission capacity. By this approach, the expected average capacity is now very close to 30.08 MMm^3/d.

F.5 ECONOMIC ANALYSIS

An economic analysis to define the quantity of stand-by units was based on Monte Carlo simulation runs performed for each gas transmission system configuration: (a) no stand-by units, (b) five stand-by units and (c) ten stand-by units. The objective was to identify the adequate number of stand-by units to provide a manageable level of risk exposure to contractual liabilities due to non-delivered capacities. The discounted cash flow was used and compared for the three configurations to identify the one that would give the better net present value (NPV). The avoided losses and liabilities were considered as revenues and the stand-by units as capital investments. No additional costs related to fuel gas plus operation and maintenance were accounted for since the units operated as stand-by units.

The evaluation results pointed to an opportunity to install stand-by units for all compressor stations as follows:

(a) No stand-by units – base case
System availability: 0.9231
Potential loss of capacity 2.28 MMm^3/d
Potential loss of revenue: 182.8 MMUS$
Potential loss in liabilities: 182.8 MMUS$

(b) Five stand-by units
System availability: 0.9761
Remaining loss of capacity: 0.85 MMm^3/d
Remaining yearly exposure: 136.3 MMUS$
Recovered capacity: 1.43 MMm^3/d
Avoided loss of revenue: 114.7 MMUS$
Avoided contractual liability: 114.7 MMUS$
CAPEX for stand-by units: 64.5 MMUS$
NPV: 164.8 MMUS$

(c) Ten stand-by units (one at each of ten compressor stations)
System availability: 0.9971
Remaining loss of capacity: 0.07 MMm^3/d
Remaining yearly exposure: 11.2 MMUS$
Recovered capacity: 2.21 MMm^3/d
Avoided loss of revenue: 177.2 MMUS$
Avoided contractual liability: 177.2 MMUS$
CAPEX for stand-by units: 129 MMUS$
NPV: 225.4 MMUS$

The results obtained for configurations (b) and (c) point to an opportunity to install stand-by units for all the compressor stations. Although the capital investment in stand-by units was not considered to be included in the original capital expenditure (CAPEX) for the gas pipeline project, they are still necessary to mitigate the risk exposure associated with the loss of revenues and contractual liabilities as a result of service interruptions. This investment in stand-by units also provides a potential increase of the gas transmission system capacity of ~4 MMm^3/d on an interruptible basis. This additional capacity will allow the recovery of the investment in the stand-by units to a certain extent.

REFERENCES

[1] Santos, S.P., Bittencourt, M.A. and Vasconcellos, L.D., 2006. "Compressor Station Availability - Managing its Effects on Pipeline Operation," Proceedings of IPC 2006 6th International Pipeline Conference, September 25-29, 2006, Calgary, Alberta, Canada, IPC2006-10560, pp. 1-9.

Appendix G

MAINTAINABILITY OF A GAS TURBINE

G.1 INTRODUCTION

The Mercury™ 50 gas turbine (manufactured by Solar Turbines, A Caterpillar Company) operates with a recuperated cycle to produce 4,600 kW with a high thermal efficiency and very low emissions [1]. Commercialized in 2003, the Mercury™ 50 gas turbine went through an extensive design, development and field evaluation program that insured a reliable and durable product that is easy to operate and maintain. This results in maximizing operation reliability and availability while lowering maintenance cost, that are critical factors in the distributed generation and cogeneration power generation markets.

Solar Turbines evaluated several turbine performance cycles to meet the product and market requirements, and the optimal choice was a recuperated cycle. The recuperation of the classic Brayton cycle gas turbine is a well-established method of improving cycle efficiency by recovering some of the turbine's exhaust heat otherwise lost in a simple cycle design. The engine flow path was changed from that of a traditional industrial gas turbine to accommodate the recuperated cycle at the lowest cost to the overall system (see Figure G.1). The new flow path and recuperated cycle also resulted in a combustion system design that could meet the low emissions requirements and allow for easy maintenance in the field.

Recuperator durability and performance issues were identified during development testing and the field evaluation phase. These durability issues did not prevent the unit from running but accelerated the performance degradation. The recuperator performance loss was primarily due to air cell creep and leakage. Material testing confirmed that the 347 SS material used in the original design did not have the required life margin to meet the turbine operating conditions. The recuperator material was upgraded to Alloy 625, a material with superior creep, oxidation and tensile strength compared to 347 SS. In addition, the recuperator size was increased to improve the performance. These two design changes resolved the issues identified during field evaluation testing. The new Alloy 625 recuperator completed extensive development testing with over 3,000 operating hours and 1,500 starts. The recuperator's predicted time-before-inspection-overhaul is now 60,000 hours.

G.2 DESIGN FOR MAINTAINABILITY

The Mercury™ 50 generator set package was designed for ease of operation and maintenance. The engine, recuperator, generator and auxiliary equipment are completely enclosed within the generator set package. Auxiliary equipment includes the fuel module, lube oil module, control system, start system, ventilation system and base and enclosure assembly.

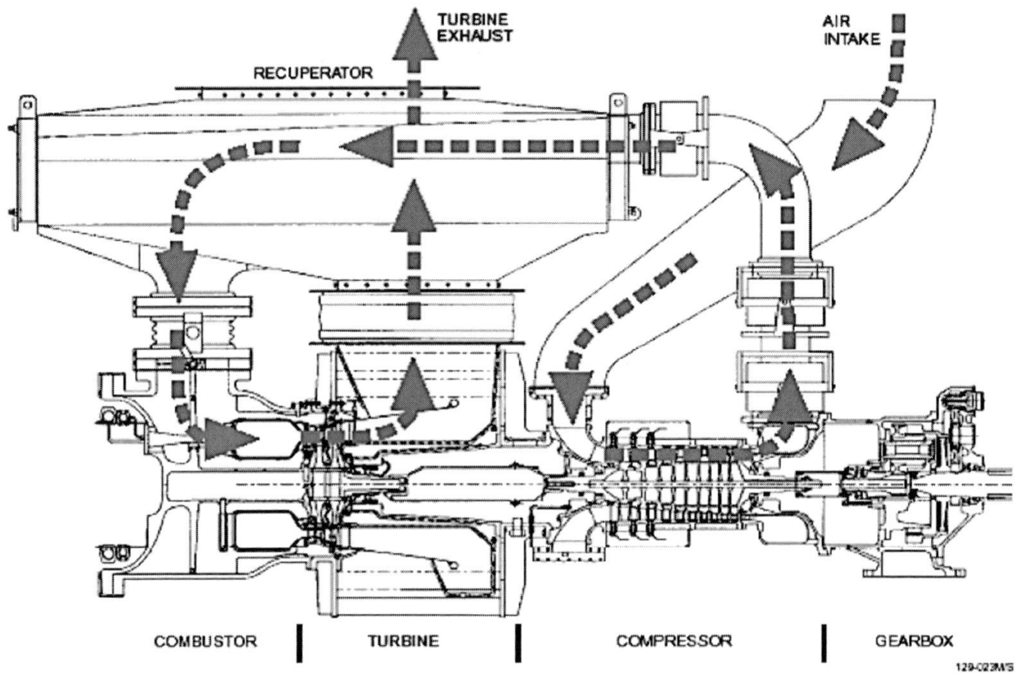

Figure G.1 Layout of Mercury™ 50 gas turbine [1]

The unique Mercury™ 50 engine layout also allows for combustion system and hot section maintenance on site, without removing the turbine.

Six Sigma methodologies were utilized to complete the Mercury 50 package design. This included gathering critical information from field evaluation customers, potential customers and Solar personnel to determine which aspects of the package were the most important. The data were then organized and prioritized using a Six Sigma tool called a Quality functional deployment (QFD) map, shown in Figure G.2. The critical buying criteria are listed along the *x* axis and shown in the table, and the overall importance rating is shown on the *y* axis. The higher the importance rating, the more critical it is to the customer. Survey feedback indicates that reliability and availability were critical and maintenance cost was very important.

In order to maximize the Mercury™ 50 gas turbine reliability and availability, the package was designed with proven components and direct service access to all package components to minimize service time. Key service features of the Mercury™ 50 package include:

1. 100% component accessibility – All package components are located no more than 18" from the skid edge to improve accessibility and reduce trip hazards within the package during service. For example, there is easy access to the nozzle case cooling and compressor discharge hub cavity system assemblies from the skid edge.
2. 100% modularity – All subassemblies are assembled in an ergonomic workstation. This allows for quick replacement of components or complete sub-assemblies, minimizing down time.

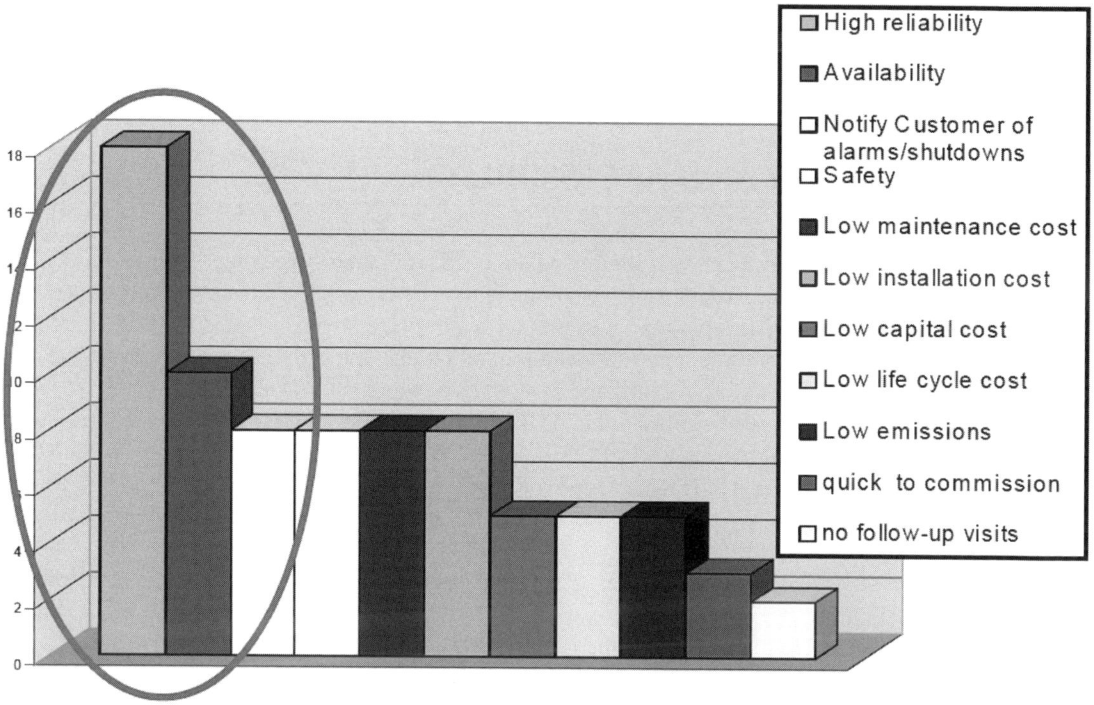

Figure G.2 QFD feedback [1]

3. Electrical wiring – Many package designs today simply run cabling along the most convenient path (Figure G.3(a)) and do not account for easy maintenance access. This may increase maintenance time and cost when repairs are required. Cable trays (Figure G.3(b)) were incorporated into the web of the Mercury™ 50 base I-beam. This improves access to the wires for maintenance.

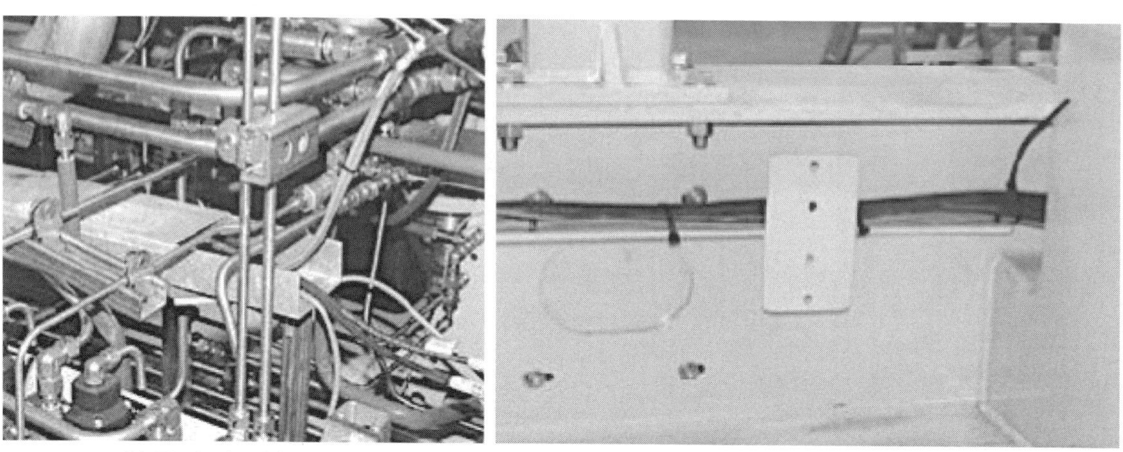

(a) Typical cable trays

(b) Mercury™ 50 cable tray

Figure G.3 Comparison of electrical cabling design for maintainability [1]

4. Large package component replacement – Lifting provisions are available for all components over 25 pounds. Field tools used to remove the turbine, generator and recuperator are designed with lifting provisions for proper installation inside the package.

G.3 SERVICEABILITY AND ACCESSIBILITY

A serviceability event was completed on the first production package to verify accessibility and serviceability of all package components. The event proved out over 90 typical service activities, such as replacement of the torch ignitor, removal and reinstallation of an enclosure door and turbine removal, using the field tools. The event led to changes that resulted in 100% accessibility for package components, which minimizes down time.

Improved accessibility to the package is gained through the enclosure design by providing doors along the wall of every side of the enclosure (Figure G.4(a)). A stable work platform is available (Figure G.4(b)) to gain access to the hot section of the turbine. This provides an area to work on the turbine hot section, including the combustor liner and injector.

The Mercury™ 50 power generation package layout reflects a special attention to availability, a key customer requirement. The most time-consuming job on a standard power generation package is the replacement of a large component, such as the turbine or the generator, or the recuperator in the case of a recuperated turbine. The package design team reviewed many concepts for large component removal and designed the package in such a way that allows easy access, which improved availability. The recuperator support structure serves as the removal railing system for the engine and the recuperator. By incorporating the removal railing system into the package support structure, the replacement of the engine and recuperator only takes two days for each component. The package design also allows for easy generator removal, which is an often-overlooked aspect of power generation packages.

In modern gas turbine engines, the most frequently serviced area is the hot section, which comprises the injectors, combustor liner, turbine nozzles and turbine blades. The

(a) Doors along every side of the enclosure (b) A stable platform for access to the hot section

Figure G.4 Examples of design for maintenance accessibility [1]

(a) Accessibility of combustors

(b) Field tooling for combustor liner removal

Figure G.5 Examples of design for hot section accessibility [1]

unique airflow path of the Mercury™ 50 turbine engine has the added benefit of making the hot section easy to access. The field service division took advantage of this accessibility and designed tooling and procedures for in situ access to these components. The injectors are accessible without having to remove the fuel manifolds, and segmented engine insulation also minimizes the amount of work and time required to access injectors (Figure G.5(a)). The combustor liner can also be removed with the engine in place. The same trolley rails that support the engine and recuperator during removal also support the field tooling used to pull the combustor end cap and liner (Figure G.5(b)).

G.4 MONITORING

Another important aspect of service is monitoring the unit operating parameters for changes that could lead to equipment failure. During package operation, the control system records operating data and returns it to a Solar Turbines customer service monitoring center for statistical evaluation. This data-logging system reviews hourly data points and notifies users of any step changes or gradual shifts that could lead to an unplanned outage. Once the field service representatives are aware of the change, they can take preventative action that can reduce the occurrence of an unplanned outage, or they can schedule parts and service to address the problem during the next planned shutdown.

Extensive analysis, component testing and development testing have been completed on the Mercury™ 50 turbine engine to verify the product reliability with over 56,000 operating hours (as of January 2005). Minimal inspection maintenance is required on the turbine, recuperator and package systems, resulting in higher availability. In addition, these turbine and package systems have been designed to allow fast component replacement when maintenance is required. The modular turbine design allows a broader range of field maintenance options. All these features allow for improved availability. The Mercury™ 50 turbine engine reliability and availability are predicted to meet or exceed gas turbines and other power generation technologies in the 5-MW size range.

The three major maintenance cost components include general maintenance and inspection, product durability, and overhaul. The Mercury 50™ design features reviewed here minimize general maintenance and inspection time. The Mercury™ 50 turbine was also designed to maximize component life, thus reducing the number of replacement parts during each overhaul. This will result in longer times before overhaul and lower overhaul costs.

REFERENCES

[1] Teraji, D., Hettick, J. and Robison, M., 2005. "*MERCURY™ 50* Product Durability, Operation and Maintenance Review", Proceedings of GT2005 ASME Turbo Expo, June 6-9, 2005, Nevada, USA.

Appendix H

EXAMPLE OF A STEAM TURBINE

H.1 INTRODUCTION

One of the major pieces of rotating equipment used in a refinery plant is a steam turbine, which is used to drive electric generators, compressors and pumps [1]. The steam turbine converts the heat energy in the steam to mechanical energy. The main mechanical parts of the steam turbine are the rotor, bearings, seals, lube oil cooler, steam throttle valve and governor. The throttle valve is used to allow the required amount of steam to enter the machine and is controlled by the governor. Another important part is the lubrication oil cooler, which is used to cool the oil that lubricates the bearing. A total of 50 steam turbines were in operation in the refinery plants under investigation. Of these 50 steam turbines, thirteen were found to have experienced frequent failures (deemed to be bad actors because they experienced more than three failures in five years) and incurred high repair cost.

H.2 PARETO ANALYSIS

A Pareto analysis was utilized to identify the most critical bad actor steam turbines based on ranking the respective number of failures and their associated repair cost (Figure H.1).

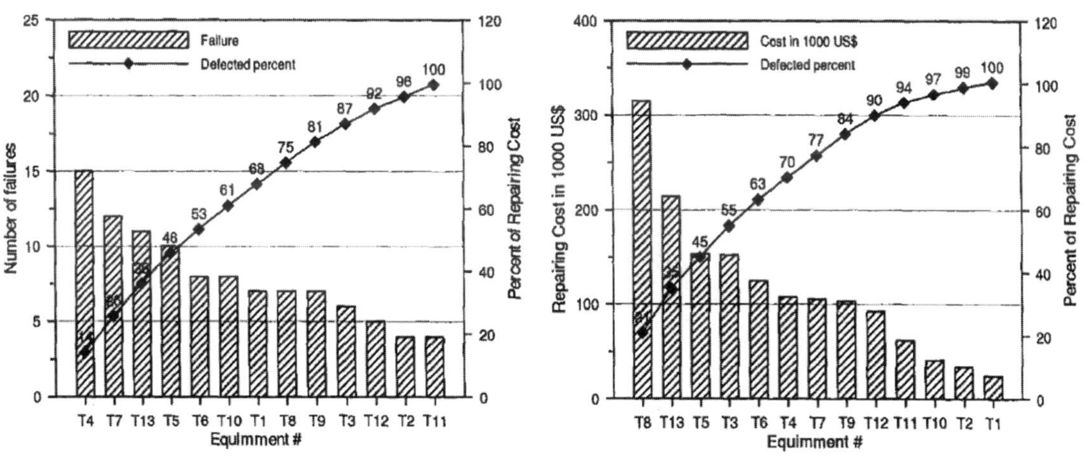

(a) Bad actors by number of failures

(b) Bad actors by repair cost

Figure H.1 Pareto charts to identify bad actors [1]

The Pareto analysis criteria for such steam turbines were used to identify the set of turbines that contributed 75% of the total number of failures and/or 75% of the total repair cost.

Failure modes for all of the most critical turbines were also summarized in Figure H.2, illustrating the frequency of different modes of failures for this group of these steam turbines. It was found that the repairable parts of steam turbine had the highest number of failures. These were governors (28%), lubrication oil coolers (25.7%) and throttle valves (24.3%). Non-repairable parts were found to have 7.5% and 4.1% of the failures for the seals and the bearings, respectively.

H.3 WEIBULL ANALYSIS

Based on the Pareto analysis, ten steam turbines were determined to be the most critical bad actors, and Weibull reliability analysis was applied with an example for two of the steam turbines shown in Figure H.3, one with a characteristic life of only 1.63 months (a) and the other with a somewhat better life of 7.3 months (b).

For the mechanical governor, the accumulated repair cost is shown in Figure H.4(a), and it continued to increase rapidly. Thus it was recommended to retrofit the mechanical governor with electronic governors to eliminate the frequent failures of the mechanical parts. Most of the repair costs for the lubrication oil coolers (Figure H.4(b)) were spent in the last ten months of the investigation period, which was five years. By referring to their history, the lubrication oil coolers had severe corrosion and could not be easily repaired. To minimize the maintenance expenditure on these coolers, it was recommended to replace them with new coolers.

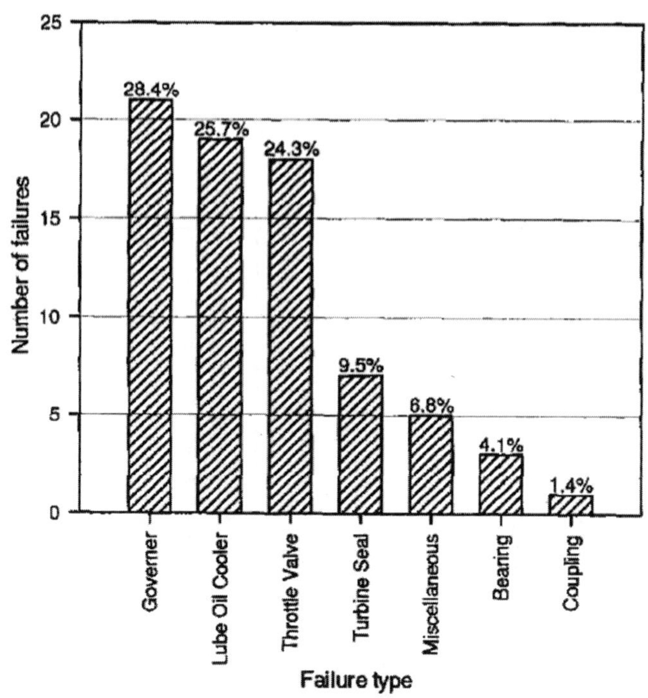

Figure H.2 Pareto charts of most common failure modes [1]

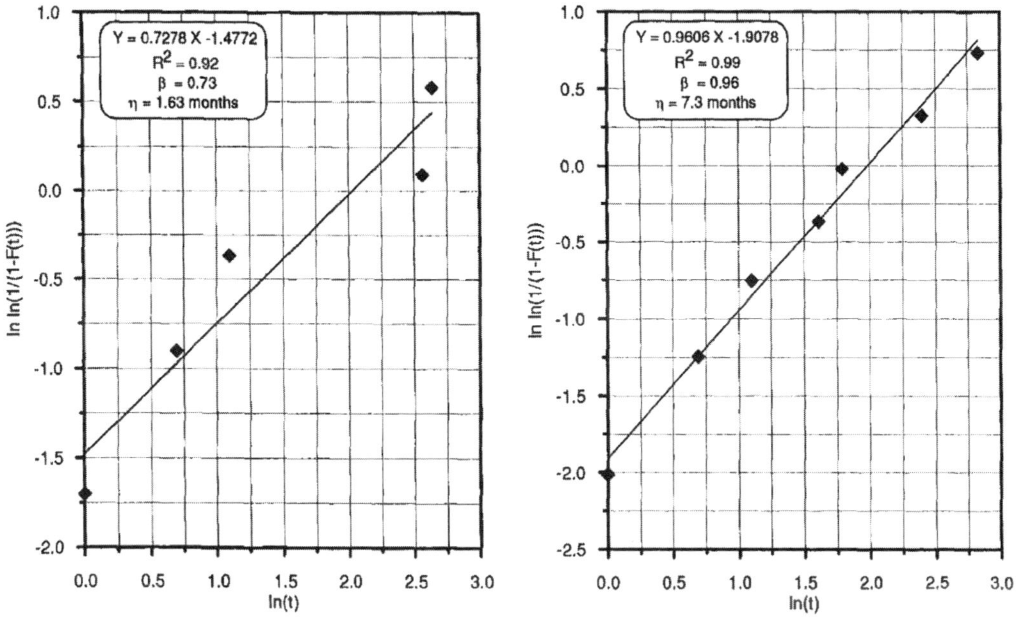

(a) Steam turbine with a characteristic life of 1.63 months

(b) Steam turbine with a characteristic life of 7.3 months

Figure H.3 Weibull analysis for two bad actor steam turbines [1]

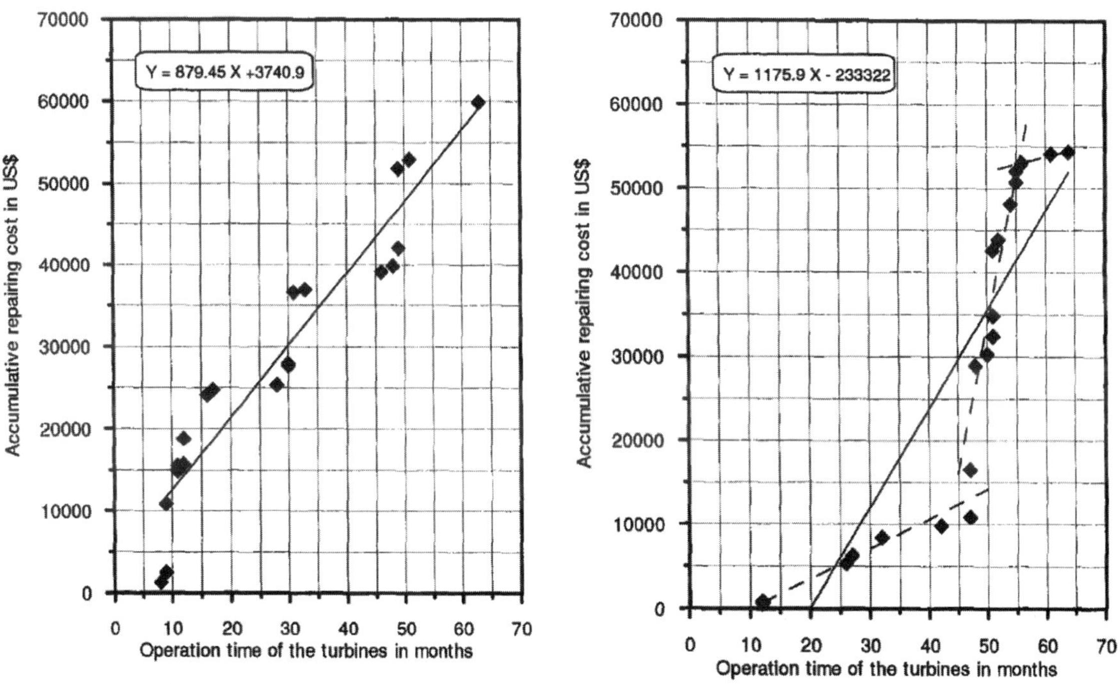

(a) Governor

(b) Lubrication cooler

Figure H.4 Accumulated costs for some repairable items [1]

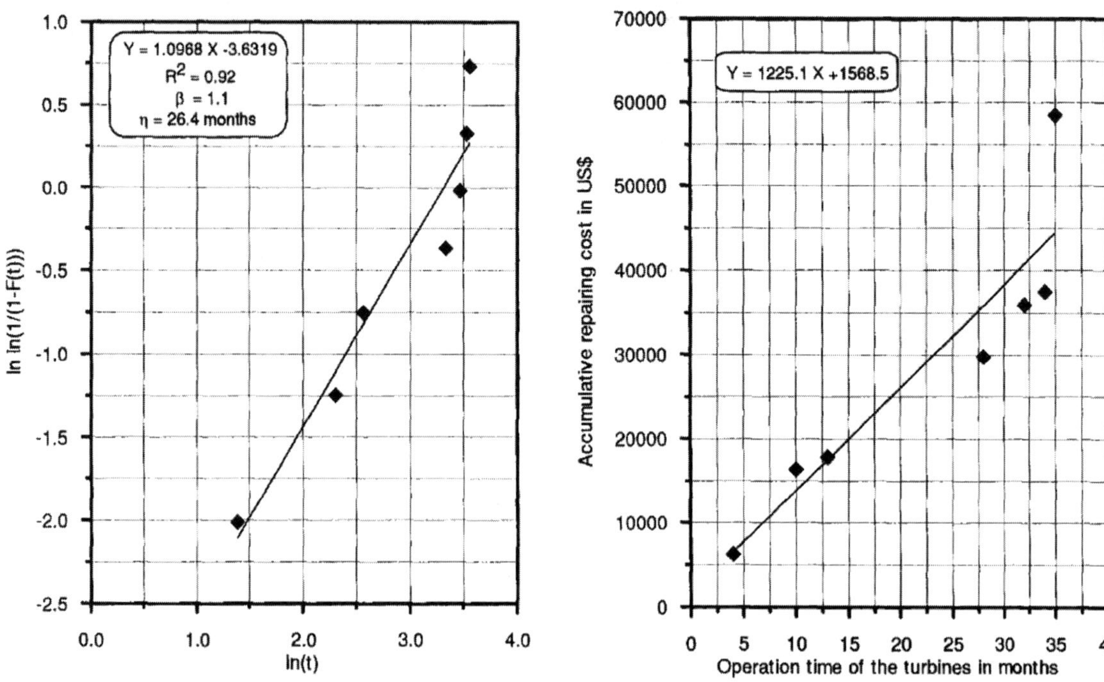

(a) Weibull curve

(b) Accumulated costs

Figure H.5 Reliability and cost data for bad actor steam turbine seals [1]

For one of the non-repairable components, the seals, the Weibull shape parameter β indicated moderate wearout, and the characteristic life was relatively short at 26.4 months (Figure H.5(a)). The cost associated with the seal replacements (Figure H.5(b)) verified that the financial impact was significant.

H.4 SPARE PARTS ANALYSIS

No design or other changes were indicated, but the data were used to ensure that enough seals were in inventory to meet future requirements. For the Weibull model, the renewal function approach was used to determine spare parts needed for a given period of time. If the time t (representing a planning horizon) is large, then $N(t)$ needed during this period, with a probability of shortage equal to $1 - p$, is given by [2]:

$$N(t) = \frac{t}{T} + \frac{1}{2}\left(\frac{1}{\beta^2} - 1\right) + \frac{1}{\beta}\sqrt{\frac{t}{T}}\,\phi^{-1}(p) \qquad \text{Equation H.1}$$

where $\Phi^{-1}(p)$ is the inverse of the normal distribution function. Considering the non-repairable parts of the previous bad actor turbines, the seals are the parts that fail most frequently. The calculation of spare parts was based on a five-year period of time (60 months). The results are summarized in Table H.1. From this table, it was found that there should be a minimum of six seals in stock to ensure 0.995 risk-free operation with no spare shortages.

TABLE H.1 Replacement seals needed for ten turbines in the next five years [1]

Type of Equipment	β	η (months)	*MTTF* (months)	*t* (months)	*Needed Parts* $N(t)$
Turbines (Seals)	1.09	27.42	26.48	60	6

TABLE H.2 Maintainability, reliability and availability parameters for the bad actor steam turbines [1]

Equipment#	*Maintainability*			*Reliability*			*Availability*
	m	θ (months)	*MTTR* (months)	β	η (months)	*MTBF* (months)	$A(\infty)$ (%)
T1	0.78	1.68	1.93	0.92	3.33	3.47	64.20
T3	0.70	4.02	5.09	0.73	7.61	9.30	64.60
T4	0.88	1.78	1.88	1.04	4.09	4.02	68.10
T5	0.84	2.77	3.03	0.85	6.41	6.98	69.80
T6	1.12	3.16	3.03	0.96	7.29	7.42	71.00
T7	1.36	2.37	2.17	1.05	4.90	4.80	68.80
T8	0.94	2.97	3.04	0.64	12.55	17.50	85.20
T9	1.28	3.17	2.94	0.94	4.46	4.58	60.90
T10	0.52	2.15	4.01	0.57	5.73	9.34	70.00
T13	1.06	2.74	2.67	1.23	6.23	5.82	68.50

H.5 MAINTAINABILITY, RELIABILITY AND AVAILABILITY

The maintainability, reliability and availability parameters from each of the most critical bad actor turbines were calculated and summarized in Table H.2. It was evident that the availability of some turbines was not very high, mainly because of the high mean time to repair (MTTR). The main factor that can be used to improve the availability is MTTR, which means the time required for repair had to be reduced. This was accomplished by providing spare parts in advance to avoid waiting time for materials, and by minimizing the actual repair activity time in the repair facilities.

In summary, the integrated use of reliability, maintainability and availability was effective in applying better maintenance strategies and implementing cost reductions.

REFERENCES

[1] Sheikh, A.K., Al-Anazi, D.M. and Younas, M., 2002. "Reliability, Availability and Maintainability Analysis of Steam Turbines Used in an Oil Refinery," Proceedings of PVP2002 2002 ASME Pressure Vessels and Piping Conference August 5-9, 2002, Vancouver, BC, Canada.

[2] Samaha, E., 1997. "Effective utilization of equipment failure history through computerized maintenance management system," ASME-ASIA '97 Congress and Exhibition, Oct. 1997, pp.2-8.

INDEX